16

10⁰⁰

D0429765

ALSO BY CHARLES PELLEGRINO

Unearthing Atlantis
Time Gate
Her Name, *Titanic*
Flying to Valhalla *(fiction)*

with Jesse Stoff
Darwin's Universe
Chronic Fatigue Syndrome

with Josh Stoff
Chariots for Apollo

with George Zebrowski
The Killing Star *(fiction)*

*with James Powell and
Isaac Asimov*
Interstellar Travel and
Communication

RETURN TO SODOM AND GOMORRAH

RANDOM HOUSE
NEW YORK

RETURN TO SODOM AND GOMORRAH

Bible Stories from Archaeologists

CHARLES PELLEGRINO

Copyright © 1994 by Charles Pellegrino
Maps copyright © 1994 by Anita Karl and Jim Kemp

All rights reserved under International and Pan-American
Copyright Conventions. Published in the United States by
Random House, Inc., New York, and simultaneously in Canada by
Random House of Canada Limited, Toronto.

Library of Congress Cataloging-in-Publication Data

Pellegrino, Charles R.
Return to Sodom and Gomorrah: Bible stories from archaeologists/
Charles Pellegrino.—1st ed.
p. cm.
Includes bibliographical references and index.
ISBN 0-679-40006-0
1. Bible—Antiquities. 2. Palestine—Antiquities. 3. Middle
East—Antiquities. I. Title.
BS621.P394 1994
220.9'3—dc20   94-2424

Manufactured in the United States of America

24689753

FIRST EDITION

Book design by Carole Lowenstein

This one is for Robert Ballard,
who took me to the Edge of Creation,
and showed me the way home.
You were right, Bob: There *are* astonishments aplenty
right down here on Earth, right under our feet.

In Memoriam

to

JANE PELLEGRINO

(May 26, 1935–February 25, 1993)

mother, friend of a lifetime—and till the very end a teacher of hyperactive and learning disabled children (and the world knows now that I was one of them).

In all your life, you hurt no one, you helped many. Your way of living is the hope of the new millennium: shining a little light wherever you could. The world did not shake the night your light faded. But it should have.

It should have.

It so happens that the words and stories passed down by my aunts made the oral transmission of history my particular interest. . . . The American Indians handed down their clan and national histories only in that way. In New Zealand, a Maori talking chief can trace royal ancestries back a hundred generations by memorizing charts over and over. It's beautiful. And of course there is a tradition that we don't think of as oral, but which existed for a thousand years, by word of mouth only, until it was reduced to writing in King David's time: the Bible. And these oral histories can be surprisingly accurate.
—ALEX HALEY

If you haven't found something strange during the day, it hasn't been much of a day.
—JOHN ARCHIBALD WHEELER

God is in the details.
—FREEMAN DYSON

Acknowledgments and Author's Note

Wherever possible, for the sake of sheer familiarity, I have (with a nod of approval from archaeologists Elizabeth Stone and Paul Zimansky) generally grouped the early Mesopotamian civilizations (of present-day Iraq) under the single heading "Babylonians." For similar reasons, I have used the modern names of places, bracketed by ancient names. Most of the ancient names used (such as Mizrayim for Egypt) derive from the original Hebrew version of the Old Testament, as translated directly into English for the Jerusalem Bible (Koren Publishers, Jerusalem, 1989, which displays the original Hebrew pages and their English equivalents side by side for easy cross-referencing). As was pointed out to me at the beginning of this project by Benjamin Mazar, the Protestant King James Version of A.D. 1611 was based upon Hebrew-to-Greek translations that had already acquired a number of errors. By a combination of chance and design, the Greek-to-English translation accumulated further mutations, with the result that in the King James Bible, Isaiah 7:14 refers to a messianic figure named Immanuel born of a *virgin*, whereas the original Hebrew refers only to a *young woman*. Further confusion arises from the lumping of diverse locations under the single heading "Red Sea" in the King James Version. Accordingly, if you want to read the early books of the Bible just as they were written down, if you really want to get close to the Hebrew tribes, you must get as close as possible to the original Hebrew. For this reason, of all the many versions of the Bible available, I have relied on the Jerusalem Bible throughout (using for comparison passages from the more familiar King James Version as necessary).

As regards my methods for assigning dates to battles and the reigns of kings mentioned in the Bible, the earliest human event that can be tied to a specific year comes from the Egyptian priest Manetho, who about 300 B.C. wrote a history of Egypt in Greek. He divided the rulers of the Nile into more than a dozen family lineages, or "dynasties," a system that all subsequent historians have borrowed. Manetho identified the third king of the Twelfth Dynasty as Sesostris III and, having access to historic records that did not survive the subsequent burning of the library at Alexandria, noted that in the seventh year of Sesostris III's reign, the star Sirius rose at sunrise during the Nile's flooding season. Astronomers can track this event back through time as easily as they can track eclipses. The rising of Sirius described by Manetho occurs once every 1,460 years, and the only such rising close to the Twelfth Dynasty happened in 1872 B.C.—which is now taken to be the seventh year of the reign of Sesostris III. The next specific date, obtained from an analysis of volcanic ash effects from the explosion of Thera in the Aegean, is Autumn 1628 B.C., during the reign of Tuthmosis III (whose contemporaries vividly recorded natural phenomena and social upheavals associated with this event). Traditionally, dates have been set around shifting pottery styles, each new style being given an arbitrary life span of about fifty years. Dates given in this book are based upon a resetting of the pottery clock around the 1872 B.C. and 1628 B.C. time lines and will therefore differ somewhat from those given in earlier works. Tuthmosis III, for example, appears to have reigned about 120 to 150 years earlier than was calculated from changing pottery styles alone.

As with all my books, I want to begin by thanking Mom and Dad, Adelle Dobie, and Barbara and Dennis Harris for the faith they showed and the encouragement they gave to a child some "experts" claimed did not have a prayer of growing up to read books, much less write them. Thanks also to two science teachers who came onto the scene a little bit later, Agnes Saunders and Ed McGunnigle. The world cannot produce enough people like you. If the seven of you could be cloned and put in every town, I'd hold out much more hope that our civilization might thrive in the third millennium.

Many people contributed to this book during wide-ranging conversations over a period covering nearly ten years. Those whose comments or insights were especially valuable include Jim Powell (Brookhaven National Laboratory), Arthur C. Clarke (University of Maratua, Sri Lanka), Father Mervyn Fernando (Subodhi, the Institute for Integral Education, Sri Lanka), Cyril Ponnamperuma (University of Maryland), Father Robert A. McGuire (Spirit Life Center, New York), Stephen Jay Gould (Harvard University), Ed Bishop of Bell Labs (the first person, to my knowledge, to summarize evolution as chaos with feedback), *the* Don Peterson (National Resource), Isaac Asimov (National Treasure), Walter Lord (National Treasure and all-around nice guy), Jerry Falwell (National Disaster), Mitch Cotter (AAAS), Daniel Stanley (Smithsonian Institution), Susan Limbrex

(University of Birmingham, Great Britain), Peter Kuniholm, Lynn Margulis, and Carl Sagan (Cornell University), Stephen Hawking (Cambridge University), Lawrence Soderblom, Jill Tarter, Mott Greene, and Kevin Pang (NASA/JPL), Michael Wise and Harold E. "Doc" Edgerton (University of Chicago), Rabbi Zola Levitt (Levitt Ministries), Eric Meyers (Duke University), Francis Crick (the Salk Institute), Joel Martin (Viacom), James Tabor (University of North Carolina), Robert Eisenman (California State University at Long Beach), Reverend "Jill," Tim White, Luis and Walter Alvarez (Berkeley), Edward I. Coher (Long Island University), Gerard R. Case, Hans Goedicke (Johns Hopkins University), George Zebrowski and Harlan Ellison (SFWA), Jesco von Puttkamer, Scott Carpenter, Fred Haise, and General Tom Stafford (NASA), Niles Eldredge and Norman Newell (American Museum of Natural History), Alan Hovard and Sir Charles Flemming (DSIR, New Zealand), John Geoffrey Leech, Alex Haley, Rabbi Seymour Freedman, Reverend John Meyer, Extinct DNA Study Group founders Alan Wilson and Clair E. Folsome, Philip and Michael Betancourt (University of Pennsylvania), Burton Rudman and Ed Pores (Archaeological Institute of America), Benoit B. Mandelbrot (New York), Rabbi Shlomo Goren, Rabbi Yeduda Getz, Benjamin and Amahai Mazar, Trude and Moshe Dothan (Jerusalem), Elizabeth Stone and Paul Zimansky (Stony Brook University), Nanno and Mrs. Spyridon Marinatos (Athens), and Bill Schutt (friend to vampire bats).

I wish to acknowledge the kind hospitality of Christchurch compound in Jerusalem and also the Israeli Army (which saved my neck on at least two occasions).

My thanks to Robert Ballard, Tom Dettweiler, Haraldur Sigurdsson, Jean Francheteau, and Roger Hekinian, my shipmates on the *Argo*-RISE Expedition, who shared with me a genuine love of volcanoes, and also to Mookau and Europa for their special, if not entirely voluntary, contribution to our understanding of the *keruvim* atop the Ark of the Covenant.

I thank Tom McAvinue, Sr. and Jr., and Joseph Pillittere II for navigating me around the core of New York Power Authority's Indian Point 3 reactor.

My thanks to Felix Limardo for Popol Vuh; to the Lunarians, for whom burning Cheez Doodles has become ceremonial; to Wayne Kaatz for pointing out a possible connection between Plato's *Timaeus* and *Critias* and chapters 27 and 28 of Ezekiel; and to James Morrow for some fascinating discussion on Lilith, immortality, the Ten Commandments, the Book of Job—and a nice subtitle.

My wife, Gloria, accompanies me on my expeditions and even, on occasion, gets me out of trouble. When we first met and a (relatively minor) gunfight began to erupt around us and I had to pull her head down to stop her from Nikoning the whole event, I knew I just had to marry *this woman*. (Most of the women I knew tended to overreact to gunfights, earthquakes, civil wars, and typhoons. What I've just said may sound funny, but it is not meant as a joke. Let me tell you something: I've seen people with far more guts than Gloria and me going into physical therapy and chemotherapy, facing up to far greater catastrophes with far worse odds of coming out unscathed.)

In keeping with my promise that they remain anonymous, several of Jerusalem's antiquities dealers, and the events that surrounded them, have been composited. Father John MacQuitty, too, is a composite of at least three unnamed Jesuits, whose theories and opinions are reproduced verbatim.

I thank my agent, Russ Galen, and my editor, Rich Kohl (Russ was right about you, Rich: You *are* Churchill), and Random House president Harold Evans for closing the wounds on "The Atlantis Fiasco" and putting all of my projects back on the launching pad. I am especially grateful for the stand Mr. Evans took against would-be censors from the Greek Ministry of Culture (and ultimately the U.S. State Department) in 1991. Special thanks to those writers and editors who stood by me and offered (much-needed) moral support, and even a little cheer, when all seemed so clearly lost: John Douglas, Arthur C. Clarke, Isaac Asimov, Stephen Jay Gould, and Stephen King. Sorry if this is beginning to read like one of those neverending Academy Awards speeches, but thanks also to Jon Karp, who pressed the launch button, and to Beth Pearson and Chris Potash in copyediting.

My thanks to the late Raymond Dart, the paleoanthropologist who once rode in the backseat of a cab with his newly unearthed Taung Child (a two-million-year-old member of the *Australopithecus africanus* clan) and who told me that his God "was not necessarily a conscious being separate from man who created the universe through a deliberate act; but rather a spiritual God that lived within us, and came into being with man." It was through his spiritual God, he believed, "that the books of the Bible, and other uniquely spiritual works, were written."

Some readers (including my editor) have already pointed out that when I quote Isaac Asimov, I may be confusing those who know he passed away in April 1992, because I tend to speak of him in the present tense. Isaac turned out to be a better friend than I realized. Only recently did I learn that even while he grew ill, and without ever telling me he was doing so or even being asked to do so, he moved quietly behind the scenes, dispelling rumors and sending his own editor (Rich Kohl) to my side while I fought off Greek antagonists, the State Department, and at least one plagiarist in what was becoming known on Publishers' Row as "all that trouble with Pellegrino's *Unearthing Atlantis.*" It is a larger story than can be told on these pages, but by speaking truth and common sense to the right people, Isaac essentially preserved my writing career, and I miss him more than I ever dreamed I would. So please forgive me if I just can't say good-bye yet.

CHARLES PELLEGRINO
Baltimore, Maryland

Contents

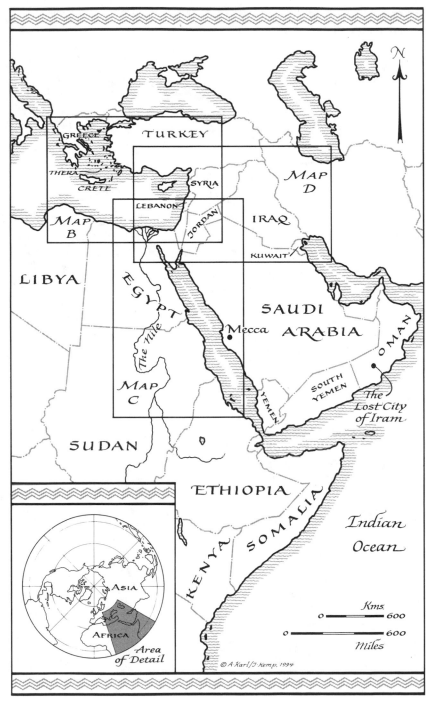

Map A

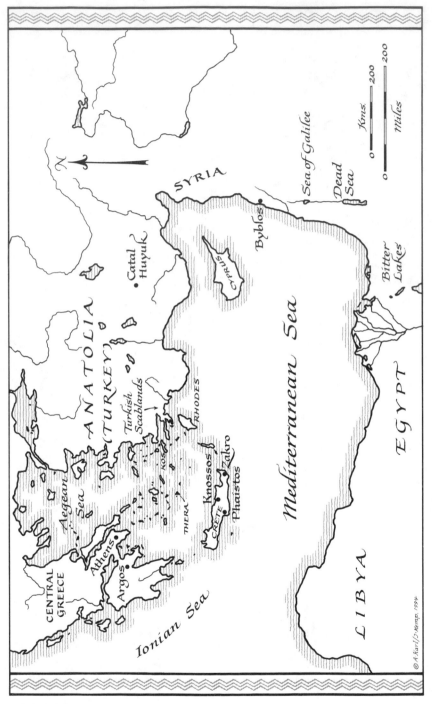

Mediterranean Sea

SYRIA

Byblos

Sea of Galilee

Dead Sea

Bitter Lakes

CYPRUS

Çatal Huyuk

ANATOLIA (TURKEY)

Turkish Scablands

RHODES

KOS

THERA

Knossos

CRETE

Zakro

Phaistos

Aegean Sea

Athens

Argos

CENTRAL GREECE

Ionian Sea

EGYPT

LIBYA

N

Kms 200

miles 200

0

0

© A.Karl/J.Kemp, 1994

Map B

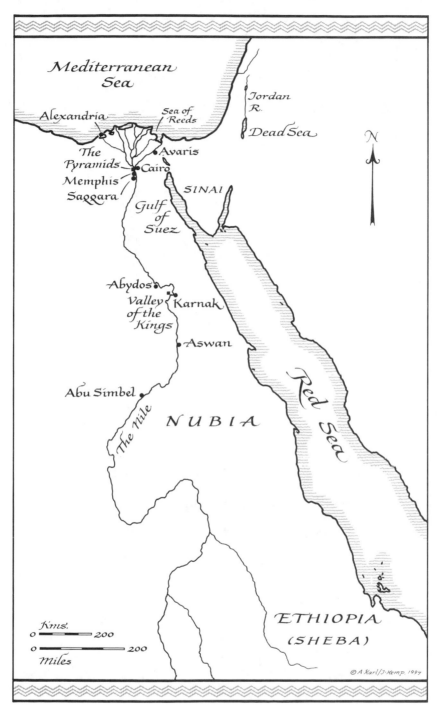

Map C

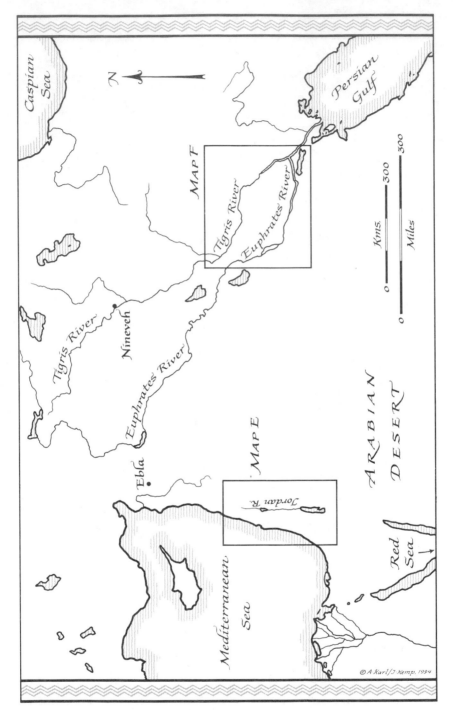

Caspian Sea

Persian Gulf

N

Map F

Tigris River

Euphrates River

Kms. 300

0

Miles 300

0

Tigris River

Nineveh

Euphrates River

ARABIAN DESERT

Map E

Ebla

Jordan R.

Red Sea

Mediterranean Sea

© A. Karl/J. Kemp, 1994

Map D

N

MEDITERRANEAN SEA

• Sidon

• Tyre

• Dan

Sea of
Galilee

• Bethlehem

• Kedesh

• Megiddo

Mt.
Gilboa

• Bet-She'an

Jordan R.

• Deir Alla

• Afeq

• Adam
(Damiya)

• Jericho

• Ashdod   Jerusalem •

Ir-Shemesh •   Qumran •

Gat •   • Egron

• Gaza

Deir el-Balah

• Hebron

Dead Sea

• Masada

The Land of Kadesh
(Northeastern Sinai)

Kms
0   30
0   30
Miles

© A. Karl/J. Kemp, 1994

Map E

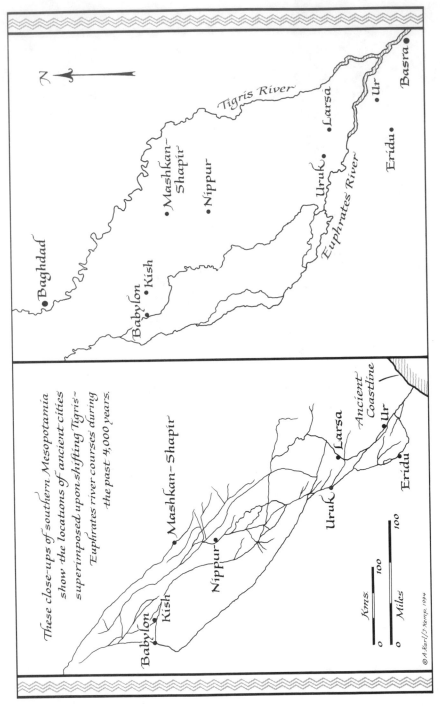

These close-ups of southern Mesopotamia show the locations of ancient cities superimposed upon shifting Tigris-Euphrates river courses during the past 4,000 years.

Mashkan-Shapir

Nippur

Kish

Babylon

Larsa

Uruk

Ur

Ancient Coastline

Eridu

Kms.
0   100

Miles
0   100

©A.Karl/J.Kemp, 1994

N

Baghdad

Babylon
Kish

Mashkan-Shapir

Nippur

Tigris River

Uruk

Larsa

Euphrates River

Ur

Eridu

Basra

Map F

RETURN TO SODOM AND GOMORRAH

INTRODUCTION:
The
Accidental
Archaeologist

When I meet the maker of the universe, I would like to be able to tell Him a little of how it works.
—JOHN D. ISAACS

All philosophy ultimately dovetails with religion—which is ultimately reducible to history. All history is ultimately reducible to biology. Biology is ultimately reducible to chemistry. Chemistry is ultimately reducible to physics. Physics is ultimately reducible to mathematics. And mathematics is ultimately reducible to philosophy.
—ED BISHOP AND C. PELLEGRINO'S FIRST LAW

THERE IS A CRACK in the Earth where a mountain was turned on its side, then swallowed whole in a spasm of quakings and compressions impressive even by the standards of today's nuclear propulsion think-tankers. Long before the first book of the Bible was committed to writing, a man in the Jordan Valley was enclosed by the mountain. You can still see his bones splayed out in the awkward attitude of sudden death, his flint tools flung about him, testimony to a geologic upheaval that was sure to echo down through twentyscore generations of oral history.

I know of a world that forced civilizations to live linearly, as though spread out along the edge of a ruler more than two thousand miles long and rarely more than two miles wide. And as if this long, narrow zone of habitability were not strange enough, it happened to weave and bump through the most geologically violent landscapes on the planet. No wonder that it became a place of "miracles."

I know of a river that stopped flowing, then changed course in a single day, allowing whole armies to cross through its basin on dry land.

I know of a puncture in the Earth more than seven miles wide, and when our seismic scans display cracks radiating out from the depression—forty miles in every direction—the mind resists coming to terms with the sheer size of it, tries to reject the idea that such destruction had required only the space of time in which a man might draw a breath, hold it reflectively, and exhale.

I know of a place where cities are buried under two hundred feet of volcanic ash: prehistoric cities with all the technological prowess of Classical Greece, a thousand years before Classical Greece was even the threshold of a dream.

I know of a sea whose waters stood up like a wall—stood up, in places, higher than the Pyramids.

I know of a volcanic death cloud that spread globular and huge, that touched the Earth and the sea with a heat sufficient to turn men into gas, that burst through the stratosphere, blocked out the Sun, turned summer into winter and man into a man-eater. In Egypt, under the pharaoh Tuthmosis III, one of history's first successful rationings commenced, and there emerged poems of darkness and pestilence throughout the land, of an Egyptian fighting force crushed under walls of water and, in the desperate economic times that immediately followed, the expulsion of Asiatic immigrants (among them merchants from Canaan) into the great eastern desert of Sinai.

The stories of Exodus and Joshua, some will argue, are strictly "fairy tales." But geology and archaeology have begun to teach us that several major Old Testament events that seem utterly fantastical to us today—among them the parting of Egyptian waters, the blotting out of the Sun, and pillars of fire in the sky—are based upon some very real kernels of truth. The rocks and the ruins tell us so.

As one who remains to this day an agnostic (no good scientist can be an atheist, for in science we must question everything, even our own questions), all of this has come as a great surprise. According to the late Alex Haley (who demonstrated most dramatically the fidelity of oral history), I was a blind fool to have been so surprised, for reasons that will become clear in the chapters ahead.

For me, the journey through the Bible lands began near a live volcano a mile and a half below the Pacific. In a very real sense, this book is mostly deep-ocean explorer Robert Ballard's fault. During the autumn and winter of 1985 I sailed with him and a crew of volcanologists to a place where a food chain based on sulfides instead of sunlight became an open window on how life got started on Earth and perhaps on other worlds as well. I was, in those days, a paleontologist (someone used to dealing with objects millions of years old) who happened to spend most of his time working in space research, and it was Ballard who, latching on to my roots in paleontology, told me to bring my head down from the clouds: "Do you not realize that we know less about the Earth we live on than about the stars in a galaxy two billion light-years away? The greatest mysteries are right here, right under our feet!"

And there I was, the closest thing to an archaeologist aboard, after one of our robots discovered the *Titanic*. Usually, on a research vessel, one can find little to do during the off-hours except read medical journals in the ship's library, but there, aboard the RV *Melville*, were miles and miles of videotape and negatives, with which it was possible to track the White Star liner's debris field backward in time and reconstruct her last three minutes.

And thus did the *Titanic* become my baptism in archaeology. Seeing how much I reveled in it, Ballard and the volcanologists said, "If you think the *Titanic* is fascinating, wait till you see what they're digging up in the Aegean. You're going to *love* that."

If there was any lost world that could outclass the *Titanic*, it was the isle of Thera and the origin of the Atlantis legend. That's where Ballard sent me, to a buried Minoan city which, according to Plato and the archaeologists, had been swallowed by a mountain, and whose homes—perfectly preserved under a cocoon of volcanic ash—were equipped with running water and bathtubs, flush toilets and central heating, yet dated back more than thirty-six hundred years. The rocks told us that a mountain had exploded with all the force of an entire nest of hydrogen bombs, but even trying to assign a date to the catastrophe became a portent of things to come, for it became immediately apparent that the isle of Thera did not exist in a vacuum and did not lend itself to tunnel vision. Thera forced us to follow a fine layer of volcanic ash under the Mediterranean Sea, into Turkey, under the Nile Delta, and as far away as the Greenland ice sheet. It forced us to acquaint ourselves with ancient Egyptian, Greek, and Chinese texts—and, yes, the Bible, too. It forced us to explore Irish peat bogs and learn everything we could about the trees entombed there, for their rings were as fingerprints running backward through time. And by drawing these diverse fields of study together, we found it possible to say with reasonable certainty that a squall was blowing west to east across Thera when all life was suddenly extinguished there in the autumn of 1628 B.C., when Tuthmosis III ruled Egypt and Jericho's City Four still stood gleaming and sharp-edged under the desert Sun.

And when all was said and done, all was far from said and done. A sedimentologist told me about a vein of Theran ash he had found sandwiched between layers of Egyptian mud, and an Israeli archaeologist told me of Minoan ruins recently discovered near the Sea of Galilee. I came to see the buried city of Thera in a rather larger dimension than the one I had begun with—which pointed more and more beyond the strict boundaries of archaeology. As I followed the Theran ash layer throughout the entire Mideast and started redating kingship from Egypt to Babylon based upon clues revealing how far above or below the sheet of ash a king's city had existed, I was living a detective story beyond the dreams of most detectives. Following the ash layer out from Thera, I inevitably began walking in the footsteps of the Hebrew tribes, asking the obvious next questions and, step by sequential step, entering the most romantic, most mystery-laden ancient worlds of all: the lost worlds of the Bible.

We all love a good mystery, and archaeology has, for me, turned out to be every bit as exciting as Agatha Christie at her best (whom, incidentally, you will be meeting in this book, along with her cohorts Leonard Woolley, Max Mallowan, and T. E. Lawrence). It is now possible, with a handful of carbon-dated mud, or volcanic ash, or a slab of stone, to illuminate day-to-day life in the world that surrounded the biblical scribes, which, in return, illuminates the books of the Old Testament. As I have often told my friends from the *Melville*, like any truly good mystery, the search for clues from the time of Abraham, Moses, and Joshua requires that we move around a lot, often to places quite distant from the "death scene," places whose connection to it may not at first be apparent. Tracking the origin and evolution of the people who wrote the Bible leads through a dazzling range of evidence (from millennia-old scars in California tree rings to stunning stone reliefs of the brilliant and beautiful queen Hatshepsut, who was forced to rule Egypt as a man), as we move from the Bible to the top of Mount Krakatoa and down along the ocean floor, through classical literature and paleontology and even into the space-time that surrounds distant galaxies. Even the soil that encloses Hatshepsut's palace in Egypt precludes studying a ruin in isolation. In the soil above the original ground level we find Roman and Christian artifacts, and higher up, a Muslim mosque. Immediately below an Eighteenth Dynasty temple are the ruins of an earlier temple. Below them, fossil clam beds and, lower still, dinosaur teeth.

As a paleontologist who brings with him all that he has learned working with relics millions of years old (including a comet-impact ash layer dating to 64.4 million B.C.) and applying it to relics only thousands of years old (including a volcanic ash layer dating to 1628 B.C.), I have learned that we must understand the land itself—how it was shaped and how it continues to be shaped—before we can understand the people who lived upon it and the stories that came from it. The rise of civilizations, the sounding of trumpets, the marching of armies over vast deserts as described in the Bible and now attested to by archaeology, the miracles and cataclysms, and even the dawn of human consciousness all were preordained by events played out in huge gyres of rock more than a million years before man existed. Hence, as I've so often told my students, "The story that unfolds may range a little deeper than you anticipated, and you may begin to feel as if you have become unstuck in time. Don't worry. If I've accomplished some modest measure of what I set out to do, that's exactly how you're supposed to feel. As a paleontologist who often surfaces into the shallows of archaeological time, that's how I normally feel. So welcome to my world."

What follows is a journey tracing man's spiritual and technological evolution from the origin of consciousness through the Dead Sea Scrolls controversy (which culminated, among other things, in an archaeological mutiny and a hand grenade attack on Jerusalem's Rockefeller Museum the day I arrived there). Very early in the journey, I decided to add one more layer of experimentation: What, I wondered, would happen if I gathered chaoticians and geologists, physicists and Jesuits, geneticists and rabbis around the common watering hole of biblical archaeology? The answer may surprise you, as it did me. While the words we used might have differed (God and infinity . . . consciousness and soul . . . DNA and the Great Attractor), we were, more often than not, saying the same thing. The result is a whole spectrum of views on the origin and destiny of man—which underscores my wish that my interpretations of what all the evidence means (which are, for better or for worse, bound to predominate in my own book) should be taken as simply that: *my* interpretations. I would never make a gross claim that they are to be taken as the one and only truth, for if the lessons of twentieth-century science are clear on any one point, it is that for all we know, there is so much more we don't know. So I urge you to look at the whole picture, wrap your mind around it, and think for yourself.*

So, with that advice, let's begin, you and I. Let's begin to raise and explore such questions as: Have scientists found Eve? Were there several pharaohs of the Oppression, and was one of them actually a woman? How did ancient Babylonian and Mayan cosmologies reveal a universe that began at a specific instant, as if from a stupendous explosion, almost exactly as revealed by modern-day Big Bang cosmology? What happened to the Arks of the Covenant and the Cross (and their duplicates)? Did a volcano part the Red Sea? Did the Jordan River really cease flowing the day Jericho fell? Why is Moses mentioned nowhere outside the Bible, and why are the most conspicuous structures in all of Egypt—the Pyramids—mentioned nowhere in the Bible? How do we know that Sodom and Gomorrah actually existed?

Our search for answers weaves a colorful tale of science and suspense,

---

*Please note that, as with all of my books, I invite criticism and discussion for the updating of future editions (for which the bringer of an important new point is always referenced and acknowledged). I still answer all my mail (except for manuscripts, which publishing house lawyers now forbid authors from reading). But please be patient if it takes awhile, for ever since Steven Spielberg made his little home movie about my recipe for cloning dinosaurs from flesh-feeding flies in amber (which, in all seriousness, ranks—along with *The Wizard of Oz*—as one of my all-time favorite movies), the pile has gotten rather higher, and I cringe every time I see the postman coming down the block with another bombload. Also, a self-addressed stamped envelope or postcard will be a very much appreciated time-saver and will receive the quickest response.

part theology, part cosmology, in which we look at ancient relics as both fascinating windows on the past and vivid signposts to the future. As a paleontologist, an archaeologist, and a sometimes rocket designer I have learned better than most people that we must vigilantly compare the past with the present if we are going to have any hope of building a future worth having. The single most frightening lesson I have ever been taught came from ancient Babylon, where cyclic collapses of civilization arose not from extreme and science-fiction-like collapses of the environment but from economic collapses brought about with chillingly simple ease at the earliest stages of environmental degradation (usually triggered by new technologies, including better irrigation channels, which at first allowed more profitable exploitation of the environment but led finally to overexploitation). Looking around and seeing an increasingly interlinked global economy and deteriorating soils, forests, and reefs in every direction, it becomes possible to believe we are on the verge of replaying Babylon's mistakes on a planetary scale.

But let me tell you something: Exploring new frontiers is always an exhilarating adventure, especially because so many of the world's best archaeological sites happen to be located in war zones. Even the scariest of the Babylonian revelations, even occasionally walking into a firefight, only adds spice to the danger of our dreams. (War zones, by the way, are not nearly so stressful as getting back to "civilization" and having to deal with lawyer types trying to claim an exclusive copyright on the two-thousand-year-old Dead Sea Scrolls or on the street plans of an ancient Minoan city. No wonder that the field of archaeology is peppered with people who will not hesitate to put a gun down a corrupt magistrate's throat and ask, "Now, what part of this don't you understand?" A hint to all city dwellers: Never argue with a man who has just walked out of the desert.)

This book is a probe not only into what it must have been like to walk with the tribes of Abraham and Moses, but also into those of us who are out in the deserts doing the probing. As such, the reader lives vicariously with people who, from time to time, will hunker down and drink from ancient goblets while war wages around their archaeological sites, and who have an uncanny way . . . of getting their way: as when one colleague, prevented from digging near a town whose politicians demanded rights of pilferage, finally, after years of battle, received carte blanche from the Greek government to dig anywhere he wished, whereupon he brought local traffic to a dead stop by digging up every road leading into the town. For an encore he undermined the foundations of city hall.

As our exploration continues, we discover that the bombs of the Iraq War were not so threatening to archaeological sites as the postwar poverty that brought so many looted artifacts to the Jerusalem black market. Together we also learn why the most important technological advance of all time was not the discovery of the wheel, or even the discovery of fire, but the discovery of death. The next logical leap was worry about the future and the discovery of time. Today most paleontologists (and some archaeologists) think in eons, from the next blue shift in the universe all the way back to the Big Bang. The biblical scribes of approximately 500 B.C. did not measure the Earth by the ages of rocks but by the standard of human life spans and the oldest tribal histories. Some biblical scholars adhere literally to the interpretations of these earliest historians, by whose yardstick the Earth and the cosmos were created about six thousand years ago. A few have suggested to me that artifacts from Jericho and other cities dating back more than eight thousand years, and meteorites 4.6 billion years older than the cities, were placed on Earth to deceive me, and that under such circumstances I cannot really prove the universe was not created only six thousand years ago. This is true, but under these same guidelines I cannot prove that the universe did not come into existence six seconds ago, with this book created in its entirety and happening to be placed in your hands during that very first chip of time, along with a false memory in your head of having already read my opening paragraph . . . of having studied a five-hundred-year history of colonization, in an America that has only now come into existence . . . of having read the plays of a William Shakespeare who never really existed at all.

Commenting on this possibility—on "the deception of the rocks"—the Jesuit scholar Mervyn Fernando said, "Einstein told us that God is subtle. I believe this to be true; but She's not malicious."

CHARLES PELLEGRINO
Sitka, Alaska

I

THE
FABULOUS
RIVERWORLDS

I

QUEST FOR ETERNITY

And God said, Let there be light. . . .
—GENESIS 1:3

*You ask me what God was doing before he created the
materials of Heaven and Earth. He was creating Hell
for people who asked questions like that.*
—SAINT AUGUSTINE

SPRING A.D. 1962

Down there on Earth, *Homo sapiens* had become adult. Suddenly and without warning, pieces of the planet were starting to detach—a phenomenon completely unknown to the world during the previous 4.6 billion years of its existence. From the Eastern and Western hemispheres, tiny magnesium sparks were coming up, every few weeks, depositing behind themselves columns of vapor that trailed back to the planet like white-hot threads. Inside one of those sparks rode an American test pilot named Scott Carpenter. Shooting out of the atmosphere at five miles per second, he became one of the first men ever to go into space, and one of the last ever to do so alone.

As he approached North Africa, night was striking across Asia. Two hundred miles below, a brownish film of dust was marching before the winds, in the direction of Thebes. With a single finger, Carpenter could draw a line across the whole of the world his high school textbooks had referred to as the Fertile Crescent. All below was brilliant white sand and sun-bleached rock, except for a faint ribbon of green. No more than thirty miles at its widest (and generally narrower than a mile), the ribbon stretched from the glittering Nile, along the shores of Israel and Syria, into the headwaters of the Tigris and Euphrates rivers, through Babylon and Baghdad to the Persian Gulf. Eight thousand years of civilization were displayed in the pilot's window. There, in the riverworlds, men had first pulled metals out of the Earth and hammered them into new shapes, pioneering a science that eventually made possible Carpenter's *Mercury* capsule. The worlds of the pharaohs, the Babylonians, nomadic Hebrew traders, all these things took on added significance . . . until the capsule turned around and pointed out, away from Earth.

Behind him was the bluish-white glow of childhood's home; ahead, it occurred to him, "There was only the blackness, only the color of death," going out forever and ever.

The works of humanity, and humanity itself, were no longer as large as Carpenter had liked to believe. If he represented the world behind him as a little ball of one-inch diameter, Mars would have been a large pea nearly two hundred feet away, Jupiter would be a mile away, Saturn two miles off, and Uranus four miles, while Neptune and the Pluto-Charon pair hovered six miles away in the frontiers of the night. And then, for tens of thousands of miles, there would exist nothing except the unexplored seas of interstellar space. The nearest star to Earth, on this one-inch scale, was forty thousand miles away, in a galaxy containing six hundred billion stars, in a universe containing at least as many galaxies.

Sooner or later every child discovers that he is not the entire universe, not even its center. This is a fundamental part of growing up—for any child, and for any civilization, though few have had this lesson driven home in quite so dramatic a fashion as Scott Carpenter. His century had opened with the conquest of the air, and he was now participating in the conquest of space. But as it had always been, and would always be, each new conquest brought new knowledge, and new knowledge often brought with it new defeats. As Carpenter threw open the doors to a wilderness and saw ahead of him new possibilities, he saw also new limitations. Explorers would certainly land upon the moon, but they could not hope to cross over to the nearest star or even to Mars in his lifetime, if ever.

And so it was from a vantage point of triumph that Scott Carpenter peered a short distance into the future and saw inevitable failure. It was the human thing to do. His species, alone among all the species that had ever trod upon the Earth, understood defeat.

More than the invention of wheels or agriculture, perhaps even more than the control of fire, the knowledge of defeat had pointed the way for human civilization. Somewhere in the remote past, humans had learned to anticipate the future in reasonable detail, and to see limitations and potential failures in every direction. With such capabilities came worry: worry that the supply of game animals might run out before the next spring season, worry that famine or drought, or both, might force the tribe to seek better pastures somewhere upstream, only to find those pastures guarded by staunch defenders, worry about attack by wolves, worry about death. The only rational escape from worry was to prepare for the worst and hope for the best. More efficient means of hunting developed, the better to assure a long-term food supply, and this in turn led to more efficient means of defense. And if one tribe discovered that another tribe's fortifications, arrows, and hand axes had become more efficient than its own, new worries drove it to match or exceed the competition's abilities. There followed a cascade of escalations that continued even into Scott Carpenter's time, when three giant tribes—Russia, China, and the United States—began dipping their arrows in deuterium, hoping somehow to enhance survival.

In the beginning there were curiosity and worry. Everything else—flint and fire and language—these were mere tools.

There is no telling precisely when an individual first grasped the notion that if all other creatures eventually died, and if the oldest, grayest people he knew always died within a few dozen lunar cycles, and if there existed no one older than the old grays, then death might not be something that happened only to everyone else. Ahead of him he began to see the end of all things, an unavoidable defeat, beyond which lay a great unknown.

If one could approximate how many spring seasons he had seen and compare these with the approximate age of the tribal elders, he could also estimate how many spring seasons remained before his own death became inevitable. The discovery of time could not have followed far behind the discovery of death. For all we know, it might even have preceded the discovery of fire.

What we do know is that by 58,000 B.C., east of the Tigris River, in Iran, the Neanderthals were digging the earliest-known graves. Bodies were placed on their sides, with the legs drawn up, and with the heads resting comfortably on the right arms as if in open defiance of the idea that with death came the end of all consciousness, as if dying had finally come to be regarded not as the ultimate defeat but as a kind of sleep, perhaps even the doorway to a new beginning. The graves contain finely crafted flint tools, charred animal bones (indicating roasted meat), and fossilized blobs of pollen (the remains of flowers scattered upon the bodies). It is not likely that people would have lowered fine tools and cooked food into a pit unless they believed that these things could somehow be used by the dead, that consciousness, or a spirit, or *something* persisted in the graves.

Fifty-three thousand five hundred years later, pyramid-shaped graves 450 feet high and weighing nearly five million tons rose along the banks of the Nile, tribute to an elite priesthood of god-kings who were obsessed with time and the notion of immortality.

Nine hundred fifty miles due east, on the banks of the Euphrates, a man was buried with objects of gold and silver, somehow meant to serve him in the afterlife. And among these objects lay dozens of sacrificial victims who, from all appearances, had followed him voluntarily to the grave. As they walked down ramps into a rectangular pit, they seem not to have been thinking of what awaited them as the end of conscious thought. There are absolutely no signs of struggle. The women wore gold and lapis headdresses that survive today as artistic masterpieces. One elegantly dressed lady—her headdress identifies her by the name Shub-ad—lay back and calmly sipped poison. The golden cup was still in her hands when the archaeologists found her. A four-wheeled wagon stood nearby, with the bones of oxen fallen between the shafts, and the bodies of their drivers beside them. Another woman—models based upon her skull structure reveal her to have been about seventeen and very pretty—was frustrated, to one degree or another, by a silver hairband during the final moments of her life. She did not have time to fasten it, so she rolled it up and slipped it into her pocket, where a man named Leonard Woolley found it, protected by the cloth from corrosion, more than forty-four hundred years later.

This was the land of Shinar, birthplace of some of the world's earliest writing, which is to say, the birthplace of historical records. We know it from the Bible as the land from which Abraham came and the land of Babylon to which the Hebrews were taken captive.

By 2500 B.C., the approximate year in which a seventeen-year-old girl descended into the Earth and drank poison from a golden cup, whole societies had grown up around the hope of conquering death, of achieving immortality. The means of living on were elaborate, often grotesque and strange, and they varied greatly even along the narrow confines of the Euphrates. According to one collection of early tablets, the Sun-god Utu taught the Sumerian Lancelot Gilgamesh that immortality could be achieved only through deeds, that nations died, men died, cattle died. The only thing that did not die was how one is remembered. The decisions that one made lived on, and the worst defeat of all was to be remembered by unjust decisions. Ignoring this lesson, Gilgamesh set forth in search of physical immortality, drawing up the plant of eternal youth from the bottom of the ocean. But before the mythical hero could drink the nectar and guarantee himself a future without end, a snake deceived him and carried the plant away. Death could not be defeated, and Gilgamesh realized that only through his deeds would he live for posterity.

Clearly, by the time of the Gilgamesh epic, and most certainly much earlier, humankind had begun to live outside the "here" and "now." The discovery of death was the discovery of the future. It should not surprise us, then, that as the first civilizations erected elaborate tombs, they also created sundials and water clocks and other means of ticking off the years, the days, and the hours. The people dwelling along the Tigris and Euphrates rivers were the most advanced mathematicians and astronomers of their day. The baked clay "workbooks" of schoolboys tell us that the Babylonians knew Euclidean geometry some two thousand years before Euclid "in-

vented" it. They partitioned the year into twelve lunar months* and introduced a week honoring the Sun (Sun Day being the first day of the week),
the moon, and the five planets visible to unaided eyes. The seven-day week,
celebrating the heavenly bodies and the seven days and nights during which
they had been created, was passed on to the Semites and later to the
Christians (along with the custom of devoting one day of seven to rest).

The Babylonians had also, by this time, invented birth certificates and
could henceforth assign numbers to the ages of their oldest citizens. Under
very favorable conditions, if disease, famine, or violence did not strike first
(and to judge from the rarity of skeletons older than thirty-three years, they
usually did), one might reach an age of seventy years. And thus do the first
birth certificates (which can be seen in Babylonian ruins) echo down to us
through the Bible (Psalms 90:10): "The days of our years are threescore and
ten."

The discovery of a limited life span is celebrated in the story of man in
the Garden of Eden, versions of which preceded the Old Testament in
Egyptian and Babylonian texts, which were probably themselves preceded
by a long tradition of recitation and epic songs, before poets began to
inscribe their oral prehistory on limestone blocks and baked clay tablets
about 2500 B.C. The earliest compilations of biblical history date from about
550 B.C., and there we read, from people who were trying to peer backward
along the stream of time as far as their interpretations of nature and science
allowed, that man began immortal. In a very real sense they were not far
from the truth.

If indeed man once dwelled in ignorance of both time and death, then
for all our forebears knew, they truly were eternal, in the same sense that
our cats, if they think at all, are unaware that they are not everlasting. In
biological terms, the explosive growth of human brains during the past
million years guaranteed that sooner or later human minds would be capable
of gaining knowledge; and with knowledge, death entered the world. In
biblical terms, a snake deceived man into eating from the tree of knowledge,

---

*The year is actually 12.37 lunar months long. Our modern calendar gets around this by
adopting the Egyptian solar calendar based upon one annual event: the flooding of the
Nile—which, as far back as recorded history goes, commenced every 365 days. The Egyptian
year, later borrowed by the Romans (and reformed by Augustus Caesar, who named the
month August after himself), was divided into twelve months of approximately thirty days
each. The Babylonian calendar followed the lunar cycles exactly, and their year was sometimes twelve months long, other times thirteen months long, and the cycle repeated itself
every nineteen years. This system was eventually adopted by the Semites (and the Greeks)
and survives today as the Jewish religious calendar.

and in an apparent variation on the epic poem of Gilgamesh and the plant of immortality, it is God who revealed to Adam (in Genesis 3:19) that he must eventually die: "For dust thou art, and unto dust shalt thou return."

By the year 2500 B.C. pyramids on the Nile, and temple-capped ziggurats on the Tigris-Euphrates, had become wonders of the world. These accomplishments can probably be traced back to and beyond early Neanderthal attempts to neutralize death with belief in an afterlife, and virtually all religions have revolved around this belief ever since. Our gods may have different names from the ones the Egyptians and the Babylonians were writing about, but still we honor them with architectural marvels, and still we measure time, though with an intensity unimaginable to Babylonian mathematicians.

We have dated the formation of the Earth's crust to precisely 4.45 billion years ago. We have peered beyond the Babylonian minute to the jiffy (the travel time of a beam of light across the diameter of a proton—or 1 billion-trillionth of a second—hence the phrase "Be back in a jiffy"). Physicist Stephen Hawking has probed beyond the jiffy to an absolute limit of smallness, at which time itself becomes timeless and cracks may appear in space—submicroscopic cracks in the universe. In certain very powerful atomic accelerators we have actually dropped particles a short distance backward in time. We have begun to design spacecraft capable of traveling so fast that their occupants will age three times slower than stay-at-home observers on Earth, and of the multiple billions who now cover the planet from pole to pole, few will be found today who have not voluntarily strapped themselves to their wristwatches, as if following instinctively some inner urgency to compartmentalize each of the day's activities, to be able at any instant to fix in time and space one's position and direction, as the hands move inexorably forward, ticking off the hours of our years—the thirty-two years of our lives that run out every billion seconds.

Man, who knows death, is obsessed with time.

So it is that we begin our journey through the Bible lands with two views arising from eight thousand years of civilization's efforts to measure, as if with compass and ruler, the Vault of the Ages.

First there is the Biblical Chronology (displayed on pages 25–29).

The ancient Babylonians, Egyptians, and Minoans of 2500–2000 B.C., asking where they and the world had come from, placed humankind at the center of creation and, tracking back as far as their new sciences would allow, placed the beginning of all things just barely ahead of the earliest events reported in their written and oral history—about a thousand years before their own time.* The edge of the visible universe, too, was perceived as being relatively near. From the Euphrates River to the Nile and the Aegean, the sky was regarded as a "firmament," a heavenly vault in which the stars were embedded. In one of Babylon's oldest legends (which came down through the Bible as the Tower of Babel story), the vault of heaven, hence the nearest stars, could be reached by building a sufficiently high tower. In the story of Jacob's dream (Genesis 28:12), it could be reached by a ladder.

The firmament itself is described in the Book of Isaiah (40:22) as a tentlike dome, under which human beings dwelled on the (apparently flat, tent-supporting) circle of the Earth: "It is he [God] that sitteth upon the circle of the earth . . . that stretched out the heavens as a curtain, and spreadeth them out as a tent to dwell in."

In Revelation, the last book of the Bible, we read (in chapter 6, verse 14) that the heavenly vault is removed at the world's end, "as a scroll when it is rolled together," suggesting that the lining of God's tent, in which all the stars were embedded, was viewed by our forebears as being no thicker than a sheet of parchment.

The firmament had other peculiar qualities. Through the desert regions of Egypt and Iraq ran three life-giving rivers, all of them originating in mountains hundreds of miles away.† The desert inhabitants knew only that the river oases existed and that rains fell very rarely into their world, like random gifts from heaven. There appears to have emerged among those people a notion that the waters of the rivers and the waters that fell from the sky were in no way related to each other and were therefore separated by the firmament, as in Genesis 1:7: "And God made the firmament, and divided the waters which were under the firmament [the rivers and the seas] from the waters that were above the firmament [the rains]."

Science has broadened our vision beyond the firmament and biblical time frames, producing a second chronology based upon cosmic distances and geologic timescales. *Apollo* and *Voyager* spacecraft have flown above the place where the firmament ought to have been, pushing back the frontiers of the night. Unless our eyes and our instruments deceive us, the universe is not embedded in the tentlike lining of a dome above the Earth, and the

*Some Babylonian texts allowed up to five hundred thousand years of prehistory. This figure was revised downward by biblical scribes about 500 B.C.

†These mountainous regions were largely unexplored. In the case of the Nile, dozens of centuries would pass before men like Sir Richard Burton, Henry Stanley, and David Livingstone discovered its source.

rains do not fall down from somewhere above the embedded stars. What appears, at first glance, to be a dome of stars is really space and distant suns spreading out thousands and billions of light-years. If objects are presumed to be billions of light-years away, and yet their light now reaches us only while covering a distance of one light-year per year, then it follows that the universe must be billions of years old.

Our more recent vision, the Cosmic Chronology (which I present on pages 29–33), has dethroned human beings as the center of the universe.* We know the triumph of new insights, yet we now see ourselves and our entire solar system reduced to a speck within a galaxy of stars that is itself a speck among billions of other galaxies. All the accomplishments for which humanity takes pride have existed for a geological nanosecond upon the Earth, and for some people the Cosmic Chronology has become a defeat even greater than the discovery of death, for we have glimpsed the insignificance of our entire civilization, of even our species in the larger pattern.

According to the latest archaeological evidence, civilization is more than eight thousand years old. As a general rule, the farther down we dig into a given layer of sediment or rock, the farther back we are looking in time. Beneath the recently excavated stone columns of Thebes, Egypt, flint choppers and hand axes were left behind by toolmaking humans nearly a half million years ago. Our people are very old, but our world is older still.

Several yards beneath Thebian hand axes lie fossil clams from a sea that existed fifteen million years ago, and beneath the clams you will find scattered teeth belonging to giant saurians long vanished from the Earth, creatures now extinct.

The revelation of extinction is what gives the Earth its history and puts

---

*It is only fair to warn you that the chronology of Earth and cosmos is still developing. New evidence may shift some of the dates even as this book goes to press. We should not be surprised, for example, to learn that the first Babylonians, the first tools, or the first living things appeared earlier than indicated. It is even possible (though not probable) that some new discovery may invalidate the entire chronology. While this will be cause for astonishment, it will not be cause for despair. It will only make the universe a bit more interesting than most scientists have supposed. The chief difference between the Biblical Chronology and the Cosmic Chronology is that the former is based on faith (answers) and is finished, while the latter is based on doubt (questions) and the picture will never be complete. A model based upon doubt is always subject to being torn down and rebuilt on the basis of new evidence, as has happened frequently during the twentieth century with the discovery of continental drift, relativistic time, and the results of *Voyager 2*'s grand tour of the solar system (which, while not entirely shattering one of my own pet theories about Triton's nitrogen seas, showed me to have missed the mark by at least one billion years). Science questions virtually everything, including its favorite models.

human beings in their place. If all the species that have ever lived came into existence with the Earth, on or about the same day, and lasted forever, we would have no perspective, no sense of time's arrow. As we dug deeper and deeper beneath Thebes, everything would be the same; we would find hand axes, clams, and dinosaurs mixed together all the way down. Instead we begin to see the stages of a lengthy history, in which dinosaurs and other creatures are segregated in specific layers of rock, and the farther back we track along the stream of time, the more unlike modern creatures the animals become.

If we begin with the notion that man is the center of the universe and the measure of all things, then what is exceptional about the dinosaurs is that their bones run down through layers of rock thousands of feet and two hundred million years deep, compared against our million years and a few tens of feet. The message from the rocks is clear: Nature is unfinished, has always been unfinished. When we examine any geologic period, including our own, all that is, is transitory. For every species that exists today, thousands more must be extinct. The thirty million species actually living on the planet must therefore represent the tiny pinnacle of a very large iceberg. Speaking bluntly, we are not the measure of all things, and the odds appear to be stacked against us in the long run.

BIBLICAL CHRONOLOGY

The Biblical Chronology depicts an Earth approximately six thousand years old, based primarily upon life spans given in Genesis for Adam and his descendants. The traditional Jewish calendar places the Creation at 3761 B.C. Christian scholars have placed it at 4004 B.C. Archaeologists and geologists have discovered that certain events described in Genesis and Exodus appear to be based upon incidents that actually did occur. Around 2800 B.C., for example, the entire plain between the Tigris and Euphrates rivers was so thoroughly flooded that all human settlements were buried beneath thick layers of mud, and Babylonian scribes began recording the legend of a man commanded by his god to build an ark and to load the animals upon it. In 1628 B.C. volcanic debris from the most powerful explosion human eyes have ever seen passed through Egypt. Independent Egyptian texts from the period describe darkness throughout the land, and plagues, providing accounts of disaster hauntingly similar to stories recorded in Exodus. Using the 2800 B.C. and 1628 B.C. time probes (now gaining currency among biblical scholars) as a guide, some of the traditional dates have been slightly reset for this chronology, adding approximately five hundred years to the biblical age of the Earth. Archaeological dates, such as the origins of the Copper and Iron Ages, provide backgrounds against which key biblical figures would have lived.

October 4456 B.C.	Creation of the heavens and the Earth.
4456–4536 B.C.	Expulsion of Adam and Eve from the Garden of Eden.
4300–4200 B.C.	Approximately eight generations after the expulsion, human conflict has degenerated from fratricide (Cain and Abel) to organized war (Lamech's victory song).
3526 B.C.	Adam dies at the age of nine hundred thirty.
3400 B.C.	Noah is born to Lamech.
3300 B.C.	"The Iceman" dies in the Alps. A copper ax entombed with him reveals that the Copper Age has begun.
2800 B.C.	Babylonian and biblical flood period begins. Severe flooding of the plain between the Tigris and Euphrates rivers occurs about this time—repeatedly (according to archaeological evidence). Genesis 7:11 places the biblical flood in the six hundredth year of Noah's life.
2800–2700 B.C.	Noah's son, Shem, and Shem's descendants occupy the Tigris-Euphrates region and the Arabian Peninsula. These are the Semitic people, whose languages include Arabic and Hebrew (the relatively modern term *Semite* is derived from the Greco-Latin form of *Shem*). According to the Bible, Shem is the ancestor of the Arabs and the people who, after the Exodus, moved into and eventually conquered Canaan and become the Israelites.
2507 B.C.	Abraham is born to Terah. Clay tablets scattered throughout the ruins of the Middle Eastern river civilizations suggest that many men of prominence began adopting the name Abraham between 2500 and 2000 B.C.
2450 B.C.	Noah dies at the age of nine hundred fifty. The Akkadians rule ancient Babylon; Sargon's lost city of Agade is built.
2300–1895 B.C.	Terah's grandson Lot escapes the destruction of Sodom.
2030 B.C.	Sin-iddinam rules Babylon. The Great Wall of Mashkan-shapir is dedicated.
1950–1925 B.C.	The sixth Amorite king, Hammurabi, rules Babylon; makes a covenant with the Sun-god; carves the prototype of Moses' laws on a stone tablet.
1895 B.C.	About thirty years after the end of Hammurabi's reign,

the city of Mashkan-shapir burns to a cinder and is never rebuilt. A Babylonian Dark Age begins.

1872 B.C. Sesostris III rules Egypt, just prior to what appears to be a gradual breakdown of authority under the first Hyksos incursions. His reign sees the construction of immense fortresses on the banks of the Nile, as if the entire country has come under siege.

1872–1800 B.C. Semitic tribespeople (the Hyksos) swarm west across the sands of Sinai and seize control of Egypt.

1770 B.C. Hittite raiders put an end to Babylon's dynasty of Amorite kings, whose empire has been disintegrating for nearly a century.

1720 B.C. Under Amose I, Egypt's oppression of the Hyksos and their kin begins.

1665 B.C. Queen Hatshepsut inherits the throne of Egypt.

1628 B.C. The year of the plagues and the Exodus (perhaps the first of several migrations/expulsions of Semitic foreigners, among them the Hebrews, from Egypt).

1628–1610 B.C. Tuthmosis III, ruler of Egypt, defeats the Canaanites at Megiddo.

1560–1550 B.C. Mycenaean Greeks are at the height of their power.

1550–1350 B.C. End of the Mycenaean palaces. The eastern Mediterranean is convulsed by economic collapse, characterized by a two-hundred-year disruption of maritime trade and chaotic migrations of peoples.

1520 B.C. The Assyrian king Tukulti-Ninurta (Nimrod) takes the throne, conquers Babylon, founds Nineveh.

1510 B.C. Akhenaton rules Egypt, attempts monotheistic reform.

1540–1450 B.C. The Philistines rise to power. Somewhere in Syria, Turkey, or Iraq, an unknown hero of the declining Hittite kingdom invents the Iron Age. Jericho's City Four is destroyed by earthquake and fire. This is the time of Joshua.

1450 B.C. Shalmaneser I rules Babylon.

1430 B.C. Ramses II rules Egypt.

1300 B.C. The population of Deir el-Balah, on the Gaza Strip, has adopted Egyptian burial customs, and is producing giant anthropoid coffins. Ramses III is victorious over the northern People of the Sea (among them the Philistines).

1150 B.C. The Chaldeans migrate from Arabia into the Euphrates region and settle in Ur.

1010–1000 B.C. The Philistines defeat the Israelites at Mount Gilboa. Saul and Jonathan are killed; their bodies are hung on the

	walls of Bet-She'an (biblical Beth-shan). David rules Judah.
1004 B.C.	David assumes rule of a united Israel-Judah; establishes capital at Jerusalem; begins construction of the First Temple.
962 B.C.	King Solomon completes the First Temple, which houses the Ark of the Covenant (apparently one of several arks containing fragments on which the Mosaic laws were first inscribed).
722 B.C.	Sargon II, ruler of Babylon, carries northern Israelites off into exile.
626 B.C.	Jeremiah begins to record his prophecies.
620 B.C.	Discovery of the Book of Deuteronomy in the First Temple.
586 B.C.	Nebuchadnezzar of Babylon sacks Jerusalem, destroys the First Temple.
586–538 B.C.	The Jews are captive in Babylon and during their captivity adopt the local custom of devoting one day out of every seven to rest. Much of the Hebrew Bible is compiled and copied during the subsequent two hundred years of tolerance and prosperity under Persian (Iranian) rule.
538–516 B.C.	Those Jews who wish to do so are allowed to leave Babylon and return to Israel. The Second Temple Period begins.
400–350 B.C.	The Book of Isaiah, assembled from many separate collections, including late additions, reaches its present form, more than three hundred years after the death of Isaiah.
300–250 B.C.	Alexander the Great conquers the Persian Empire. The Septuagint, the first translation of the Hebrew Bible into Greek, is written.
200 B.C.	The Dead Sea Scroll biblical texts are begun.
165–150 B.C.	The Book of Daniel is written; the Book of Psalms reaches its present form.
63 B.C.	With Pompey's capture of Jerusalem, Israel falls under Roman rule.
17–6 B.C.	Christ is born during the reign of King Herod (40–4 B.C.), who, in an attempt to stop the Messiah while still a child, orders all male children of two years and under to be killed (Matthew 2:16).
A.D. 15–22	Christ is crucified.
A.D. 46–62	Saint Paul journeys throughout the eastern Mediterranean, beginning the spread of Christianity.

Writings attributed to Paul (compiled almost five hundred years after his death) constitute approximately one third of the New Testament.

A.D. 70 Titus sacks Jerusalem and destroys the Second Temple. The Dead Sea Scrolls are hidden.

COSMIC CHRONOLOGY

The Cosmic Chronology allows us, in only a few minutes, to sweep our eyes over the history of the world, from end to end, in accordance with the best evidence now available from newspapers, ancient texts, archaeological sites, fossils, radioactive dating of rocks and meteorites, and astronomical observation. Starting with the turning point of the 1992 International Space Year, the chronology probes back in time (first by one year, then by two, four, eight, and so on), in much the same way that archaeologists and paleontologists, digging into the Earth layer by layer, view time as if they were leafing backward through the pages of a book.

A.D. 1992 International Space Year; implementation of the Space Cooperation Initiative; Antarctic ozone hole reaches New Zealand; human population nearing 6 billion.

A.D. 1991 *Voyager 2* spacecraft leaves the solar system; strategic defense shield funding reduced almost to extinction; first medical applications of genetic manipulation of human cells (to treat immune dysfunctions) signal that human beings are on the verge of dictating their own evolutionary destiny.

A.D. 1989 A thaw in the Cold War; Berlin Wall crumbles; first hesitant steps toward East-West demilitarization; biomap project begins; world population reaches 5 billion.

A.D. 1985 First interstellar rockets designed; matter-antimatter engine concept tests proposed; scientists confirm rising CO_2 levels; Antarctic ozone hole detected; President Ronald Reagan proclaims Strategic Defense Initiative (an international space weapons system known as "Star Wars").

A.D. 1977 Departure of *Voyager 2* spacecraft on grand tour of the solar system; means for global nuclear annihilation is in place; world's first "test tube" baby, Louise Brown, is conceived in England; world population reaches 4 billion.

A.D. 1961   Yuri Gagarin becomes first man to orbit Earth; Berlin
                   Wall erected; increased Cold War tensions signal an era of
                   nuclear proliferation.

A.D. 1929   New York Stock Exchange crash ushers in a worldwide
                   depression that culminates in global war and the first
                   atomic bomb; the Belgian astronomer (and Jesuit priest)
                   Georges Lenaître, tracing the motions of the galaxies back
                   in time, proposes that all the substance of the universe
                   was once compressed into an infinitely dense state, from
                   which the universe as we know it emerged at a specific
                   instant, as an outburst of energy called the Big Bang;
                   world population reaches 1.7 billion.

A.D. 1865   Exploration of Antarctic coast begins; germ theory of
                   disease adopted; widespread use of fossil fuels
                   commences; world population reaches 1 billion.

A.D. 1737   Reemergence of the experimental method in science;
                   invention of the steam pump; colonization of the
                   Americas; world population reaches 760 million.

A.D. 1481   Renaissance in Europe; voyages of discovery from Europe
                   and Ming Dynasty China; printing presses and handguns
                   proliferate; world population reaches 500 million.

A.D. 969   Widespread use of steel in China; Viking colonization of
                   Greenland; Arab Empire explores Indonesia and the
                   Philippines; world population is stable at 300 million
                   people.

55 B.C.   Roman Empire; first use of hot-air balloons in Nasca,
                   Peru; Great Wall of China (first of many construction
                   periods) fails to prevent conquests by nomadic herdsmen;
                   Ptolemaic astronomy; world population reaches 300
                   million.

2103 B.C.   The height of Minoan civilization; the Pyramids at Giza
                   are still new, cased in glass-smooth surfaces of polished
                   white limestone and capped with gleaming gold plate;
                   world population between 150 and 250 million;
                   development of astronomy; early dynasties of Babylon,
                   which will remain a large village until Hammurabi, the
                   sixth king of the Amorite Dynasty (1947 B.C.), builds it
                   into a world metropolis that, despite the rise and fall of
                   civilization, lasts more than a thousand years; subsequent
                   to northern Akkadians' moving in on Sumerians
                   (Babylonians) on the Tigris and Euphrates rivers and,
                   around 2500 B.C., subjugating them, scribes in Babylon
                   look back with nostalgia and wishfulness to a time when

their people were free, to a golden age in the land they called Eden, which means "plain"—the plain between two rivers east of Israel (identified in Genesis 2:10–14 as the Tigris, the Euphrates, and their branches).

6299 B.C.   End of most recent glacial period; proliferation of agriculture along the Tigris, Euphrates, and Nile; use of fire-baked clay in towns along the Tigris; Çatal Hüyük, a city of five thousand people, flourishes in Turkey, as does Jericho in the Jordan Valley; there are fewer than 30 million people in the world.

14,391 B.C.   Ice sheets two miles thick cover Europe and North America; giant beavers and sloths roam Montana; nomadic herdsmen inhabit the Nile Valley; invention of pottery, lamps, ropes; extensive cave painting in Europe; world population probably below 8 million.

30,775 B.C.   A small population of anatomically modern humans has, during the past hundred thousand years, spread across Europe, into Siberia, and sailed across the Exmouth Trench into Australia; invention of blade tools; Neanderthals rapidly dwindling or extinct.

63,543 B.C.   A warm interglacial period; Neanderthal man flourishes from western Europe to central Asia and Afghanistan; African presence of anatomically modern humans whose stone tools are, at this time, inferior to Neanderthal's.

129,079 B.C.   Warm interlude between Ice Ages; Neanderthals range from China to Europe; more modern-appearing humans, probably numbering no more than a few thousand, begin leaving a handful of fossils in Tanzania, Africa; *Homo erectus,* probable direct ancestor of both Neanderthals and anatomically modern (thinner-boned) humans, is apparently extinct.

260,151 B.C.   Ice maximum; *Homo erectus* and "Archaic" *Homo sapiens.*

522,295 B.C.   Another Ice Age; *Homo erectus* elephant hunters; control of fire; building of huts; speech has probably originated by this time.

1,046,583 B.C.   *Homo erectus,* apparent descendant of an Asian lineage of *Homo habilis,* inherits the Earth, leaves stone tools from China to Thebes, Egypt.

2,095,159 B.C.   Beginning of the Pleistocene Epoch; invention of stone tools by *Homo habilis,* who spreads from Kenya to Java and shares the African continent with the australopithecine man apes.

4,192,311 B.C.	Pliocene Epoch; a dry Mediterranean canyon has refilled; desertification of equatorial Africa; australopithecine man apes emerge (apparent ancestors of the *Homo* lineage); chimpanzees still thrive in Africa, resulting in coexistence of ancestral (chimp) and descendant (australopithecine) stocks.
8,386,255 B.C.	Miocene Epoch ends; African plate collides with Spain, building a natural dam across Gibraltar and triggering the Mediterranean dryout; the Nile River spills over a cliff forty times higher than Niagara, then slowly carves out a valley two miles deep; the moon is three hundred miles closer to Earth; the Hawaiian Islands do not exist yet; Australia and India are four hundred miles south of their A.D. 1992 positions; the Atlantic Ocean is eighty miles narrower; *Dryopithecus* (also known as Proconsul), probable ancestor of apes and men, has spread across Europe and Asia.
16,775,223 B.C.	Warmest part of the Miocene Epoch; forests line Antarctic coast; the African continent, drifting north, makes its first physical contacts with Eurasia, allowing the breakout and diversification of dryopithecine apes.
33,552,439 B.C.	Oligocene Epoch; the first cats and dogs; protomonkeys appear on the island continent Africa; Egypt is a shallow sea, in which the shells of sea creatures accumulate to form the limestone beds from which the Pyramids will eventually be built.
67,016,871 B.C.	End of the Cretaceous Period; the last dinosaurs; the Mediterranean is an open ocean larger than the Atlantic; global cooling, falling sea levels, and mass extinctions are under way; diversification of birds and mammals is also well under way.
134,215,735 B.C.	Early Cretaceous Period; dinosaurs dominate; birds and mammals are restricted to small niches (mostly insect feeders and scavengers); first marsupials; first honeybees (and the first flowers) have appeared by this time; Europe, the Americas, and Africa merge to form a supercontinent.
268,433,463 B.C.	Permian Period; the first dinosaurs; mammal-like reptiles (probable ancestors of mammals) dominate but will decline almost to the vanishing point during the Terminal Permian Extinction (about 245,000,000 B.C.), which will erase 96 percent of all species known from fossils; global Coal Age forests have reached an all-time peak following the Gondwanaland Ice Ages of 290,000,000 B.C.

536,868,909 B.C.	Middle Cambrian Period; first vertebrate animals; Canada's Burgess Pass formation contains virtually the only fossils we have from this time, which reveal exotic invertebrate water fauna with few modern parallels; except for windblown knots of DNA, the landmasses are without life.
1,073,739,821 B.C.	Pre-Ediacaran Time; the first wormlike creatures have probably appeared in marine communities; seaweeds flourish; interstellar space is two hundred times as dense as it will be in A.D. 1992 (each cubic yard contains at least twenty atoms), and is twice as warm.
2,147,481,645 B.C.	Bacterial reefs in Australia preserve multicellular threads of algae; a significant oxygen atmosphere begins to develop on Earth; formation of the lunar maria.
4,294,965,293 B.C.	Protocell Era; the Earth's crust is barely formed, yet the invention of multicellularity and sex (by protocells) has probably already occurred.
8,589,932,589 B.C.	The solar system does not exist as yet; the materials of Earth and human bodies are still being fused from hydrogen and helium in the cores of very large stars, whose lives culminate after just a few million years in supernovas that blast silicon and carbon through interstellar space.
17,179,867,181–20,000,000,000 B.C.	The Big Bang, the current recession of every galaxy from every other galaxy suggests that we all are caught up in the aftermath of a stupendous explosion. The Big Bang may mark the beginning of the universe or the end of a previous universe about which all information was destroyed in the cauldron of creation. Tracing the motion of the expanding universe backward in time, we glimpse a state in which the cosmos was infinitely dense and infinitely hot, with photons (particles of light) emerging from every direction at once. As near as we can tell, in the beginning . . . there was light. . . .

A BRIEF HISTORY OF TWO CHRONOLOGIES

YEAR	AUTHORITY	PROPOSED ORIGIN OF EARTH
2500–2000 B.C.	Unknown Babylonian authors	500,000–60,000 B.C.
500 B.C.	Unknown Hebrew scribes	Approximately 3800 B.C.
A.D. 1642	John Lightfoot, professor of Greek	9:00 A.M., September 17, 3928 B.C.

A.D. 1658	James Ussher, archbishop	October 23, 4004 B.C.
A.D. 1869	Thomas Huxley, Charles Darwin, biologists	300,000,000 B.C.
A.D. 1897	Lord Kelvin, physicist	40,000,000 B.C.*
A.D. 1907	B. B. Boltwood, geologist	1,640,000,000 B.C.
A.D. 1947	Norman D. Newell, paleontologist	3,000,000,000 B.C.
A.D. 1977	Gerald Wasserburg, physicist	4,600,000,000 B.C.
A.D. 1990	Thomas Gold, planetary geologist	4,600,000,000 B.C.

*Lord Kelvin based his estimate on a proposed cooling rate of the Sun since the solar system's formation and argued fiercely against Huxley and Darwin's old-Earth hypothesis, which was based upon estimated life spans of fossil species. At that time, the only imagined energy source for the Sun was frictional heating left over from the solar system's accretion. Nuclear fusion, a still-ongoing process that allowed for a sun even more advanced in age than Huxley and Darwin's fossil record, was as yet undiscovered.

2

THE
CRESCENT
OF
FIRE

*The doors of Heaven and Hell are adjacent and
identical.*
—NIKOS KAZANTZAKIS

THE ASWAN HIGH DAM, TIME PRESENT

Twenty million years lie under my feet. Twenty million years of Egyptian history that began with a seafloor raised up into sunshine and breezes, and rains that transformed this part of North Africa into a garden; and in that garden, according to paleontological accounts, our prehuman ancestors once stood. But the trees and the streams vanished so long ago that their very light has left the galaxy and is now lancing out toward Andromeda. They are not gone without trace, however. If you travel forty or fifty miles into the desert and dig down deep enough, you will find hippopotamus bones, hinting at watering holes and green pastures where none presently exist.

Also underfoot, like a capstone on all those vanished ages, lies one of the great marvels of twentieth-century civilization. In cross section, the Aswan High Dam looks like a step pyramid grown out of control. Two and a quarter miles long and a half mile wide at its base, it contains a volume of stone sufficient to accommodate all sixty of Egypt's Pyramids, with enough left over to get a start on the Great Wall of China. Aswan is one of the few man-made objects clearly distinguishable from space, and with this dam Egypt has at last harnessed the mighty Nile. There is enough stone here to guarantee that the structure will last four thousand years or more. Broken and weatherworn, it will probably come to be regarded as one of our own pyramids, somewhere in time. By then mankind may very well have rebuilt the entire planet to serve his interests. Standing alone in the center of the Nile, atop a technological miracle, I begin to suspect that such accomplishments will not speak well for our species. Perhaps a truly advanced civilization would leave no mark at all upon the world.

To the south, a man-made lake covers most of the horizon. Below me, the river is much as it was before the dam, barely a mile wide and running from this point almost as straight as an arrow toward Thebes and Cairo.

Here the land rises as vertical cliffs from the riverbed. The water table does not come up to the ground surface, so the desert begins only two feet from the river's edge—little splinters of milky white rock stretching uninterrupted for hundreds and hundreds of miles. The silence is unnerving. There are no animals in this world except for the fish in the lake and the insects in the trees. And the only trees are those maintained by a roadside sprinkler system that runs east to west over the dam. Aside from the dam-top oasis, there is not a live tree or human in sight.

There are no distracting elements in the landscape, as in the woods of New England. In the desert there are only naked hills and gullies and the peculiar interplay of light and shadow beneath a thoroughly cloudless sky. I wish I possessed the sort of camera that could capture this light, but no such machine exists. I can only try to record it in my memory's eye. Every angle of rock seems to stand out so sharply that without the blue sky overhead I could easily believe the planet had no atmosphere. An Egyptian archaeologist has told me that once a foreigner sees the desert light he can never be the same man he was before, for it has a mysterious presence that is not merely perceived by the eye. She says it touches the soul as well. I don't know. That may hold true for the wondrous colors of desert dawn, but in midafternoon, when temperatures reach 126°F, the brilliant white light is, to me, the color of death. I squint a hundred miles east, and I wonder how I could possibly make it that far, if I had to, on foot and alive. Yet somehow Moses and his people are said to have gone out into that hell and survived there for forty years.

Don't ask me how they did it. The air wants to soak the moisture out of my mouth as soon as I open it. To someone like me, adapted to steamy Long Island, collapse from dehydration would come after only a few hours out there. For years I have been accustomed to sipping seltzer water with every meal and barely paying attention to it. Here I must carry drinking water with me always, and the water from the tap usually has reed fragments and other Nile decay products floating in it. To avoid ingesting parasitic organisms, I must boil my water, but this habit insults my Egyptian hosts, so I must boil it in secret, whenever I can, with the result that my daily supply is rationed. It is late morning, and the thermometer has just topped 110°F. All of a sudden, for the first time in my life, I notice that water tastes delicious.

No, this land was not always so cruel to living things. Even in geologically recent times, during the height of the last Ice Age about sixteen thousand years ago, there were cloud banks overhead, and the rains, for more than five thousand years, came regularly to these hills. From an airplane you can still trace the paths of dried streambeds. Ice Age climate must actually have been pleasant on the Egyptian desert, turning the land into savanna. The same thing was happening far to the northeast, in Iraq's land of two rivers. But by 4000 B.C. the northern glaciers were in retreat, and a great dying began. The grasslands were still disappearing when settlers along the Tigris and Euphrates rivers invented writing and started capturing their oral traditions on baked-clay tablets. It is possible that retreating water tables displaced whole populations and that the displacements are recorded in the first books of the Bible as the wanderings of Abraham and Semitic tribespeople, the very same tribes whose descendants wrote the Old Testament and whose perspective on the history of this land is preserved better than any other.

All the standard history books refer to this portion of the world as the Fertile Crescent: down here the life-giving Nile, up there the fields of Israel, Lebanon, Syria, and up a little higher the Euphrates River, which arches southeast through the heart of Iraq. Most schoolchildren are taught to regard the crescent as a three-thousand-mile stretch of tranquil farming communities evolving slowly into pyramid- and ziggurat-building nation-states. It wasn't that at all.

The incessant turmoil we read about in the Old Testament—the burning of cities, the putting of children to the sword, the marching of many feet—this is closer to the truth. The Fertile Crescent is a myth. We would be more correct to call it a Crescent of Fire.

Try to imagine the Nile and Euphrates rivers, one in the east, the other in the west, each producing patches of vegetation weaving through a thousand miles of desert. In some places the patches are no more than a few inches in diameter, in others, as many as twelve miles, but no wider. Now imagine a desert corridor between the two rivers—again almost a thousand miles long—running from the headwaters of the Euphrates in the north to the Nile Delta in the south. This warm, wet path is, like the river corridors, very narrow, ranging from its widest parts in Israel and Syria (generally no more than thirty) to only a few feet across on the northern shores of Sinai. From end to end—from the Nile through the Jordan and the Euphrates— the inhabitants find themselves cultivating an oasis thousands of miles long with an average width of less than ten miles. Civilization becomes linear, and this very linearity dictates that the people will naturally break up into competing fragments or kingdoms. Along this giant winding path of fertility, different cultures are spread like markers along a ruler, and if they create armies and try to move anywhere at all, they must move through one another. Though many have tried, no one can establish a universal empire for very long, and it has been during those times when empires were established, as when Egypt conquered all the land within reach of its horses, that competition between economic and industrial centers was snuffed out, and technological stagnation inevitably set in, followed by decline.

Competition between diverse cultures, sustained and force-fed by geographical partition within and between the rivers, virtually guaranteed that civilizations would rise here and that their development would be greatly accelerated. In and around this crescent originated the world's first libraries, street lighting, and residential houses whose walls concealed plumbing and whose bathrooms were equipped with showers and flush toilets. Here were metallurgy, herbal medicine, the towering stone columns of Thebes, and a strange little toy, weighing barely more than an ounce, that seems to incorporate detailed knowledge of wing and stabilizer design. When tossed, it glides easily through the air. Technological marvels there were, everywhere; but of stability and peace within and about the riverworlds there would be none, forever.

3

THE
MAKING
OF
THE LAND

There is a grandeur in this view of life, with its several powers, having been breathed by the Creator into a few forms or into one; and that, whilst this planet has gone cycling according to the fixed laws of gravity, from so simple a beginning endless forms most beautiful and most wonderful have been and are being evolved.
—CHARLES DARWIN, *The Origin of Species*

We are convinced that masses of evidence render the application of the concept of evolution to man and the other primates beyond serious dispute.
—PONTIFICAL ACADEMY OF SCIENCE,
  Vatican City, 1983

WHEN THE EARTH was already ancient, of an age most of us seldom encounter in our thoughts, events that might ultimately prove to be of galactic and even universal importance were taking place in Africa.

During the warmest part of the Miocene Epoch (about 20 million B.C.), as lush forests penetrated into northern Alaska and even the Antarctic coastline sprouted evergreens, the cosmopolitan dryopithecine apes spread across the savannas of Ethiopia and Kenya to the shores of the Egyptian sea. The changing climate, probably the result of increased solar radiation (from a Sun that had been behaving strangely for some fifty million years), had transformed equatorial forests into semiarid plains, across which a form of life the world had never seen before was advancing like a wave.

Four million years earlier an obscure bamboo descendant had begun scattering its seeds to the wind, and the first grasses now carpeted the treeless ground. In time the descendants of grass—including wheat, corn, barley, and common lawn grass—would, in their own way, enslave the descendants of dryopithecines. They were going to spread everywhere the toolmakers went, until they became as a living membrane over the surface of the planet, manicured, watered, bred, and fertilized.

The dryopithecines walked on all fours. Looking somewhat like large monkeys without tails, they stood somewhere between monkeys and apes, and probably closer to apes. Casts of the insides of their skulls reveal small but nevertheless recognizable frontal lobes. They were the newest and brightest things around, but they were a long way from dominion of the Earth. Even the lowly snake held sway over them, though it, too, was a relative newcomer, in terms of geologic time, having crept out of the Earth only yesterday.

The cobra's ancestors had lived underground during the reign of the last dinosaurs. They were, in those days, barely larger than the ratlike mammals who huddled with them beneath the Earth, but they were far better adapted to a burrowing lifestyle. Any genetic changes that had damaged or even reduced the legs to relics of shoulder blades and hipbones did not harm the first snakes. Quite the opposite: The stunting and eventual loss of limbs improved their streamlining. There were further adaptations to a subsurface environment. Efficient burrowers can easily damage the delicate eardrum, and hearing is not of much use underground (vibrations are better detected through body surfaces in direct contact with the ground, as is the case with snakes, which *feel* sounds), so mutations that eroded the eardrum were more blessing than curse, and the snakes encountered by *Dryopithecus* were, like all snakes, deaf.

The ancestral snakes of the Late Cretaceous Period (about sixty-five to one hundred million years ago) probably preyed upon insects, worms, and burrowing shrewlike mammals. Early in the Cenozoic Era (the Age of Mammals), muddy lake bottoms began preserving the skeletons of snakes more than ten feet long. In megayears past, the saurians had crowded their brethren out of the sunlight. And now that lost reptilian tribe was coming back—changed. When it ventured aboveground, the dinosaurs were either vanished or fading, and the "rats" had begun radiating and diversifying into vacant niches, evolving toward the varied mammalian forms we know today. This newer, more varied menu of prey species (combined with an absence of predatory dinosaurs) may be precisely what tempted the snakes aboveground, in turn triggering their own radiation and diversification. By 20 million B.C. pythons and cobras had grown to such lengths that they could easily kill a dryopithecine and swallow it whole.

Twenty million years later, human infants exhibit an untaught fear of snakes and iguanas, but not of kittens, rabbits, or chicks. Evidently our DNA carries a deep-rooted, instinctive message: Reptiles are our enemies.

The dryopithecines, too (who appear to have been located near the trunk of our family tree), must have been born with an innate fear of snakes. It persists among infant chimpanzees, gorillas, and baboons as well as humans. This shared genetic memory probably goes back past the dryopithecines to a time when our ancestors—furry and ratlike—were perpetually underfoot, living in the nooks and crannies of a world whose chief denizens usually possessed scales and were often terrifyingly large.

"The pervasiveness of dragon myths in the folk legends of many cultures is probably no coincidence," says Cornell University astrobiologist Carl Sagan. "The implacable mutual hostility between man and dragon, as exemplified in the myth of St. George, is strongest in the west. (In chapter 3 of Genesis, God ordains an eternal enmity between reptiles and humans).

But it is not a western anomaly. It is a worldwide phenomenon. Is it only an accident that the common human sounds commanding silence or attracting attention seem strangely imitative of the hissing of reptiles?"

The dragons of our myths and our nightmares apparently reside in the deeper, more archaic parts of the brain programmed by our genes, those same parts that tell us how to walk and eat and to be afraid of the dark without ever being taught. A genetic memory should not persist for tens of millions of years unless it imparts some sort of survival value. Like the snake's eardrum, it should have become vestigial soon after it ceased to serve us.

By the time *Dryopithecus* appeared, the last of the pack-hunting monsters known to paleontologists as "terrible claw" had been dead for ages. There were no more velociraptors or tyrannosaurs in Eden, but cobras had taken their place.

Had the snakes remained hidden in the Earth, had they never spread across the continents and evolved into newer, deadlier forms, it is possible that lingering memories of an ancient hostility between reptiles and mammals would have, in a fate similar to that of the human appendix (which was rendered obsolete by the control of fire and the invention of cooking), degenerated by dryopithecine times, and children would not now be both fascinated and horrified by fantasies of slitherings underneath their beds. But on an African savanna in 20 million B.C., with cobras lurking in the grass, and crocodiles camouflaged to resemble logs at the edges of water holes, an easy way for a dryopithecine to die was to forget the fear of reptiles. So dark shapes rasped and hissed in the savage brain, and in the first books of the Bible, the serpent emerged as a symbol of power and evil; and in some small way, the dinosaurs were alive-seeming still.

The most ironic feature of the coevolution of serpents and humans is that we mammals are at least partly responsible for the persistence of unpleasant memories. It was the very diversification and success of our ancestors that created a need to preserve images of hostile reptiles. By encouraging the emergence and diversification of snakes, the dryopithecines and their contemporaries had, in essence, become the creators of their own tormentors.

〰〰〰〰

By 18 million B.C. huge gyres of solid rock, circulating so far below the African continent that the sheer weight of overlaying material caused them to flow like a warm, superdense plastic, had raised Egypt out of the Mediterranean Sea. Grasses, snakes, and dryopithecines covered the new land almost as quickly as it was pushed up.

The Red Sea did not exist as yet. The Arabian microcontinent was still welded solidly to Africa. It extended north from Egypt, through Israel to Syria, and east across Iraq. Turkey and Iran were foreign shores, on the other side of a narrow sea. For tens of millions of years the African-Arabian continent had stood apart from the rest of the world, and the Mediterranean was an open ocean much like the Atlantic. Floating atop a sluggish but constantly convective stew, the landmass was (and still is) drifting northward at about the rate human fingernails grow. Two or three inches per year may not sound particularly dramatic, but over the course of thousands of years it adds up to more than two hundred feet. In a million years the continent moves forty miles; in twenty million years it can drift halfway across an ocean.

Dryopithecus was living on a giant raft of land, bound on a collision course with Eurasia. The first jolts were felt in east Africa. The Earth buckled there, rose like a dome, then cracked open and produced rift valleys studded with volcanoes. Pressures from below pushed layers of rock above the basement until they stood forth as a new mountain range. The rise per century was minuscule, barely greater than the erosive forces trying to tear the blocks down, so that even over the course of a dryopithecine life span, if one of them had ever cared to keep track of changing landscapes, he would not have noticed that it was happening, even as it happened before his eyes. In a million years mountains grew ten thousand feet into the clouds, stopping trade winds in their tracks, collecting their moisture, and sending forth streams that, as they increased in age, merged to form the Nile River.

By 17.5 million B.C. Italy and Greece (formerly Mediterranean islands) were being driven like mighty nails into southern Europe, pushing up the Alps, whose highest peaks are today strewn with fossil cockleshells from the Mediterranean seabed. South and east of Greece the Earth's skin wrinkled and cracked under the mounting pressure, erecting Crete, the Cyclades, and the volcanic isle of Thera. What had once been the Mediterranean Ocean was presently being squeezed so severely between two continental plates that by A.D. 15 million it would be entirely eliminated.*

Around 17 million B.C. the dryopithecines broke out of Africa, crossing over on Arabia's first physical contacts with Turkey and Iran. As they spread north into Europe and east into China, the collision continued unabated

---

*Under the stubborn persistence of the African plate, the Nile Delta will ram Cyprus and Turkey, and Libya will burrow under Greece and Italy, raising Rome, Athens, and Çatal Hüyük higher than the Himalayas.

behind them, plowing up mile after mile of basement rock and building mountain ranges on the plains of southern Iran and eastern Turkey, along the fringe of what, during an era of desertification, would come to be called the Fertile Crescent. Rain on the mountains began to feed the Tigris and Euphrates rivers, which, together with the Nile, would forever command the attention of dryopithecine descendants.

The Nile, in particular, underwent astonishing transformations. In the west Morocco rammed Spain amidships. The Earth crumpled there, pushing up a dam that cut off the Atlantic. The Mediterranean was (and is) two miles deep, and its rate of evaporation was (and is) so rapid that if the flow from the Atlantic were blocked, the sea could dry out completely over the course of seven centuries.*

The story of the great Mediterranean dryout is recorded in the geology of Africa's oldest river and can be easily read if one looks closely enough. Under the Nile itself are the remnants of a deep valley to rival the Grand Canyon. River silts began covering it up as soon as the Gibraltar dam broke open and the Atlantic spilled in, but oil geologists drilling through thousands of feet of mud have located the solid bedrock of the Nile Canyon's floor. It lies nearly two miles beneath the city of Cairo.

For a brief time, for perhaps two or three thousand years after the dryout, the Nile poured over a cliff forty times higher than Niagara, but within a half million years, at a rate of inches per day, it had chewed back the bare limestone, slashing the Earth from Cairo to Aswan. The river ran east of Karnak in those days; the slash bypassed Karnak's limestone fields, left them intact for stonecutting beings, who were then only a distant potential in dryopithecine descent. In the river valleys north and east of Karnak, and two miles below it, reeds, papyrus, and tall stands of palms were sprouting. At a depth of two miles the atmosphere piled up 50 percent thicker than at Karnak. The more richly oxygenated air was probably healthier for creatures living in the canyon, but dense air trapped more of the Sun's energy, and beyond the shade of the palms, the oases must have been unbearably hot—up to 140°F on August afternoons. Isolated as it was from the rest of the planet, somewhat like Tibet's mythical Shangri-la, the Nile Canyon

---

*Samples of alternating layers of mud and salt beneath the Mediterranean floor tell us that the sea has dried out several times. More than a million years of fossils lie undiscovered on the drowned Mediterranean basin—especially in the remnants of the Nile Canyon. The lost Nile probably ranks among the most exciting paleontological treasure chests anywhere on Earth, but the field of deep-sea paleontology has not been born yet and must await the further development of robots now being built for the emerging science of deep-sea archaeology.

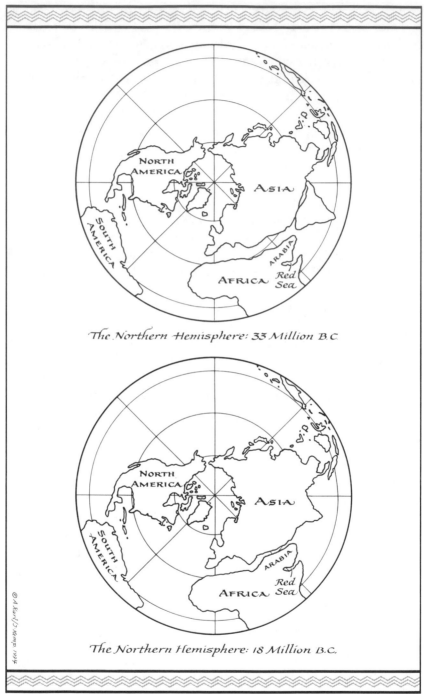

The Northern Hemisphere: 33 Million B.C.

The Northern Hemisphere: 18 Million B.C.

© A. Karl/J. Kemp 1994

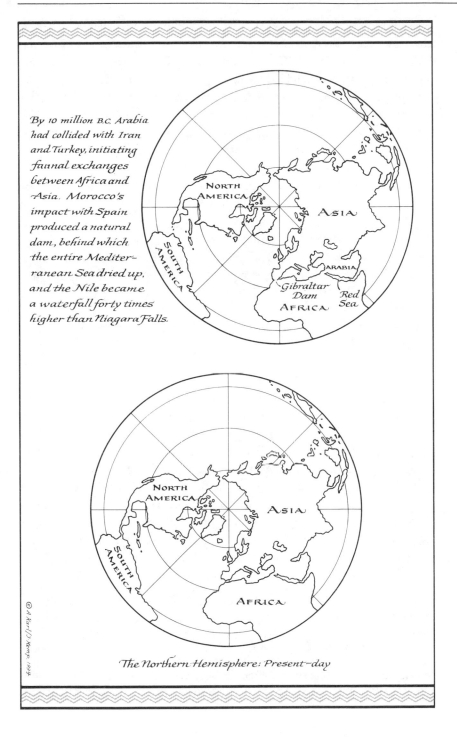

By 10 million B.C. Arabia
had collided with Iran
and Turkey, initiating
faunal exchanges
between Africa and
Asia. Morocco's
impact with Spain
produced a natural
dam, behind which
the entire Mediter-
ranean Sea dried up,
and the Nile became
a waterfall forty times
higher than Niagara Falls.

NORTH
AMERICA

ASIA

SOUTH
AMERICA

ARABIA

Gibraltar
Dam   Red
AFRICA   Sea

© A Karl/J. Kemp, 1994

NORTH
AMERICA

ASIA

SOUTH
AMERICA

AFRICA

The Northern Hemisphere: Present-day

was among the most remarkable habitats the world had ever seen. In the thicker air, ordinarily harmful mutations that reduced the lengths of wings took on new meaning. Even stubby-winged birds, even today's gliding snakes and squirrels would have been graceful fliers on the canyon floor. We do not know if some branch of the dryopithecine clan lived there or if we owe some part of our origins to the canyon, but we do know that the Nile flowed north under Cairo, carrying river silts over Mediterranean salt flats as far away as Crete (the river's banks can still be seen today on the actual bottom of the Mediterranean Sea). On Crete the fossil bones of dog-size hippos, shrews, and elephants provide only random snapshots of a few creatures that happened to venture up from the basin. Wandering into alpine bogs, they drowned and fossilized when the islands of Crete and Thera were forest-crowned mountains whose foothills lay at the northern-most reaches of the Nile oasis. There can be no doubt that the inhabitants of the Nile began to evolve in strange, new directions, but they and their world lay in the path of the Atlantic Ocean's inevitable breach of the Gibraltar dam, and about 6.5 million B.C. there came a deluge of such proportions that the biblical parting of the Red Sea might easily have been dwarfed by one of its eddies, as a raindrop is swallowed by the ocean.

As new mountains and rivers were born, as the Nile Canyon drowned and silted over and as the continents shifted into positions resembling a freehand sketch of a modern map of the world, the dryopithecines under-went equally dramatic transformations. The stage was now set for an instance of migration and diversification that would produce the most extraordinary lineages seen on Earth since the rise of the dinosaurs. Within a million years of the dryopithecine breakout, they had extended their range from Africa to Spain and Portugal in the northwest, to China and Thailand in the Far East. South African, European, and Asian dryopithecines were as separated by sheer distance as by the new mountain ranges and deserts forming between them. Slowly, at about the rate mountains are erected, genetically isolated populations began to drift in different directions.*

*A million years after the dryopithecine breakout from Africa, new creatures began depositing scraps of skull in the fossil-bearing strata of northern Pakistan. Seen from a paleontological perspective, in which adjacent layers of mineralized mud and bone will mark off, at most, an interval of fifty thousand years, a million years is but a fortnight. In a layer of African sediment dating from about 17 million B.C., one finds only dryopithecine bones. In the very next layer, in Asian bone fields only a million years younger, *at least* two new branches have been added to the dryopithecine family tree. They have been named *Ramapi-thecus* and *Sivapithecus* (after the Hindu deities Rama, the preserver, and Siva, the de-stroyer). The sivapithecines very closely resemble the modern Asian genus *Pongo* (the

In 4 million B.C. reptiles still held dominion over the best parts of the Mediterranean world. Slowly, and with crushing force, a global deterioration of the climate had begun to build giant ice sheets and to deflect trade winds in odd directions. Rain patterns changed. Then, as drought swept across east Africa and savannas degenerated into deserts, the cobras, deadliest of the snakes, found their territory reduced to a strip of land only twelve miles wide. For more than two million years they continued to hold sway over a narrow band of palms and tall grasses that twisted with the Nile through a thousand miles of drifting sand.

Then, as they slid and weaved along paths that they had used for tens of thousands of generations, as they prospered on good hunting for mammals gathering around water holes, occupied the limestone crevices near Karnak, and multiplied into living masses of snakes, a newer, more terrifying creature came into their world and instantly challenged their authority.

The toolmakers had no poison glands and could not strike as quickly as a cobra; but they carried sharpened sticks and could loop hemp snares at the ends of long poles, and in time they would master fire and cook cobra meat in their hearths. Charred cobra vertebrae buried in prehistoric campfires suggest that one of man's first and most symbolic acts in the Nile Valley was the killing of the snake.

As a symbol of power the cobra would, in ages to come, be respected above all other creatures of the Nile. Its likeness would be found on the walls of Thebes, on its altars, and on the headdresses and golden death masks of pharaohs.

Across the river from Thebes, in the Valley of the Kings, frescoes in the tomb of Sethos I (Seti) tell a strange story. Snakes on the wall have been painted with legs and scaly feet. They grin menacingly, and the accompanying hieroglyphs indicate an ancient knowledge that snakes were not always without legs. According to the Sethos I inscriptions, "the serpent's forebears possessed feet." It seems reasonable to conclude that the Egyptian priests and scholars must have dissected snakes and that the vestiges of limbs

---

orangutans). *Ramapithecus*, once thought to possess a mixture of ape and human characteristics, seems to have been related only to *Sivapithecus* (to judge from the structure of its skull), making it part of a genus (or group of species) that came in various shapes and sizes and whose only surviving branch is the orangutan. Elsewhere in Africa, Europe, and Asia other branches continued to split and meander. By 4 million B.C. they had begat gorillas, chimpanzees, and the australopithecine apes. The australopithecines, in turn, branched into three major new directions (and probably at least a half dozen others not yet known from fossils), of which we appear to be a lucky side branch, one of only a small handful of still-living twigs on a once-luxuriant dryopithecine tree.

Three views of Tutankhamen's golden death mask. The striped flaps extending along the sides of the pharaoh's face (which were worn only after his ascent to power) were an article of clothing designed specifically to mimic the shape of the cobra's head. Extension of facial flaps is the cobra's way of commanding attention and letting an adversary know that he is ready to strike if provoked. By copying the cobra's threat display in his headgear, the pharaoh was conveying a similar message—a very clear, nonverbal command for respect. When viewed in natural daylight, the god-king's death mask is revealed as the most haunting work of sculpture ever created by the hand of man. The artist used the reflective surface of the gold itself to produce facial lines that change as our point of view changes. Viewed from left to right, the pharaoh's expression changes from serene, to sad, to regal.

did not escape notice. Indeed, the builders of Sethos I's tomb were anatomically correct when they painted hind legs on snakes, precisely where the remnants of hipbones are located.

The Sethos I inscriptions connect the snake's ancestors with an evil curse, which was cast upon them for one offense or another when the world was still very young. The nature of the offense is not described, but the punishment is clear: Their legs were taken away, and henceforth they were obliged to crawl upon their bellies.

This is, of course, very reminiscent of God's punishment of the snake for its deception of man in the Garden of Eden. In Genesis 3:14 the Lord God says unto the serpent: "Because thou hast done this, thou art cursed . . . upon thy belly shalt thou go. . . ."

The Sethos I account of the transformation of snakes predates the Bible by a thousand years.* Not only is it so similar to the Genesis account as to render a common origin extremely likely (especially when considered in the light of such additional Egyptian parallels as the four beasts of the Apocalypse before the biblical Apocalypse), it is surprisingly faithful to the scientific account of snake origins: They evolved from four-legged lizardlike ancestors.

In Genesis 3:15 God proclaims eternal enmity between the descendants of Eve and the descendants of the snake. Then, turning to the woman, he says: "I will greatly multiply the pain of thy childbearing; in sorrow thou shalt bring forth children."

The story of Eve plucking the fruit of knowledge from a tree, of passing along to her descendants an expanded intelligence that would one day beget spears and rockets, and of God punishing her for this by causing human birth to be the most difficult in the animal kingdom is, like the story of the snake's missing feet, oddly akin to the scientific story of origins.

The most conspicuous trend in vertebrate evolution during the past hundred million years has been an increase in brain size (and the largest infant skulls—man's—have indeed produced the most painful births in the animal kingdom). Comparing changes in skull shape, and sweeping our eyes from end to end through Late Cretaceous time into the present, we can easily believe that the whole Earth has been trying to forge large brains.

Once the mammalian diversification began, brains enlarged in surges along lineages leading to kangaroos, cats, horses, whales, and dryopithecine apes. It was a pattern that, sooner or later, seemed bound to create thinking, toolmaking beings.

*When scholars finally succeeded with the first translations of Egypt's mysterious hieroglyphs during the nineteenth century A.D., they were astonished to learn that ancient Egyptians had been speaking to us all along through the Bible. On the walls of tombs and in containers of papyrus scrolls were stories of Isis and the child of immaculate conception, of a Theban trinity (Isis, Osiris, Horus), of baptisms and resurrections. Farther east, Babylonian texts, including the Gilgamesh epic, also contained curious resonances, as when Gilgamesh, like Adam, was robbed of physical immortality and gained knowledge through the deceptions of a snake.

〰〰〰〰

And so, around 1.5 million B.C. the first toolmakers occupied the Nile, the Tigris, the Euphrates, and the land between the three rivers.

It was not a hospitable land, that day when our ancestors arrived. A tribe could wander east from the Nile or west from the Euphrates for weeks and never find enough water to survive. For twenty million years Egypt, Arabia, Israel, and Iraq had lain across a fracture point in the uneasy marriage of continental plates. It was a land of earthquakes and violent climatic upheavals. In millennia past, deserts had been overrun by swamps, and the swamps had subsequently shriveled into deserts. Hundreds of feet below the Arabian wastes, underground streams flowed as vast as seas. Wherever the desert floor dipped below sea level, the waters bubbled forth as springs and whole forests of palms sprouted on landscapes otherwise as lifeless as the sands of Mars.

It was a cruel, desolate land, subject to nature's wildest whims. In just one million years canyons two miles deep had come and gone, and even as the toolmakers moved in, a mountain near the river Jordan was torn apart and tipped on its side. Within the mountain, where a huge fissure had opened and snapped shut—a fissure through which construction workers of the twentieth century A.D. would eventually cut a road—lay the crushed skeleton and stone tools of a man who, in the hour of his death, had witnessed geologic manifestations of which his distant descendants could scarcely dream.

The toolmakers were the inheritors of 4.6 billion years of continental wanderings, upliftings, and degradations. Everything they set out to accomplish would be either aided or hindered by what had happened to the land during those vanished years. Here and there dissolved gold had percolated up through volcanic steam vents and condensed into crystalline veins. In the deserts of Sinai, mountains of copper had been pushed up from somewhere far below. They were slowly being weathered into blue-green sediments and redeposited over a wide area. Ancient seabeds had risen across Egypt, exposing outcrops of marine limestone from which Theban columns and the Sphinx would be cut. The Red Sea and the Bitter Lakes looked much as they do today. They were a crack in the world, where a new ocean had begun to form, nudging Egypt and Israel in opposite directions. The Gulf of Aqaba, the Dead Sea, and the Sea of Galilee were mere pits along the crack's eastern branch. From the south of Israel, north to Turkey, human activity was confined to a narrow strip of land pressed between the Mediterranean Sea and an equally formidable sea of sand. The strip formed the only corridor between Asia (the Tigris-Euphrates) and Africa (the Nile), and from the start its pastures must have been fiercely contested. Beneath the corridor, and beneath the sands to the east, lay seas of oil, which would one day be valued even more than green pastures, even more than gold.

Toward the end of the third millennium B.C. mighty cities were going to rise up along the Nile, in Iraq's land of two rivers, in Syria, and on the Aegean. As these powers swelled and inevitably pressed against one another, the place where they collided would be Israel. From that moment armies, not just continents, would meet and clash. While the stage was being set, even before the actors existed, geography had dictated that human history would be every bit as violent as the geologic history that had preceded it. The shiftings and groanings of the Earth itself had set the stage for a focus of human forces.

4

ALL
ABOUT
EVE

For some reason, we are powerfully drawn to the subject of beginnings. We yearn to know about origins and we readily construct myths when we do not have data. Most of us know that the Great Seal of the United States pictures an eagle holding a ribbon reading e pluribus unum. *Fewer would recognize the motto on the other side (check it out on the back of a dollar bill):* annuit coeptis—*"He smiles on our beginnings."*
—Stephen Jay Gould, Brookhaven lectures, 1984

FOUR DAYS SAILING out of Aswan, following the flow of the Nile north, I have entered the archaeological zone of Thebes. Night has blotted out the cornfields, which spread about ten miles east, to the pink and white cliffs that enclose the Valley of the Kings. Few lights burn in the nearby town. Standing on the foredeck with a cup of hot tea, I can see nothing except the sky. Not since the Ballard expeditions to the East Pacific Rise—more than a thousand miles from the nearest street-lamps—have I seen such a sky. In the cities it is the ground that is brightly lit, and the sky is simply black, with a solitary star burning through here and there. In the middle of this ocean of sand, with only the Nile oasis cutting through it, the land surface, from horizon to horizon, is a dark floor, and the constellations are drowned out by a mass ensemble of light covering every inch of the heavens. Even in the galaxy's dust lanes, there are no patches of sky that are truly dark. Where the ship's mast rises against the backdrop of stars, its silhouette is as sharply defined as the edge of a knife. City people never see this, but it is how Moses and Pharaoh saw the sky. No wonder they thought of it as a solid dome, a mighty firmament.

I would like to stand in the ruins of Karnak tonight and view the obelisks and the giant stone pillars as perhaps they should best be viewed: silhou-etted against the vault of heaven. But that will not be possible. Someone has built a light show into the ruins. Now the buildings sing and speak and blush radiant colors into the night, erasing from view a starscape that has, for thousands of years, enhanced their beauty and which, taken by itself, in absolute silence, without multicolored lights and the distractions of an actor's narration, might actually produce in us a genuine feeling for deep time.

Near this place, forty-five centuries ago, almost a thousand years before Tuthmosis III and the Exodus, the Egyptians had developed a sophisticated society, with surplus resources and manpower that allowed nonfood producers to operate within the community. Already there had emerged astronomical measurement, literature, advanced geometry, and civil engineering projects, including water-lifting machinery for irrigation and hydropower (to turn paddle wheels and shafts, possibly to assist in the grinding of grain). Many modern observers smugly dismiss the know-how of ancient Egyptians by pointing out that they appear never to have discovered such simple machines as the pulley, and thus the standard method of construction in these parts was to haul building materials up earthen ramps. But the critics are not doing their homework, not observing closely enough. Why fashion a tool if perhaps there was no need for it in the first place? Karnak's written records tell of cedar being shipped from as far away as Crete and Syria. Almost all that wood, imported at great expense, was reserved for fine furniture and royal barges. The most readily available building material, found everywhere along the Nile, was limestone. It had the very convenient quality of being easy to carve, when damp, and then turning as hard as concrete when dry.*

At some places along the Nile (including the Great Gate of Karnak), structures were, for one reason or another, abandoned in mid-construction. Within decades, sand dunes had rolled over them, sheltering and preserving the buildings, and the means by which they were built. The mud-brick ramps are still there, at Karnak's gate, teaching us that they were used to raise the city stone by stone. Upon completion, entire structures were buried under ramps, whose bricks were then carted away, presumably for recycling.

It was a brilliant engineering strategy, no less clever than the construction of the Aswan High Dam. That pulleys appear to have been absent does not give us cause to believe that the Egyptians could not have invented them. Pulleys were simply unnecessary without decks of wooden scaffolds. Since there was no surplus wood, they built earthen ramps instead. It was the optimal solution to a common engineering challenge. The Egyptians were, by every architectural standard, an extraordinarily clever and advanced people.

The oldest limestone structure in Egypt—indeed, the oldest-known freestanding structure in the world—is the massive Step Pyramid of King

---

*Powdered limestone is a main component of concrete. It is clear from ancient repairs on relief sculptures that the Egyptians had learned how to make a paste out of lime, similar to concrete, though they appear to have used it only to fill in accidental chip-outs. Since they lacked dump trucks and highways, it was, in those days, just as easy (if not easier) to prefabricate pillar segments and other building blocks and haul them to the construction site than it was to cart loose sand and powdered lime (which weighed as much as limestone) to the site, build expensive wooden molds for concrete walls, and bring water there, too.

Djoser (Zoser), near Cairo, just three hundred miles north of Karnak. It dates from about 3000 B.C. In terms of engineering techniques, the pyramid's ancestry goes back a mere two hundred years earlier, to the cutting tools and methods used to build permanent irrigation canals, which had been growing in complexity for a thousand years. The canal systems required, and very likely forced, the development of mathematics (particularly geometry), so that distance, area, and cubic volume of water transport per hour could be planned in advance of construction. Sometime during the fourth millennium B.C. someone (most likely a nomadic tribesman) discovered copper strewn between the hills of Sinai, on the other side of the Red Sea. The metal could be drawn easily from the rocks with the application of a little heat, and the copper was used to create newer, more powerful limestone-cutting tools, which facilitated the expansion and elaboration of canal systems. Alloyed with tin, copper became sufficiently strong to hold a sword edge, and the Bronze Age could not have lagged very far behind the Age of Canal Building.

The earliest evidence of social organization in Egypt goes back almost to 3500 B.C., to fragmentary hieroglyphs that describe the division of Thebes and its surroundings into rectangular "water provinces," ruled over by men called *adj mer* (diggers of canals). It was these first engineers, many of them farmers, who perfected plows and irrigation systems and set in motion, for better or for worse, the flourishing of civilization.

A community will survive as long as there is food to sustain it. From the moment that canalization produced food supplies exceeding those necessary for day-to-day survival, men's hands slowly became free. The community could begin to support members who left the fields, devoting their time to other concerns. The first of these were probably the men who maintained and designed the irrigation systems. In no time at all, it seems, they were organizing others to do the job. They became administrators.

By the time of Egypt's first written records, the *adj mer* had determined, from the rising and setting of certain stars, precisely when the annual rise and fall of the Nile would occur. They monopolized this knowledge, keeping to themselves the ability to predict the best times to plant and harvest. They began to live as kings, and as reed baskets daubed with clay gave way to permanent fire-baked containers, and as the world's first granaries overflowed with pots, a need to identify the ownership and contents of the pots developed. The first hieroglyphs (picture words) come from the time of the *adj mer*. They are signatures on pots and tablets bearing lists of pots and totals of the objects contained in them—the etchings of prehistoric accountants and tax collectors.

Ever since the *adj mer*, community structure has been reversing itself from a predominance of farmers to a predominance of workers who have moved away from the fields. Increasing food production continues to swell the Earth's population (though the planet cannot sustain such growth much longer), and yet few of us ever meet a farmer. Most of us are freed from the task of food production, free to explore, commute daily to steady jobs, or even to do nothing at all. Here, near Karnak, and east along the Tigris-Euphrates, new technology produced prosperity in the Crescent of Fire. Surplus food became goods for barter with other communities, a circumstance that created nonagricultural niches for merchant-middlemen, shippers and shipbuilders, bookkeepers, and eventually lawyers. Some communities inevitably looked with envy upon those more prosperous than themselves. Members of each community—more and more of them—were free to develop metal weapons and to contemplate invasion against those whom they envied or defense against those who envied them. Along the narrow confines of the riverworlds, arms races became inevitable. If a community, upon achieving agricultural prosperity, did not develop some form of military organization very quickly, someone upriver, or downriver, would likely move in and take that prosperity away. One of the sad hallmarks of man's existence, as described in the beginning of the Bible and seen in strata dating back to prehistory, is that we were killers. In time, as prosperous human communities spread over the Earth and became nation-states, arrows and spears evolved into rockets. In America they have mounted artificial eyes atop a missile and hurled it past Neptune. On the Tigris they have threatened to mount a pot of poison gas atop a missile and hurl it toward the Jordan and the Nile. We stand today on the receiving end of a cascade of consequences that gained momentum six thousand years ago, near this very spot, with the discovery that a digging stick could redirect the flow of water.

But humanity's cascade point does not begin and end with canal builders in the Nile Valley. It must be pushed back deeper still. Beneath Karnak, beneath the remnants of the first crude mud ridges and water basins, excavators have found flint hand axes a half million years old.

One of these was delivered to me via the Royal Ontario Museum Expedition. Its edges have softened with age, but when it was new, it was fearfully sharp. Its owner probably protected his hands with a thin sheet of animal hide, or perhaps even an ancient leather glove. The "ax" is roughly the shape of a spearpoint and about as long as a man's hand. There are indentations along one side that seem to have been carved specifically for grasping by the thumb and fingers of the right hand.

We will probably never know exactly what this tool was used for. If one

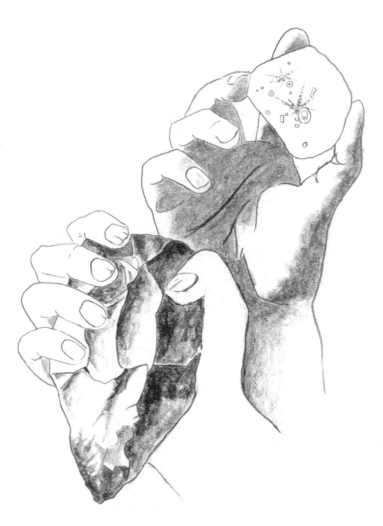

A Homo erectus *hand ax (bottom), found in a stratum below the ruins of Karnak, dates back nearly a half million years. Humans have been drawing rocks from the Earth and reshaping their world ever since. The Genesis Stone (top), a meteorite that fell in Allende, Mexico, has reshaped man's view of the home planet. The stone contains the decay products of aluminum 26 and other short-lived elements, ejected from an exploding star and stirred into the matrix of the solar system as it formed 4.6 billion years ago. Today, all over the world, nuclear reactors are coaxing from uranium atoms faint echoes of the supernovae that created them.*

wanted to chop firewood (and the remnants of campfires tell us that fire was indeed under our ancestors' control by 500,000 B.C.), climbing atop a ten-foot-high boulder and dropping a heavy rock onto the center of a log propped between two other rocks would have done the job quite nicely, with a lot less effort than chipping away with a sharpened flake of flint.

With its pointed end projecting down from the user's fist, the hand ax was a good tool for defleshing prey or digging up edible roots. It would also have been useful for preparing a hide and was certainly a formidable weapon. My personal guess is that it was a multipurpose tool, and therefore priceless, always in its owner's possession.

The flint cutter belonged to that branch of the human tree grouped under the heading *Homo erectus,* which appears to be the only tribe whose bones were entering the fossil record during the time of the Egyptian hand ax. All the older branches, including *Homo habilis* and at least three twigs on the australopithecine limb, had disappeared with the flourishing of the toolmakers, who spread during a geologic fortnight out of Africa, over Asia, and across the Exmouth Trench (probably in the world's first boats) into Australia. Cooked australopithecine bones in the remains of several *Homo erectus* campfires suggest that in some parts of Africa, at least, cannibalism was the flavor of choice. We do not know exactly what role killing played in the ascent of our ancestors, but wherever the flint cutters went, other, older manlike tribes either had recently died off and been replaced by them or were systematically and forcefully displaced.

Homo erectus's body was more robust than a modern man's. Standing five feet tall and weighing about 150 pounds, he was a stocky figure with huge shoulders. Bone circled the eyes, and the front of the skull was conspicuously narrower than yours or mine. He walked erect and probably had a smooth skin that was dark brown and produced no great amount of body hair. To judge from the shapes of all known *Homo erectus* hand axes, he was, like most of his descendants, right-handed. He must also have mastered a language. His jaw, throat, and tongue were physically capable of articulating sounds, and it seems unlikely that the control of fire and the ax maker's skill could have been passed down from generation to generation without at least a rudimentary vocabulary.

The man or woman who prepared to strike a rock that day about a half million years ago, in a muddy stratum beneath Karnak, had all the essential capabilities that the *adj mer* and the Aswan engineers would possess, including an innate sense of proportion, design, and beauty. The flint cutter had selected a large core stone from which, using a hammerstone and rotating the core as she struck, twenty or more knifelike flakes could be made to leap. When the cylindrical core was used up and could no longer produce promising flakes, it was discarded. Several large flakes were also discarded on the pre-Karnak ground surface without being worked on further, and these tell a story every bit as important as the finished hand axes now prominently displayed in museum collections. They tell me that a flint cutter once held a core stone and freshly struck flakes in hand, then peered a short distance into the future and knew from experience that no matter how much additional effort was applied, neither the core nor the flakes would yield the shapes desired. The stones imply memory of past failures and how they differed from successes; they imply a foreknowledge of what certain slivers of flint, cut in certain ways, were bound to become. Like you and me, the owner of the hand ax was capable of anticipating the future in reasonable detail, and had begun living outside the present. There is no evidence of communal burial among the tribes of *Homo erectus,* no evidence that the toolmakers had discovered death or begun to break time up into measured segments, but already, by 500,000 B.C., the seeds of those twin catastrophes were present, and had begun to take root.

〰〰〰〰〰

The Bible tells us that from dust we came and to dust we shall return. Dust to dust. The imagery is haunting and poetic; and to those of us who spend our lives poking about in the cellars of time and occasionally pulling skeletons out of the Earth, the inevitability of dissolution is both immediately apparent and mythologically consistent. But it is impossible to believe that our *Homo erectus* forebears, or Adam and Eve, or even our mitochondria arose from the dust of the Earth. Unless you happen to believe in miracles. Or unless you happen to be an evolutionary biologist.

Charles Darwin saw at the cradle of life "a warm little pond." He wrote to a friend in 1871:

> It is often said that all the conditions for the first production of a living organism are now present, which could ever have been present. But if (and oh what a big if) we could conceive in some warm little pond, with all sorts of ammonia and phosphoric salts, light, heat, electricity, etc., present, that a protein compound was chemically formed ready to undergo still more complex changes, at the present day such matter would be instantly devoured, or absorbed, which would not have been the case before living creatures were formed.

Abundant heat and ever-expanding accumulations of mineral deposits near deep-ocean volcanic vents are just one source of Darwin's "warm little ponds." When I met him in Sri Lanka, the biochemist and evolutionary biologist Cyril Ponnamperuma pointed out that life must have evolved in such places through a process of natural selection that began with the exhalations and regurgitations of the planet itself, with volcanically heated water and the dust of the Earth. He was saying, in effect, that if we track our ancestry back far enough, we had better be prepared to think a little more kindly of dust and clay. Sometimes the process of scientific discovery is the disquieting feeling of climbing a mountain, only to find rabbis and priests, Buddhists and Islamic scholars sitting at the top and saying, "See! We knew all along."

A century after Charles Darwin wrote his letter, those of us who puzzled most over life's origins began finding remnants of what had once been hydrothermal streams inside certain crumbly, claylike meteorites. Called carbonaceous chondrites, these splinters of once-larger bodies were made largely of dust and carbonaceous soot. Residues of aluminum 26 and other radioactive elements allowed us to track the warm, wet interiors of asteroids back to 4.6 billion B.C. Alongside meteoritic salt veins and microgeodes full of pyrrhotite crystals we found ethanol, purines, globs of protein, and *porphyrin molecules*—actual precursors of chlorophyll and hemoglobin. There, before our eyes, went carbon, hydrogen, nitrogen, and oxygen, teaching us that the origin of life's building blocks on Earth was not an event so improbable as to be a miracle that happened only once in the entire universe.

A particularly dramatic example of the ease with which the carbon atom binds to produce compounds of higher organization was demonstrated during the Second World War. For years the Germans had been synthesizing hydrocarbon fuels on a commercial scale. With burning rubbish used as a heat source, mixtures of simple gases like methane were passed at high temperature over bentonite (a type of clay) and metallic ore catalysts (which accelerated the formation of chemical bonds) in closed pressure-cooker-like containers. The process did, however, have its shortcomings, primarily because of the production of unwanted by-products; these included arginine (and other amino acids, sometimes clumping into globs of protein), purines, fatty acids, and the occasional porphyrin molecule. Had it occurred to the German engineers that they were possibly simulating reactions between lava, dust, and a "primordial broth," they might have gone to Sweden (as two American scientists did ten years later, with a variation on the same procedure) to claim their Nobel Prize in biochemistry.

Biology is the production of order from chaos; and it seems that the carbon atom, impelled by just a little energy, is a most splendid organizer.

If this view of our genesis is correct, then the miracle we call life was drawn from a disarmingly simple bag of tricks: We are the result of the most likely chemical reactions undergone by some of the most abundant elements in this part of the universe.

So here we stand, you and I: symphonies of chemical activity written on DNA and performed by protein. As such, every cell in our bodies is a living, breathing museum, for by all accounts, the composer in our lives was the Earth itself, in whose waters carbon and volcanic dust came together to produce something new. Pulsing with life's surge, the very fluid in which our mitochondria, flagella, and other cell organelles thrive is almost an exact replica of seawater. The concentrations of sodium and zinc, cesium and cobalt in our tissues are no different from those found on the high seas. When some of our remotest forebears liberated themselves from the oceans, they apparently took the oceans with them onto the plains and into the mountains, stubbornly adhering to the chemistry of their origins, as if to trap the Pacific in our veins, like a living fossil, until the end of time. And in each of those miniature Pacifics that lives on inside us, cell nuclei and mitochondria carry the seeds of the evolutionary process, as did the most competitive protocells (and later the bacteria) more than three billion years ago. Every form of animal life on Earth—cat and flea, elephant and intestinal amoeboid—shares a common ancestry, somewhere very low down the evolutionary tree, with every other creature. We know this because the cells of mice and men, cockroaches and lobsters (and what is a lobster but a giant seagoing cockroach?) are all filled with and powered by the same organelles: mitochondria.

And how did this begin? Put simply, there is another side to the story. According to Lynn Margulis, a biologist specializing in primordial organisms at Yale University, our mitochondria (which provide the oxidative energy that burns the nutrients we take in, drives every organ in our bodies, and sends us out to doughnut shops and supermarkets) might not be entirely ours, if we trace their evolutionary history back far enough. On close inspection, they look like and behave like bacteria, hinting that bacterial cells might simply have entered our remotest forebears and stayed aboard.*

*If you think Margulis' ideas about mitochondrial origins are rather farfetched, you're in for a surprise because such get-togethers are probably occurring all the time. In 1980 I had an opportunity to observe firsthand how linkages between organisms unrelated to one another can begin, when the waters around Wellington, New Zealand, became infested with Portuguese men-of-war. I was prepared to avoid these pain-dealing creatures, but no one in the South Pacific had ever been told to beware of the common sea slug. Everyone knew

Just look at the skin and veins of your right hand, and think for a few seconds about the mitochondria down there: organelles that might originally have been nonself yet now are the reason for every breath you take. Without oxygen, the mitochondria would die within minutes, ceasing to generate power for your cells. While live the mitochondria, you too shall live. When fall the mitochondria, you too shall fall.*

Mitochondria have their own stores of DNA, quite distinct from the genetic blueprints contained in our cell nuclei. They are indeed like bacteria living inside us, replicating independently of our cells whenever it suits them, living their separate lives yet sustaining us and under our control in some strange way. The point of this odd tale of symbiosis is that the very nature of mitochondrial DNA—living with us as we evolve yet in its own way separate from us—makes it the perfect place to look for historical records, to seek out files on where we came from and what we once used to be. The mitochondrial library allows us to reconstruct the branching lineages of the hand ax makers at Karnak, leading us ultimately to a woman called Mitochondrial Eve, the apparent ancestor of us all, and to make very good guesses about where she came from and what she looked like.

∿∿∿∿∿

The Book of Genesis is a story about origins, and the riverworlds have, since their birth, been a hive of origins. History has always surrounded them, and because the rivers are so firmly embedded in the ebb and flow of civilizations (and even species), they are the ultimate archaeological and paleontological window—a window from which we can, if we pay very close attention, step back and take a grandstand view of nature, as nature, even at its strangest and most thrilling, has hitherto revealed itself to man.

---

that sea slugs could not sting—that is, until they began feeding upon the men-of-war. One afternoon I came upon and was promptly stung by one of nature's surprises. The nematocysts, or stinging cells of the "jellyfish," were able to survive in the bodies of the sea slugs, which coincidentally possessed the right equipment for transporting the stingers to their skin surfaces, thus creating, in only a few days, something that had never existed on Earth before: an organism expressing the characteristics of both the sea slug and the man-of-war. If Lynn Margulis is correct, we, too, were colonized by strangers who, three or four billion years hence, just happen to be working in our best interest.

*When toolmaking humans discovered that gold could be extracted from rocks, one of the chemicals they used, a simple carbon compound bound with potassium, proved to be a very fast-acting poison. No longer popular with gold miners, it was adapted by twentieth-century civilization for use in gas chambers and as suicide pills in certain high-risk occupations. Known as potassium cyanide (KCN), it acts by attacking the mitochondria directly and shutting them down.

As I write, new tools are continually broadening our vision, allowing us to accomplish feats that only a few years ago could not be discussed openly in polite scientific circles because they were the stuff of science fiction. Today we can routinely copy and read mitochondrial genes or even the entire human genome (and soon, I am sure, we will be reading the genetic codes of creatures long extinct, among them the dinosaurs). As when Neil Armstrong and Buzz Aldrin made those first hesitant steps on the moon, the reality of scientific achievement is once again catching up with the fiction. And—oh—the things we have begun to see.

A comparison of the mitochondrial DNA of every race on Earth has focused our attention on the tribe of a single woman who lived somewhere near the Nile River about the time a flint cutter fashioned my Karnak hand ax. At least one camp of anthropologists has condemned the theory of "Eve" as being simultaneously "against religion!" and "too close to the Biblical account to be true and therefore the product of a [religious] conspiracy masquerading as science." If such contradictions appear jolting, be warned that they are but the first in a series of outcries that are bound to increase to hysterical pitch (as even newer tools become available), and that may never wholly cease. Rarely has something so small as a mitochondrion fomented so much "loud discussion."*

It has been said that paleontologists in general, and paleoanthropologists in particular, seem to make up for a lack of fossils with an excess of fury. At least one cynic notes that in human paleontology the consensus often depends on who shouts loudest—with plenty of accusations of incompetence and fraud flying about. This is not to say that humans studying humans are necessarily crazier than the rest of human society, although I must admit that I know of no other field of endeavor that can boast papers

---

*I have seen one discussion of fossil evidence degenerate from name-calling to food throwing and the total destruction of two laboratories. This very book is being published a bit late because of a fight with the Greek Ministry of Culture (and ultimately with the U.S. State Department) initiated by a colleague overseas who objected to my redating of volcanic ash from the island of Thera (he was joined by a prominent Egyptologist who threatened suicide over this very same issue). These extreme situations should not be taken as evidence that scientists are terminally crazy. Such behavior is, as in most segments of the human family, limited to a relatively small portion of the population. Yet for some reason outbursts of passion and fury do indeed seem to be most heavily concentrated amid those of us who stroll through the cellars of deep time. Perhaps the human brain was never built to think in millennia and eons, and perhaps a fair percentage of us who make this particular leap of thought are doomed to be driven a bit mad by our gains. I don't know, and neither it seems do my fellow paleontologists, anthropologists, and archaeologists. All I can say for certain is that if you are thinking of studying science and you want to avoid the certifiables, take up botany or particle physics. I've never met a botanist or particle physicist I did not like.

written from prison, where the author is serving time for feeding poisoned candy to the judge.

Suffice to say that when you go out into the world and talk about Mitochondrial Eve and the dawn of human intelligence, you are entering a realm in which passions run very high, largely because the subject at hand is nothing less than the descent of man.

The mitochondrial picture of human origins began to take shape in the year 1987. Berkeley biochemist Alan Wilson, assisted by University of Hawaii geneticist Rebecca Cann and Penn State geneticist Mark Stoneking, had followed a trail of mitochondrial DNA and found evidence that all members of all races share mitochondrial genes so astonishingly identical that, geologically speaking, there has been almost no time at all for mutation and differentiation to take place.

Mitochondrial DNA has one feature that makes it a particularly useful tool for the interpretation of ancestral histories. Though the mitochondria infiltrate both sperm and egg, the business end of a sperm is all nuclear DNA (the mitochondria are located where they are needed most, at the base of the sperm's tail, to power the all-important propulsion system), so it is the mother's mitochondria, and hers alone, that ride down to the first diploid cell of the newborn, dispersing their progeny through cleavage into succeeding generations of *Homo sapiens*. We can therefore follow trails of linear descent, unencumbered by the fractured, reconstituted, and criss-crossing pathways followed by a blending of both fraternal and maternal genes to produce, for example, A, B, AB, or O blood types. While the nuclear genes governing such characteristics as blood type may vary widely among a mother's children and grandchildren, the only differences between the mitochondrial DNA of a child and that of its mother—or a great-grandmother fifty generations removed—are the result of random mutations.*

Studies of diverging primate pathways, wherein genetic and fossil dates are in agreement, suggest that the gorilla split from a dryopithecine line ancestral to chimps and australopithecines between ten and seven million years ago, and that over the course of any million-year interval, the mito-chondrion genome will mutate at a rate of 2 to 4 percent.

---

*Mutations may occur through accidents of division or when a cosmic ray particle chips off a piece of DNA. In the case of the mitochondrion, mutations serious enough to cause grievous harm quickly result in the death of the organelle and are not likely to live on in the mother's egg or in any other nucleated cell. There are, however, several rare, nonlethal forms of epilepsy and blindness known to be inherited only maternally, and these are now understood to be hereditary mitochondrial diseases. Rates of aging, too, may be tied to the mitochondrial throughput of the mother.

The first figures from the Wilson group (a variation of less than 1 percent between all people whose mitochondria were tested) suggested that when the Earth was inhabited by perhaps fifty thousand humans—between 290,000 and 140,000 B.C.—the mitochondrial mother of us all came out of Africa, crossing with her people over the bridge between riverworlds (Israel) and eventually winning for her tribe an entire planet. By 1990 Wilson team member Linda Vigilant had refined the mitochondrial clock. Comparing the mitochondrial DNA of Africans* via a new technique called polymerase chain reaction (which allows researchers to snip off a chosen segment of DNA and make millions of copies for analysis, instead of painstakingly comparing random snippets cut one at a time by enzymes), she was able to scrutinize a section of mitochondrial DNA that seems to mutate faster than the rest: about 8.4 percent per million years. The advantage of a faster mutation rate is that it should, in theory, allow us to discern smaller increments of time.

Africa's fossil record suggests that (following the dryopithecine/ancestral chimp split of ten to seven million years ago) chimps split from an australopithecine line ancestral to humans about 5 million B.C. Vigilant's rapidly mutating length of DNA (called the control region) seems to confirm the date arrived at by paleontologists, yielding a divergence of 42 percent (between chimps and humans)—which, at 8.4 percent per million years, adds up to about five million years.

When Vigilant compared control regions among modern Africans, she found a divergence of only 2 percent. According to the timetable written on our mitochondrial genes, African mitochondria began diverging from a single ancestral cell less than 250,000 years ago.

Clues inscribed in stone and bone lend support to this interpretation of the mitochondrial clock. Near the top of *Homo erectus*'s stratigraphic range (the youngest and hence the highest level of rocks in which this fossil "species" occurs), a nearly complete skull from Dali, in China, was unearthed in 1981. The Dali skull dates from approximately 200,000 B.C. (give or take fifty thousand years) and has been classified as "an archaic type

*One flaw in the initial 1987 study involved the sampling of mitochondrial DNA from African Americans and its subsequent use as a basis for comparison of genetic excursions among Africans, Europeans, and Asians. Owing in part to the conditions of American slavery up to the nineteenth century, in which African women were frequently forced to bear children by masters of European origin, such comparisons might have lost some small measure of their meaning. It should be noted, however, that it was the female victims of these attacks whose mitochondria were passed on and that since the abolition of slavery, social conditions in America have been such that mitochondrial contributions by European American women were, until the latter part of the twentieth century, exceedingly rare.

of *Homo sapiens.*" It could represent a morphologic excursion that originated at the fringe of *Homo erectus*'s range. Later, driven home to Africa
by advancing glaciers, this newly evolved Asian stock might then have given
rise to our ancestral African stock. Alternatively, Dali's people could have
descended directly from our mitochondrial mother's tribe, having already
migrated and expanded out of Africa by about 200,000 B.C. (or the Dali
skull could simply represent an evolutionary "side branch" that vanished
utterly and without heir). No one knows for sure.

What we do know is that blade tools considerably more sophisticated
than *Homo erectus* hand axes, including delicately carved spearpoints, made
their first appearance in Africa about 100,000 years ago. The oldest, thoroughly modern human skulls are also known from African sediments, dated
between 130,000 and 100,000 B.C. (followed by modern remains in Israeli
strata dated to about 100,000 B.C.). We know that the Neanderthal tribes,
which retained much of their ancestral *Homo erectus* robusticity, spread
across Europe, the Mideast, and Asia before or during the last interglacial
period (a warm period extending from about 130,000 to 80,000 years ago).
Then, as new tools and less robustly boned humans spread out of Africa
about 45,000 B.C., fewer and fewer Neanderthals left traces of themselves
and their tools in the fossil record. The last known Neanderthal specimens
date from about 38,000 B.C. in Australia and 34,000 B.C. in Israel.

In Israel's Skhul Cave, a skull dating from about 35,000 B.C. has a
Neanderthal face tucked into a modern cranial vault. If this curious "hybrid" left (female) descendants, then some modern humans, especially
those of Mideastern ancestry, should possess mitochondrial genes expressing a greater variation than that seen between modern tribes or "races." At
this writing no such genes have been found, suggesting that (A) if the Skhul
"hybrid" left numerous female descendants, their mitochondria are genetically indistinct from modern races (a very unlikely event) or (B) the more
divergent Neanderthal mitochondria vanished without heir (or have just not
been found yet), and if you could trace the branches of our mitochondrial
mother and the Neanderthals back through time, they would join at a *Homo
erectus* ancestor more than 250,000 years ago. According to the latter
interpretation (which may or may not hold up under the weight of future
evidence), the Neanderthals did not branch off from a portion of our family
tree lying somewhere between our mitochondrial mother and us. They are
simply members of a more ancient side branch from the *Homo erectus*
line—cousins a quarter million years removed, perhaps, but not great-
grandparents.

If we assume that the Wilson team's mitochondrial clock is telling the right time, the picture is quite different for today's races: All of us, from Alaskan Eskimos to Australian Aborigines, can trace our mitochondrial DNA to one woman in the recent geologic past (so similar are the mito-chondria of modern humans). At first hearing this may sound impossible, and the name given by Wilson to our mitochondrial mother—Eve—once triggered an avalanche of media coverage that made its way into books and became self-perpetuating dogma and, for the most part, portrayed her almost as accurately as that other paleontological oddity, the Irish elk (which was neither exclusively Irish nor an elk).

That all the tribes of man might be traced back to a single woman should not be taken to mean that according to the scientific story of Genesis, there once existed only one ancestral female on all the Earth, with no other humans living before her or beside her.* It does mean that sometime between 250,000 and 140,000 B.C., when the Earth was inhabited by perhaps fifty thousand human descendants of *Homo erectus,* one woman (or a few women from an inbreeding tribe with the same mitochondrial type) among those fifty thousand became the catalyst from which began a cascade of branching lineages, including Israeli and Palestinian, Roman Catholic and Northern Irish Protestant, white supremacist and native South African.

"This principle rests upon a well-established mathematics," says Harvard paleontologist Stephen Jay Gould. "Its conclusions are firm, though sur-prising to those (most of us, alas) who do not understand the nature and power of random processes. For example, in a purely random system even for large populations begun with fifteen thousand unrelated females, we can calculate a fifty percent probability that, eighteen thousand generations later, all members of the population would be descendants of but one female among those fifteen thousand."

Since mitochondrial genes are passed down to us only through our mothers, in a manner analogous to the way family names are traditionally handed down to us by our fathers, the Wilson team findings seem a perfectly sensible (in fact, inevitable) outcome of the laws of probability.

For example, two American presidents spaced nearly a half century apart (Franklin Delano Roosevelt and George Bush) can trace their ancestry directly to John Tilley, who sailed to New England aboard the *Mayflower.*

---

*It should be noted that not even the biblical story of Genesis makes this claim. The tale of Adam and Eve seems to be derived largely from an oral history of one tribe's origins. In the Bible there are unmistakable references to the existence of other people besides Adam, Eve, and their two sons, as in Genesis 4:15–16, when God puts a mark upon Cain's head, to protect him from the other tribes. Cain then goes out into the world and finds a wife, who bears him a son.

Neither of these presidents nor hundreds of other descendants of Tilley carry his surname. This does not negate their ancestry; it merely means that there were many female lines involved, including Mrs. Tilley—who, while failing to pass down her own family name, did manage to pass down her infinitely more important mitochondria, with the result that President Roosevelt's control region (though sided with the Democrats) was in virtually every way identical to President Bush's.

A study of genealogies on Pitcairn Island, which was settled in A.D. 1790 by 9 mutineers from the HMS *Bounty* and their 19 Tahitian wives, bears similar results and offers a deeper understanding of the mathematics involved. By 1856 their numbers had swelled to 193, but already the probabilities had begun to catch up with the children of *Bounty*. Several family lineages had passed without male heirs, and their names simply went extinct with the marriages of their daughters. Nearly half the original names had disappeared in barely four generations. Notably all the surviving names were British. Though every Pitcairner carried the genes of Tahitian great-grandmothers, only the names of their great-grandfathers had passed down to each generation of newborn (which means that surnames alone, taken at face value, would appear to bar an Eurasian ancestry, and mitochondrial data should be viewed with this in mind). By A.D. 1960, 1,500 Pitcairners sharing three last names were living on two islands. Had this population continued as an isolated, inbreeding tribe for another century or two (this process has been interrupted beyond repair by the arrival of rapid, global transport), all the descendants of *Bounty* would eventually have shared the same last name (most probably Christian), and a visitor to Pitcairn or Norfolk islands might conclude (wrongly) that all the inhabitants had descended from a single eighteenth-century "Adam"—Fletcher Christian.

We must be similarly cautious about jumping to the conclusion that "Eve" was the only woman on Earth around 200,000 B.C. She represents merely a very young twig on a very old bush, with DNA common to all her contemporaries (just as Fletcher Christian possessed DNA and physical features common to all humans). She had a branching lineage of *Homo erectus, Homo habilis,* and the australopithecine man apes behind her, and four hundred million years of vertebrate evolution behind them.

One of the most important pieces in the mitochondrial puzzle is the variation seen in the control regions of native Africans. African diversity is greater than that seen between the descendants of Mrs. John Tilley and the Tahitian wife of a *Bounty* mutineer—which is to say, greater than that seen between European and Asian lineages. If Africans are more different from each other than Europeans are from Asians (or than either stock is from the

African lineage), the implication is that Africans have been around longer than either Europeans or Asians, their mitochondrial genes having had more time to accumulate random mutations and generate diversity.*

If our mitochondrial tree began diverging out of Africa less than 250,000 years ago, Eve's tribe, from which all the present races diverged, had already shed the robust bone structure of her *Homo erectus* ancestors.† Her forehead was higher, her cranium was more thinly boned, and her people were (relative to *Homo erectus*) as lithe as any humans living today. If, as some scientists (including University of Michigan anthropologist Milford Wolpoff) have argued, the mitochondrial clock is keeping the wrong time and Eve's tribe flourished at least a half million years earlier, then our most recent common ancestor had a thick ledge of bone over her eyes and looked like *Homo erectus*.

If we accept the former interpretation of the mitochondrial clock, then a tribe that was already very human-appearing spread over the Earth and developed the regional diversity whose most outward expression is man's present range of skin color. If we accept the latter, then an extra, highly improbable step must be inserted: All *Homo erectus* tribes already existing in Africa, Europe, and Australasia between 1 million B.C. and 500,000 B.C. independently and by extraordinary coincidence underwent the same reduction of *Homo erectus* robusticity, identical in almost every aspect of eye socket shape, forehead height, and jaw structure. Then, on the heels of this remarkable convergence—in which all *Homo erectus* branches (except descendant Neanderthals) converted independently into modern humans—Asian people diverged into a race that was characteristically Asian, the Africans became African, the Europeans European.

While it is true that nature often molds similar shapes from different ancestral stocks, this phenomenon tends to be the result of function dictating form. Thus a creature that must swim rapidly through the water will not

---

*One can only smile indulgently when considering how this news will be received by America's homegrown sheet heads and other "my race is better than yours" types, who frequently use distortions of the Bible as a means of claiming racial superiority. Let them think upon this: If our mitochondrial forebears were African, and if (according to the Bible) God created man in his own image . . .

†A decrease in robusticity of bone structure brought with it a shortening of the human jaw, which produced new conditions never faced by *Homo erectus* or his robust Neanderthal descendants. Whatever advantages a lithe skeleton produced, there are indications of imperfect adaptations to our thin-boned existence. Our mouths no longer have room for the wisdom teeth, which only rarely grow in successfully and even then with great difficulty. More important, our skulls and our legs are easily broken.

be shaped like a television set (lest it expend too much energy trying to move and thus fail to survive). It will be torpedo-shaped, possessing fins in just the right places, with the result that the skeletons of dolphins, tuna fish, and the extinct ichthyosaurs all look very similar. During the time of Karnak's flint cutters, there was no known physical challenge requiring a common solution (convergent pathways) in *Homo erectus*. Moreover, nature's influence on the human form was vastly reduced from the "moment" *Homo erectus* began to make tools and control fire. If a redesign of the human nose was required (say, narrower nasal passages to warm inhaled air more efficiently during an era of glacial advance), an animal-skin scarf would have met the new demand in only a few minutes and been genetically less complex than remodeling the noses, in parallel, of at least three diverging and geographically separated tribes of *Homo erectus*. Far more likely that a newly shaped nose was inherited from a common, more recently emerged tribe of thinly boned *Homo sapiens* that branched from the *Homo erectus* tree, established a large central population in Africa, and then, as it swarmed over the Earth, continued to diverge into the present races.

If the Wilson team's interpretation of the mitochondrial clock is correct, then the *Homo erectus* flint cutters who were living at Karnak, Peking, and Java between 500,000 and 200,000 B.C. died off without contributing their more archaic, more divergent mitochondrial genes (and by implication any other traits) once Eve's people became established and began to spread. After a parting of branches at some robust/thin-boned junction, genetic communication between the tribes of Eve and the surrounding tribes of *Homo erectus* must have ceased, and in time all *Homo erectus* tribes, along with Europe's Neanderthals and other archaic, thickly boned peoples, seem to have been swept aside by the more gracile Africans. Wherever the daughters and granddaughters of Eve went, only their mitochondria survived.

And how can it be that my Karnak flint cutters, after more than a quarter million years of existence, were simply swept aside? What was really happening in the world when the mother of us all emerged? What can science really teach us about where she came from, where she went, and how she became the seed for all the different races we know today? In the first books of Genesis (whose accounts of geology and genealogy were borrowed directly from ancient Babylonian science) the tribes of Eve and Noah spread out of the Crescent of Fire to all points of the compass and, in their isolation across vast distances, became the Arab, Israelite, Ethiopian, and Far Eastern races. This was the best guess that Babylonian logic could offer, and for one reason or another, it comes extraordinarily close to the modern story of Mitochondrial Eve.

The modern story is, of course, a bit more complicated than the Babylonian account. It is painstakingly assembled (amid arguments and counter-arguments) from pieces of a scientific jigsaw puzzle in which the emerging picture of man's evolutionary history upon this planet begins (as one observer has put it) "to look like chaos with feedback." On this score, too, the biblical and scientific accounts agree, but as science allows us to peer more deeply than ever before into our chaotic origins, the details sometimes become difficult to sort out. Yet if we look closely enough, there is a simplicity to (and beauty behind) the whole scheme—so much so that it seems an inevitable outcome that great masses of carbon should now be standing on two feet and asking, "Where did I come from?"

Australian National University anthropologist Alan Thorne argues against the idea that modern humans originated in the startlingly recent geological past and then spread from Africa and the riverworlds to invade the rest of the Earth. He bases his argument on the theory that modern Aborigines in Australia can be traced back in time, through a chain of closely related ancestors, to Java Man (a *Homo erectus* who lived more than 750,000 years ago). The oldest Australian skulls date from about 40,000 B.C. and possess forward-facing cheekbones (a feature generally associated with modern Asians), thick cranial bones, and low, sloping foreheads characteristic of Neanderthals. The full, "classic" Neanderthal pattern is not present, however, and skulls that succeed these first Australians, beginning about 38,000 to 35,000 B.C., possess relatively thin cranial bones and high unsloping foreheads and, while retaining forward-facing cheekbones (as do most present-day Aborigines and Asians), are fully modern.

Thorne suggests that over the course of more than a half million years, a Southeast Asian *Homo erectus* tribe gave rise to the "Australian Neanderthals," who in turn evolved directly into more modern-appearing Australian Aborigines.

The only way to Australia was across the Exmouth Trench, and this suggests that by about 40,000 B.C. our forebears had learned to build simple boats. I have never seen any cause for believing that one group of founding Neanderthals set out to sea, discovered a new continent, started a colony and that no one else arrived after them—just as there is no cause for believing that the Vikings, once they landed in America, were destined to remain the only seafaring people able to do so. Migrants must have continued to sail down from mainland Java and Borneo throughout the last glacial period, a scenario that places Australia on the tail end of African and Asian migrations and (contrary to the Thorne scenario) transforms the continent into the world's most valuable tool for viewing changes occurring in Asia, the Mideast, Africa, and Europe. Viewed in this way, the Australian

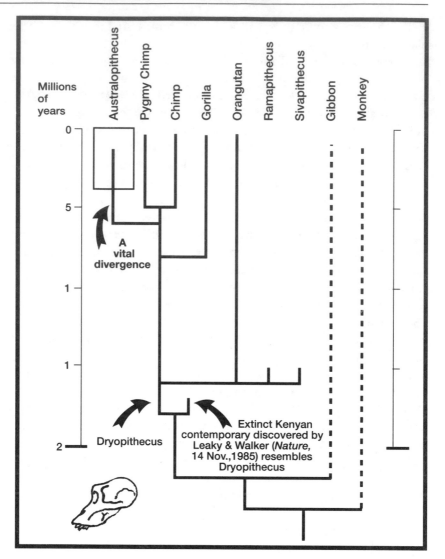

Figure A. This simplified family "tree" combines clues from the fossil record with the genetic history written on blood proteins and mitochondrial genes (whereby the molecular clock may be used, somewhat like the steady decay rates of carbon and argon isotopes, to date branching lineages). According to this interpretation, dryopithecine apes (probably several species) radiated out of Africa into Asia and Europe about 17 million years ago. Three major branches, leading to sivapithecines and orangutan-like creatures, split off from the lineage ancestral to gorillas, chimpanzees, and humans some 16 million years ago. The gorillas split next, between 10 and 7 million years ago, while protochimps and protoaustralopithecines parted between 7 and 5 million years ago. (Figures and data compiled by C. Pellegrino.)

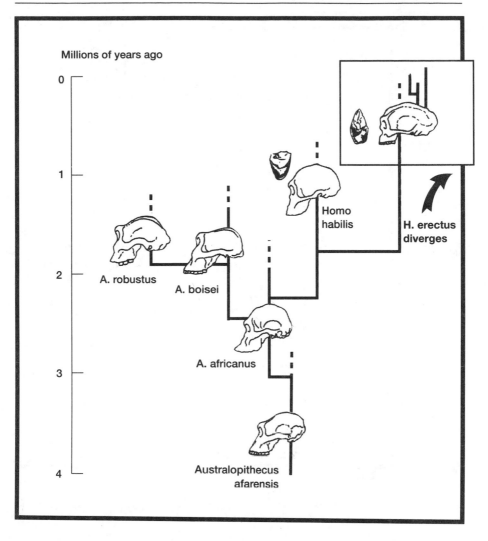

Figure B. This close-up view of the box in the upper left corner of Figure A shows a vital divergence of Australopithecus afarensis *("Lucy," or a still undiscovered species that looked much like her) into more varied and generally larger-brained austra-lopithecines and toolmakers of the genus* Homo *(including* Homo habilis, *a name that probably encompasses several distinct "races" or "tribes" standing somewhere between "Lucy" and* Homo erectus).

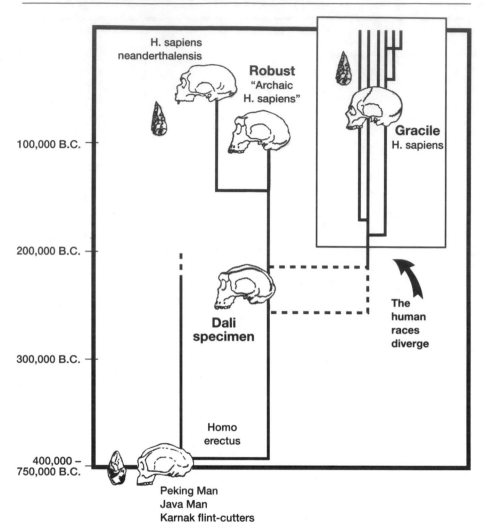

Figure C. Much as Figure B zoomed in on a small corner of Figure A, this third figure in the set provides a close-up view of the approximately 750,000-year span that includes the divergence of Homo erectus *to produce the Neanderthal tribes and "Eve's people." Seen in a sequence of increasing magnification, the tree of life acquires a fractal aspect—in which each piece, once enlarged, looks much like the piece that preceded it. At every level, the tree produces candelabralike shapes, concealed within candelabras, within candelabras.*

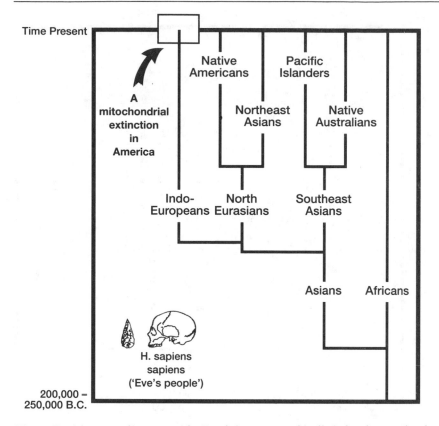

Time Present

A
mitochondrial
extinction
in
America

Native
Americans

Pacific
Islanders

Northeast
Asians

Native
Australians

Indo-   North
Europeans Eurasians

Southeast
Asians

Asians   Africans

H. sapiens
sapiens
('Eve's people')

200,000 –
250,000 B.C.

Figure D. After spreading across the Earth into geographically isolated areas, local variations of facial features, skin color, and hair color began to distinguish Mitochondrial Eve's descendants. Ultimately, the toolmakers invented rapid transport, pulling the old geographic barriers apart. In America, especially, the varied tribes began to converge after thousands, even hundreds of thousands of years of isolation. When the matriarch of one immigrant family (the McAvinues) landed at Ellis Island in 1921, an ancestral stock of Scottish and Irish had been marrying Scottish and Irish for as far back as recorded history. There was shock when the first American-born generation began marrying other Americans descended from Abraham and the southern European tribes. A generation later, when European and Asian lineages converged after more than 100,000 years of separation, no one even shrugged. The McAvinue lineage also carried a gene complex that now supports the theory of a mitochondrial "Eve" whose people came out of Africa. Expressed as a rare immune disorder called ankalosing spondylitis *(which can turn soft tissue into bone), the gene complex is* known to exist in European and Asian stocks *(including Native Americans) but not in native Africans. The implication is that this troublesome bit of nucleic acid is a latecomer to human evolution, one that originated after an Asian-African split but before the Asian-European divergence. The disease seems to be telling us that Africans were ancestral to Asians, who in turn gave rise to Europeans.*

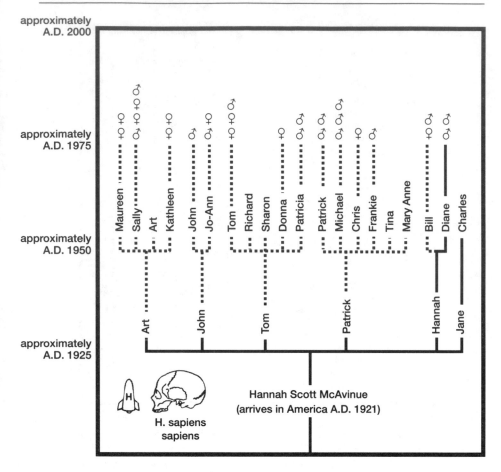

Figure E. A mitochondrial lineage dies in America. As one branch of an Irish immigrant family spread halfway across the continent, a mitochondrial ancestry going back millions of years became extinct within a single century, ceasing abruptly with the male mitochondrial heirs of Jane and Diane. The extinction illustrates the statistical nature of mitochondrial descent, in particular the rapidity with which any individual woman's mitochondrial lineage is likely to vanish, failing an unbroken chain of female carriers giving rise directly to at least one female carrier in every generation. The pattern for inheritance of the family surname (carried by males) is similar. In the next generation after Patrick and Hannah McAvinue, four out of six children carry the surname. By the third generation, ten out of twenty offspring have a potential for carrying on the name, and by the fourth generation, the ratio is down to eight out of twenty-eight. Though such family traits as musical and artistic ability will continue to be inherited, both the surnames and the mitochondrial signatures of the first immigrants will probably be diluted out of existence within two hundred years.

record is not one of thick-boned humans evolving locally into moderns, but of a constant influx of people who were undergoing a loss of archaic robusticity somewhere beyond Java. This record perhaps reflects episodes of outward migration (with Australia on the receiving end, somewhat like an oversize catcher's mitt), wherein one breed—a newer, less robust one— somehow outperformed the others and won for itself an entire planet.

Support for this view is beginning to emerge from the Wilson group's further studies of Aboriginal mitochondria.* So far, fifteen mitochondrial types have turned up in the cells of native Australians, suggesting that boatloads of settlers arrived at least as many times. Their common mito- chondrial mother was not the bride of Java Man or even the first Australian Neanderthal (whose descendants appear, like Neanderthals everywhere, to have vanished without leaving mitochondrial heirs†), but part of an influx that, it now seems, came out of Africa through the Crescent of Fire.

The characteristically Asian cheekbones preserved on the fossil skulls of Australian migrants can also be found on early "modern" human fossils in Africa and Israel—this forces us to ask how essentially Asian these and other regional traits can be—and complicates an interpretation of what Eve's people really looked like. We should probably avoid the *Newsweek* syn- drome, in which major magazines, on first reporting the Eve hypothesis, depicted the common ancestor as a contemporary African. That the widest mitochondrial diversity seen anywhere occurs among modern Africans should, if anything, put to rest this popular misconception, for the Africans have had close to 250,000 years (a longer period, according to the mito- chondrial clock, than any other race) to develop the features that distinguish them today, and the chances are that a modern African looks no more like Eve's people than does a modern Asian. It is more likely that Mitochondrial Eve's face bore hints of all the races she was about to spawn: Asian, African, and European.

*Alan Wilson, the founder of the Mitochondrial Study Group, died during the writing of this book and prior to the mitochondrial discoveries in Australia. Even in his absence, I have continued to name the Study Group after Wilson, in homage to him.

†The lesson of the Pitcairners' clearly Asian features, despite the fact that there exist no male heirs whose surnames would suggest an Asian ancestry, should not be forgotten when interpreting Australia's mitochondrial clock. While it does not seem likely that inbreeding between modern-appearing migrants and Australian Neanderthals ever took place (as indi- cated—*so far*—by the absence of a mitochondrial deviation significantly greater than the fifteen minor deviations already known for native Australians), we cannot altogether elimi- nate the possibility, at least on rare occasions in which lines of female descent failed to survive.

When a family line diverges in, say, five different directions, while it is true that no line will ever again look exactly like the ancestral line from which all groups started, there will nevertheless exist two inevitable extremes. One group is bound to diverge most from the starting point, and another group least. Australian Aborigines seem to have retained in their skulls more ancient features than the others, including thicker cranial bones and other vestiges of our ancestral robusticity.

If Aborigines do indeed represent the least divergent group, then they (and related Asian lineages) may look more like our ancestral African than do today's Africans.*

To understand how this might have come to be, we must suppose that Java Man, and my Karnak flint cutter, were always in motion, always in transition, much like the ancestral tribes described in the stories of Noah and Abraham. But Eve's people spread across distances unimagined by the writers of Genesis, sometimes developing regional traits as far away as Australia, Tibet, or Iceland—and never, ever standing still. The picture, seen in the time-lapse frames provided by a geologic perspective, was like taking a handful of pebbles and tossing them into a pool of water. Up there, in Asia, a tribe may generate outward-spreading ripples of nomads that, sooner or later, bounce up against ripples spreading from other pebbles. Some ripples are weaker than others and disappear beneath the surge of the strong—winners and losers in nature's extinction lottery.

Chaos with feedback . . .

Taken alone, the mitochondrial clock tells us only that a large, powerful ripple of relatively thin-boned humans spread out of Africa. It reveals nothing definitive about where the pebble that set the ripple in motion came from or what it looked like. If the pebble was flung into Africa less than 250,000 years ago and sent forth modern humans to become the only surviving limb on a once-luxuriant *Homo erectus* bush, then we are forced to explain such matters arising as the occurrence of shovel-shaped incisors and forward-facing cheekbones on the skull of a Chinese *Homo erectus* dating back a quarter million years *before* the African ripple. These features clearly predate Eve's people, yet they now characterize modern Chinese people and their Native American "cousins."†

---

*European lineages, and some African ones (including the dense-brush-dwelling huntsmen known as Pygmies), might have diverged, in terms of physical appearance, more from Mitochondrial Eve's starting point than all others.

†Teeth, though they decay with such ease during life, are remarkably resilient after we are dead and no longer need them. Our teeth are in fact so hardy that they tend to outlast our skulls and tibiae. Many fossil species are known only from teeth, and some paleontologists

The incidence of even one shoveled incisor, or one pair of forward-facing cheekbones on a Chinese *Homo erectus* skull (combined with Asian features stamped on some of Africa's and Israel's first "modern" human fossils), is at least suggestive of a genetic potential for these features predating and being directly inherited by Eve.

In other words, though the ripple that directly gave rise to us spread out of Africa, the pebble that caused the ripple could easily have been tossed from China, or somewhere in that general direction.

During the approximately 250,000 years since Eve, the African population has diverged more from its ancestral stock than the ripples that went out through the Nile to China, Australia, and Europe. Since we are first meeting Eve's people as fossils when they have already become a large, central population, the probability of finding the actual founders (whose numbers must only have been in the hundreds, or even the dozens) is disarmingly small, though not quite zero. It is not impossible to believe that the quarter-million-year-old skull of a gracile woman (that is, a thin-boned member of "Eve's" tribe) with prominent cheekbones, shoveled incisors, and other distinctly Asian features will eventually be found in Africa (bearing in mind, however, that shoveled incisors were of low frequency, even among Asians, before 40,000–38,000 B.C.).

While the trickle of jigsaw pieces from fossil beds reveals a fossil record apparently full of gaps, some, if not many or most, of the gaps may arise from small founding populations unlikely to be preserved (that is, they are an artifact of the jerky mode of speciation). The gaps may thus be more real than apparent, meaning that while the fossil record is indeed imperfect, we should perhaps be paying closer attention to the imperfections of our questions about the record.

As for what the record of teeth, bones, and mitochondrial genes tells us about the fate of yesterday's child—about my Karnak ax maker and all her robust contemporaries as far away as Java—their noble lines, it seems, became victims of a Pleistocene holocaust carried out by Eve's people over tens of thousands of years. It might only have been a matter of moving into and somehow maintaining control over the best hunting grounds, perhaps by aid of better technology, swifter running ability, greater social organization, or some combination of these three and one or two other advantages as yet unimagined. Most anthropologists I've spoken with would like to believe that the lost tribes were merely squeezed out by the newcomers into

---

have been heard to complain that much of our fossil history seems to be nothing more than the mating of teeth to produce slightly modified descendant teeth.

less productive highlands, where they quietly died out under a series of adverse climate changes.* I think they are in denial. What less productive highlands were there for the flint cutters of Karnak to go to? There was only the desert. To stay in the Nile Valley, or to be squeezed out of it, was the difference between life and death. In this place, at least, there must have been a terrible fight. Any who doubt a capacity for slaughter among the children of Eve need only consider such descendants as Stalin (who executed more than two million of his own countrymen), Pol Pot (who built pyramids out of human skulls), Hitler (more than 10 million killed in concentration camps), Custer (an American Hitler), Radovan Karadzik (a Serbian Hitler), and the settlers of Tasmania (the only people in modern history claiming to have exterminated an entire race to the very last mitochondrial heir). If, as seems feasible, Eve's progeny wiped out all rivals in their path, then perhaps the Eve hypothesis is gravely misunderstood and wrongly placed, and should be named instead after her murderous son, Cain.

〰〰〰〰

As punishment for eating the fruit of a tree that gave man the knowledge of good and evil, God (in Genesis 3:16) passed a curious sentence upon Eve: "I will greatly multiply the pain of thy childbearing; in sorrow thou shalt bring forth children."

You may recall from Chapter 3 that God's curse is eerily akin to the scientific story of origins. Childbirth is more difficult for human beings than for any other known species, a price apparently paid to accommodate the brain's tripling in size during the past two million years. The head of the newborn is the largest part of the body and the first to emerge. In sorrow and pain women do indeed bring forth children. What is curious about the Genesis account of our species' uniquely unpleasant childbirth is that biblical scribes linked it directly to the gaining of knowledge, and it is in fact the growth of the human brain that made all knowledge possible.

According to those who have most carefully studied the Bible (and pieces of its earliest stories still surviving in Egyptian and Babylonian writings), the Book of Genesis is an apparent blending of two documents, commonly referred to as the J-document and the P-document.

---

*Berkeley anthropologist Tim White is not among these. In 1982 he found a partial *Homo erectus* skull bearing unquestionable marks of scraping with stone blades—"the first known scalping," he says. "It appears to be intentional defleshing, but it is impossible to say why it was done."

The J-document is the more ancient one and seems to borrow heavily from the epic of Gilgamesh and other early legends that had originated with the Sumerians (Babylonians) in Iraq's Tigris-Euphrates Valley, and eventually spread out along trade routes to Israel and other distant lands. Into the J-document passages, hints of an earlier belief in multiple gods occasionally intrude, as in Genesis 1:26 ("And God said, Let us make man in our image, after our likeness") and Genesis 3:22 ("And the Lord God said, Behold, the man is become as one of us, to know good and evil").*

The P-document was compiled about 550 B.C. by Hebrews held captive in Babylon, on the Euphrates River. At that time, the legendary Hanging Gardens of Babylon are said to have rivaled the Pyramids in sheer beauty. There, the P-document, too, picked up a distinct flavor of the Tigris-Euphrates, including Babylonian views of cosmic origins based upon nearly three thousand years of thought dating back to the founding Sumerian tribes. Consequently, in Genesis 2:14 the Garden of Eden is placed squarely in the land of the Hiddekel (known to the Greeks as the Tigris) and Euphrates rivers.

The Babylonians were, in terms of their understanding of engineering, every bit as sophisticated as the Egyptians, the Minoans, and the builders of the Indus Valley civilization. They had, by about 2500 B.C., begun to lay the foundation of what we might call science, keeping records of changes observed in nature, including the shifting positions of stars and planets over long periods of time. They did not know what those specks of light in the night sky were, for they did not possess telescopes and spectrographs, so the Babylonians thought of them simply as lights splattered across a dark, paper-thin firmament, placed there almost as an afterthought by their creator (as in Genesis 1:17: "And God set them [the stars] in the firmament of the heaven to give light upon the earth"). Tracking the generations of all living creatures backward as far as they could (and lacking the tools of

---

*The tribes of Israel were probably polytheists in their early days, as were most tribes of most people. The story of Moses' smashing of the tablets when his followers reverted to Minoan-style bull-worship indicates well enough that there was disagreement between monotheists and polytheists even among Hebrews, even during the time of the Exodus from Egypt (about 1628 B.C.). By the time the eleven chapters of Genesis were being compiled in Babylon, monotheists predominated. and the history of the Hebrew tribes was retold from a monotheist, Mosaic point of view. Nevertheless, the pluralisms of the past crept through from ancient texts and, once included, apparently became too sacred to remove. A similar occurrence is known from the first translations of the New Testament into Greek during the second century A.D. The Hebrew name Joshua was incorrectly translated as Jesus, and by the time Saint Jerome prepared the Latin version of the Bible, about A.D. 400, the erroneous name had become too widely established to be altered.

radiometric dating combined with a jigsaw of fossils from all over the world), the Babylonians imagined a cosmic egg from which the Earth and the heavens began. They saw a primordial garden from which the tribes of man emerged, and the garden lay between the Tigris and Euphrates rivers, on a flat world created about sixty thousand years ahead of their time (this span was later shortened to six thousand years by Hebrew scribes). We who can speak with Scott Carpenter and hear firsthand that the Earth is a sphere, who can look out upon islands of stars known to reside billions of light-years away or down through layers of fossils billions of years old need not feel superior to the Babylonians. They had only the evidence of their eyes to read from, and after the Dark Ages had buried almost all traces of the subsequent Minoan and Ionian Enlightenment, the rest of the world re-established and held stubbornly to these same beliefs (a young, flat Earth, with stars nearby and embedded in a solid firmament) until about A.D. 1500.

While it is true that if the biblical writers had available to them the scientific knowledge we now possess, the early chapters of Genesis would have been written quite differently, we nevertheless come across descriptions that, in some essentials at least, ring strangely familiar. In many instances Babylonian logic (aided, much like Mayan logic, by advanced mathematics) was not far off the mark. The universe truly does appear to have had a specific beginning, and while many readers find the biblical description of a six-day Creation fantastical, the scientific story of Genesis is no less fascinating: All the major events making possible our existence can be traced back some twelve to twenty billion years ago, to the first three minutes of the Big Bang, and logically the tribes of man should have been traceable, if a complete record of family lineages had existed, back in time to either one or very few ancestral mothers. Though the Babylonians lacked the tools to peer beyond where their tribal mother came from, they did suspect that there was a time before man had learned to redirect the courses of rivers, a time in which he lived in ignorance of good or evil, or death, and his savage brain was smaller, perhaps, and childbirth was less painful, and in his innocence he was more akin to the beasts of the Earth.

Somewhere along the line in the development of human life emerged two patterns: a high-energy feeding strategy and a direct connection to our mothers throughout fetal growth. From a paleontological perspective, these traits could have contributed much to the explosive growth of human brains and suggest that woman has given us far, far more than her mitochondrial genes.

Human children require more parental care than any other mammalian species. Our enlarged brains account for much of the burden. Human fetal growth is characterized by an acceleration in the development of both brain and body, as the British anatomist Robert Martin of University College has observed: "Human infants have brains and bodies twice as big as you'd expect, given the length of gestation. This must be extremely costly energetically. A high-energy feeding strategy was essential for its development."

The human brain quadruples its volume between birth and adulthood. Other primates merely double their brain volumes. No primates (including humans) grow new nerve cells after birth. Postnatal brain growth is limited to tissues that support the nerves and to making new nerve connections. In spite of our accelerated fetal development, we are born with essentially fetal brains, rendering human infants the most helpless creatures in the world. By contrast, a horse is born with enough supporting brain tissues and sufficiently advanced nerve connections to permit standing and walking within hours, while a chimpanzee will outperform a human child in every way during the first year of life.

The burden of support falls upon our parents, and a simple comparison of skulls and pelvises spanning the past 2.5 million years permits us to make a good guess at when a marked increase in parental care became necessary.

The explosive growth of the human brain late in our phylogeny had a wide range of consequences, from anatomical digressions of skulls and teeth to the discovery of time and death. Incomplete closure of the newborn's skull was very likely one of several inelegant solutions to the problem of ballooning brains, and I suspect that this necessary accommodation may in some way account for the eventual survival of thinner-boned over thicker-boned humans. If this is the case, then perhaps changes in only one or two regulator genes will explain the geologically sudden branching of gracile *Homo sapiens* (Eve's people) from a more robust stock.*

As the human brain changed, so, too, did the female pelvis. The skeletal

*American Museum of Natural History paleobiologist Niles Eldredge and Harvard paleobiologist Stephen Jay Gould have argued that the growth and development of one body part might often be linked to the development of other body parts. In this case both reduction of the thick ledge of bone over the eyes of *Homo erectus* and a thinner tibia may arise from an overall thinning of bone necessary to accommodate incomplete closure of the newborn's skull—which results in a ripple effect: a thinning of bone that resounds through the whole system (which may in turn offer multiple advantages, such as greater running speed and endurance, thereby rendering increased odds of survival to the bearer of such genes). "The principle [that a change in the tempo of growth in one body part will influence others] is as obvious as your big toe," writes Gould, "which appears to have developed for no better reason than to keep pace with your thumb. Repeated parts of the body are not fashioned by the action of individual genes. There is no gene 'for' your thumb, another for your big toe. . . . It may be genetically more complex to enlarge a thumb and *not* to modify a big toe than to increase both together."

opening became substantially wider in the region of the birth canal, but enlargement of the pelvic girdle could go only so far before eventually rendering women incapable of walking. The redesigned pelvis is an imperfect, jerry-rigged accommodation to large brains, whose only virtue is that it works—sometimes.

Here we encounter a phenomenon I have called selective balance, whereby requirement A (a need for the widest possible birth canal) is balanced against requirement B (a need to walk efficiently). In this case, requirement A, acting in one direction and carried to an extreme, would be harmful (if one makes the pelvic opening too wide, women cannot walk). Requirement B, acting in the opposite direction and carried to an extreme, would also be harmful (if one makes the pelvic opening too narrow, women cannot give birth to large-brained babies).

Nature forces a chancy sort of balance, in which bigger brains and mobile mothers are accommodated at the expense of a higher incidence of mortality during childbirth. It is not an easy bargain. Prior to the advent of twentieth-century medical technology, childbirth was the leading killer of women between the ages of fifteen and thirty.*

According to Robert Martin, the brains of modern infants are about as large as pelvic engineering will allow (meaning an upper limit of approximately 350 cc at birth). If we apply the typical primate pattern of postnatal doubling of brain volume, we yield an adult brain volume of 700 cc.

Support for Martin's prediction comes from comparisons of different limbs on our family tree with Nariokotome Man, the most complete early hominid skeleton known at this writing. He was a Kenyan representative of *Homo erectus* dating from about 1.6 million B.C. Nariokotome Man's adult brain volume is estimated at 900 cc, and the opening in his pelvic girdle is consistent with the delivery of babies possessing 300 to 350 cc brains.† By the time Nariokotome Man appeared, about 1.6 million years ago, Martin's 700 cc limit had already been reached, and any further brain growth had to

---

*Cesarians are just one product of large brains, having all but removed this statistic from memory by allowing the delivery of those babies that would normally be unable to pass through the birth canal. Contrary to popular legend, the cesarian is strictly a twentieth-century invention. Julius Caesar was not born that way, and the procedure is not named for him, nor he for it.

†Although Nariokotome Man (specimen KNM-WT 15000, discovered in 1984) represents a male youth (twelve years old, plus or minus one year), pelvic openings are not known to vary greatly between sexes of modern humans, and it has been inferred by Kenya anthropologist Richard Leakey that the male and female *Homo erectus* likewise had similar sacral diameters.

be assigned to postnatal life. Nariokotome's head size at birth was nearly one third the adult size; hence the brain became less developed at birth (compared with chimps and gorillas, whose brains merely double in volume after birth), its owner was more helpless, and the demands upon the parents rose proportionally. To determine when this uniquely human characteristic of extending the greater part of brain growth into infancy might have begun, we must track back past *Homo erectus,* closer to the trunk of our ancestral tree.

The australopithecine man apes, whose most famous representative is a female skeleton named Lucy, had adult brain volumes varying between 400 and 500 cc, requiring only the typical primate doubling throughout life from an initial volume of 200 to 250 cc (suggesting that australopithecine birth was not particularly painful). The vital shift must therefore have occurred somewhere between *Australopithecus* and *Homo erectus.*

Homo habilis, the next closest branch to *Homo erectus,* had an adult brain volume of 700 cc, and though no fossil pelvises are known from this clan (actually, there appear to have been at least two distinct *Homo habilis* branches or "races"), a presumed postnatal doubling of brain size places its newborn squarely at Martin's 350 cc boundary. Judged from skulls, *H. habilis* was the first primate brain outside the australopithecine lineages to show a substantial increase in volume, and the first primate to feel Eve's curse *(in sorrow thou shalt bring forth children).*

Perhaps the most significant drawback of big brains is that they are energetically expensive and can therefore be nourished only by voracious feeders. A quarter of all blood pumped from the heart goes directly to the brain, an organ accounting for less than 2 percent of the body's mass. No other brain in the animal kingdom makes such demands upon its owners. In a very real sense, our bodies can be viewed as parasitized vessels providing nutrients, and mobility, and sensory pleasures to an organ that—alone among the Earth's offspring—has grown to know the difference between good and evil.

Giant brains can be afforded and evolved only under certain favorable circumstances, such as a high-energy feeding strategy and correspondingly abundant food sources. A large body also helps. Living primates range in size from the mouse lemur, weighing in at about two ounces, up to the six-hundred-pound mountain gorilla.

If we plot primate metabolic rates against body weights, we derive a slope of ¾. This tells us that, as monkey- and apelike creatures increase in size, body weights increase 25 percent faster than metabolic rates. There appears to be a special relationship between increased size and energetic efficiency (expressed as a relatively slower metabolism among the larger primates), which is consistent with what we know about primate dietary habits. The largest apes feed on leaves, and leaves are relatively low in energy content. Monkeys tend to be fruit eaters, and fruits provide more nutrition per ounce than leaves. The smallest primates tend to subsist on high-protein seeds and insects. A notable deviation from this trend (the largest primates eat leaves; smaller primates eat fruit; the smallest primates eat high-protein foods) occurs in the primate branch closest to the australopithecines. The mid-size chimpanzees have a varied, high-protein diet that includes fruit, insects, seeds, and meat.

If this does not sound particularly dramatic, then go ahead and plot the ratio of brain weight to body weight. Robert Martin has done this, and the result again is a slope of ¾. (As primates increase in size, body weights increase 25 percent faster than brain weights.) The fact that both comparisons—metabolic rate with body weight and brain weight with body weight—produce the same slope hints at a relationship between them.

"It is the mother's energetic potential that determines the brain size of the developing fetus," says Martin. "After birth, brain growth then follows a trajectory already set" (by the energy throughput of the mother's metabolism).

Generally, humans are among the largest of all primates, and large size runs parallel with improved energy conservation (beginning simply from the standpoint of heat retention). In addition, we subsist on a generalized, energy-rich diet that expands in many directions to include fruits, vegetables, meats, and seafood. Together an efficient metabolism and a high-energy feeding strategy might have fostered the development of large brains. But it was not enough.

It is often said that our greatest evolutionary step was the one taken on two feet. It freed the hands for carrying objects and making tools, and that in turn encouraged the further development of the brain. I'm afraid this explanation is at best only partly true, for if it is wholly true, it raises a mystery that troubles me: Why have other creatures with free, manipulative limbs, including kangaroos, elephants, and octopi, not developed along the same pathway as humans? Why is it we who cover the earth with our farms and our cities, and shepherd kangaroos and elephants into zoos, instead of the other way around?

I believe the answer lies within the energy requirements of big brains, and hence our greatest evolutionary step was not that taken on two feet but the one taken within the womb. Stone hand axes, the pillars of Karnak, a rocket filled with nerve gas, and the Aswan High Dam all may be the result of a high-nutrient throughput by large-bodied, energetically efficient mothers who were constantly consuming the most energy-rich foods in their surroundings and were connected to their developing young by umbilicals (and later by nutrient-rich milk glands). This may explain why despite hundreds of millions of years of biologic experimentation, civilized species did not arise on Earth sooner. Kangaroos and elephants were around long before australopithecines arrived on the scene, and both had free, manipulative limbs.* But both were grazers. Both had low-energy feeding strategies.

Octopi, despite their manipulative limbs and a high-energy feeding strategy, have never, given nearly two hundred million years to do so, developed beyond the intelligence level of dogs. The story might have been different were they not exclusively egg layers. All the nutrients the developing fetus will ever receive are (and always have been) deposited in the yolk at conception. There is no continuous energy throughput by the mother. Precious little is left over for the construction of large, nutrient-guzzling brains.

I have always wondered why certain dinosaur lineages, such as *Ornithomimus* (the man-size "ostrich dinosaurs"), despite the fact that they walked erect, possessed hands, and appear (like today's birds) to have had efficient metabolisms and the necessary high-energy feeding strategies, and despite tens of millions of years of opportunity, never developed brains large enough to carve out a flint hand ax. Perhaps we need look no farther than their eggs for an answer.

The Babylonian story of Eve, who passed down knowledge to mankind, who first bore her children in pain, seems somehow to anticipate the scientific story of how intelligent life originated on Earth. Of course, I may only be indulging in mankind's vice of seeking a symbol or a dawning reality in every coincidence. Scratch any house cat hard enough and you'll scare out a flea—right? I don't know. Some of my more spiritual friends insist that there are no coincidences.

Not only by our large brains and by our bipedal gaits but by our

---

*Arguably the kangaroos would seem to have had an important advantage over primates. They were spared the pressures of selective balance between width of the pelvic opening and skull size, for the latter part of fetal development was assigned to a pouch. Theoretically there were no limits on kangaroo brain size, yet when the explorers James Cook and William Bligh (whose loss of nerve in Hawaii resulted in Cook's death and whose later misadventures aboard the *Bounty* led to the Pitcairn lineage) first set foot upon Australia, the kangaroos looked and behaved much as they had for millions of years. When finally Cook asked the Australian Aborigines, "What are those odd, hopping animals?" they answered in their native tongue, "I don't know," which is pronounced *kangaroo,* and we have been calling them that ever since.

umbilicals, our stomachs, and our wide pelvises did we set forth toward a civilized existence, following a line of destiny set in motion by woman.

So here we sit, by either destiny or chance: subduers of the Earth. In Karnak tonight the stars are very bright, but north in Cairo, and in most of the world's cities, they are obscured beneath lenses of filth a mile thick and often more than six hundred miles wide. Parts of the Mediterranean and the Atlantic are turning black, and even as I write, the white limestone of the Pyramids is tarnishing brown under the residuum of a carbon-burning civilization, while in Iraq Saddam Hussein waves plutonium fists at us.

Looking back across time, I am able to glean a lesson and a simple proposal, if not a symbol, from our mitochondrial mother. Perhaps we should push for a United Nations resolution requiring those nations that are most vigorously fouling the planetary nest, and those in possession of nuclear arsenals, to be governed only by women with young children. Two million years of reinforced instinct cannot possibly lead us wrong. Women with young children live, more than anyone else, in the future. They understand, without having to be told, how far civilization has come and how high the stakes are.

5

HER
MURDEROUS
SON

*Unfailingly, humans pity their ancestors for being so
ignorant and forget that their descendants will pity
them for the same reason.*
—EDWARD R. HARRISON, at the New Zealand
   National Observatory, 1982

*Oh, Lord, in a world so rich and lovely, why can your
children find nothing better to do than to dig iron from
the ground and work it into vast, grotesque engines for
blowing each other up? Is it because Abel's next-door
neighbor was Cain? Is it because if my enemies make
deadly engines then I must do it better or die? Maybe
the vicious circle . . . will never end.*
—HERMAN WOUK, *The Winds of War*

*One day, somebody should remind us that even though
there may be political and ideological differences between
us, the Vietnamese are our brothers, the Russians are
our brothers, the Chinese are our brothers; and one day
we've got to sit down together at the table of
brotherhood.*
—THE REVEREND MARTIN LUTHER KING, JR.

And they shall beat their swords into plowshares.
—ISAIAH 2:4

*History teaches us that he who beats his swords into
plowshares usually ends up plowing for those who kept
their swords.*
—ANON.

THE WEST BANK OF THE JORDAN RIVER, MIDNIGHT, SEPTEMBER 22, A.D. 1991

A woman carving a hand ax. That is how it began.

The bus from the Galilee to Jerusalem is stopped near a barbed-wire barrier. There has been gunfire up ahead. A helicopter sweeps the fields to the south and west with a powerful search beam, and we know now that we are going to be milling about here for a very long time. In the glare of floodlamps, the young Israeli soldier with the assault rifle has the most shockingly blue eyes I have ever seen. "They came across the river and cut through the barriers," she explains, offering me a fresh can of Coca-Cola. "Terrorists. They killed a farmer who had settled here only months ago. So they got one of ours. Now we have got two of theirs."

"Captured?"

"Shot. Gone to Allah."

Everything about man has changed, Einstein once observed: "That is, everything except man's way of thinking." Here in the fertile valley, a quarter mile below sea level on the floor of the newest crack in the continent's surface, war, humanity's Pandora, seems more a fact of everyday life than on the planet above. One night not very long ago, as Saddam Hussein's Scud missiles were lancing overhead, Israeli television resumed its regularly scheduled broadcast of a concert after a brief announcement that there had been explosions and citizens should put on gas masks. At the very same moment, in Europe and America, every channel became instant wall-to-wall coverage of "Israel Under Siege." It's all part of the psychology of living near the Jordan River: that you can see a man on television playing the flute while a rocket attack is under way. And thus do the soldiers and the passengers behave as if our interrupted journey is no more an inconvenience than getting caught at the end of an unexpectedly long line at the bank, with perhaps the slightest trace of blood sport thrown in. They mutter, complain, pass Cokes and candy bars around, joke, kick idly at the dirt, groan, and wait.

Away from the glare of the lamps, the moon, though only in partial phase, appears brighter and clearer than I've ever seen it before. The faint ejecta rays from the crater Bruno are actually visible without the aid of binoculars or telescopes. Toward the south, about a quarter of the way up the sky, a compact galaxy of stars shines brightly enough to be a minor rival to the moon. The cluster is crescent-shaped, with its two horns pointing Earthward, and below them no stars shine at all. The crescent caps a vivid dark mound, above which rises the backdrop of the Milky Way, below which is only black, and I realize that the object is not a galaxy at all but the mountaintop city of Jerusalem.

So here I stand, in the green stretches of Israel, at the crossroads of conquering armies. In times of peace as in times of war, crossroads are hotly contested. Up there, in the sky, the electric lights of Jerusalem burn atop still-uncounted layers of residua left by eight thousand years of marching feet. As far back as archaeologists can see, the city has always been enclosed by walls. By about 3000 B.C. the ancestors of the Israelis who now rule behind those walls were already present in this land. In those days they were the people who dwelled outside the city walls, nomadic Hebrew traders and herdsmen living free under the desert sky. It is they who most likely wandered upon the copper fields of Sinai, who later came across outcrops of tin and discovered that it could be blended with copper to produce a newer, stronger compound that would hold a sword edge. It is these nomadic shepherds, too, who are credited by ancient scribes with the discovery of caffeine's sleep-inhibiting properties, having noticed that their sheep walked around all night after eating berries of the coffee plant. Much as settled city life allowed for organized storage and mass marketing of ground coffee, it also facilitated the mass production of bronze weapons. Most archaeologists believe it was the city dwellers who invented the Bronze Age, but this scenario seems no more likely than one crediting them with the discovery of coffee. If small patches of tin, coffee plants, and copper ore are in the wilderness, the greater probability is that a wandering people will find them, and learn how best to use them, before a sedentary people. The nomads took with them only what was worth carrying and brought to the cities only what was worth trading. I suspect it is they who perfected and sold the first bronze tools, and they who sold the science of metallurgy, from village to village.

The nomads used to call this land Canaan, so named for a Semitic city-building people who called themselves Canaanites. By 1650 B.C., and possibly many centuries earlier, the Canaanites had invented an alphabet in which twenty-two characters could be strung together to spell out the spoken word. This at once rendered obsolete the Egyptian hieroglyphic method of representing each word with its own, separate pictograph. Within the Canaanite cities were silos filled with grain, cisterns for trapping rainwater, shelves filled with records of grain and water supplies, furnaces for smelting bronze, and a few man-size storage jars imported from the seafaring Minoans who, from the islands of Crete and Thera, ruled much of the eastern Mediterranean. According to Benjamin Mazar, the great-grandfather of Israeli archaeology, the oldest alphabet finds are rarer and even more fragmentary than 1650 B.C. Minoan jars, but when we find them, their mere existence tells us that, at least as far back as the time of Moses and Joshua, someone in these parts was keeping records. From the period nine hundred years later we begin finding the earliest-known fragments of biblical texts, such as the plastered wall at Deir Alla, only a few miles from here, inscribed with red and black letters spelling out, in Hebrew, the prophet Balaam's vision of cosmic disaster: "and they said to Sha[ma]sh [the Sun]: Sew shut the skies with your cloud! Let there be darkness and no shining . . . for you will provoke terror by a cloud of darkness."

There is little doubt among the Mazars, or any of the other native archaeologists I have met, that the seer is the same Balaam mentioned in the Book of Numbers, whose oracles (Numbers 23–24) refer to "the dust cloud of Israel . . . a meteor comes forth from Israel; It smashes the brow of Moab."*

"There came a period," explains Benjamin Mazar, "in which there was the real first development of writing—alphabetical writing—in Hebrew and Phoenician.† This was about the tenth century B.C., the period of David and Solomon. There comes a later period in which writing became common not

---

*The Moabites were a tribe known to have been firmly in possession of the territories east and south of the Dead Sea, in what is now Jordan, before and after the time Tuthmosis III (whose reign marks the probable time frame of Exodus). The writing on the wall of Deir Alla seems to echo several prophetic texts in the Bible, notably Isaiah 13:10–13 and Joel 2:1–3: "The stars and constellations of heaven shall not give their light; the sun shall be dark when it rises; and the moon shall diffuse no glow. . . . A day of darkness and gloom; A day of densest cloud; Spread like soot over the hills. . . . Nothing like it has ever happened. And it shall never happen again."

†Phoenicia was a kingdom that, during the eleventh century B.C., encompassed Syria, Lebanon, and parts of Israel. The Phoenicians' extinct Semitic language utilized script that eventually was incorporated into Greek, Roman, and all other Western alphabets. The term *phonetics,* describing the study of speech sounds and their production, is derived from the word *Phoenician.*

only in Israel, not only in Phoenicia, but also in Greece, the Aegean, and parts of Asia. And this is the eighth century B.C., the time of the Deir Alla prophecy."

The writing on the wall could not have been a unique event. Many other such walls must have crumbled to dust and will never be discovered, and plastered walls are so much more easily preserved than paper scrolls, which must also have existed by that time. According to Benjamin Mazar, Deir Alla marks the period, between the tenth and eighth centuries B.C., in which the earliest versions of the Old Testament evolved from oral tradition to something less mutable (in a figurative and sometimes even literal sense, they became cast in stone).

"In this country, three thousand years ago, these Books of Genesis, Exodus, Joshua, and Numbers were old stories, which were well known to the public, and somebody put it in writing, eventually. And this is the prehistory of the Jewish people, of the Hebrews. The main idea was to write the Hebrew history, starting with Abraham or whoever or whatever, and to tell in a few chapters what happened before this time, before King David built his Temple in Jerusalem."

As one reads the Old Testament verses, a poetic, almost musical quality is immediately apparent. On every continent and in every time frame right up through the Middle Ages into present-day Africa and New Zealand, there have always been what Benjamin Mazar calls "these song stories." They were the next best way of memorizing history in the absence of writing, and there existed singers who traveled from town to town, entertaining the inhabitants and passing their oral traditions to the next generation of singers. The song stories were powerful mnemonic tools, so persistent and often so immutable that the word *unbelievable* sometimes does not seem big enough, as was demonstrated so dramatically when Alex Haley visited an African griot who, in A.D. 1974, recited events on a family tree whose tribal roots ran back more than two hundred years.

"So the old songs were being passed from generation to generation," says Mazar. "And when the temple builders of Jerusalem wanted to write what had come before David's time, there were these legends, which were still being passed on from grandfather to grandfather. Someday it would be interesting to know if these legends were general legends—for they were also well known in Egypt, Babylon, and Syria—or if they were original Israeli legends."

Since Israel was a crossroad, a bridge between the riverworlds of Egypt and Babylon, the possibility of interchange between diverse cultures clearly bears interesting implications. God's punishment of Sodom and Gomorrah has a distinctly Babylonian flavor, and portions of the Mosaic laws are virtually identical to those passed down by the Babylonian king Hammurabi, while the story of the infant Moses being set adrift in a reed basket has counterparts in both Egypt (the pharaoh Akhenaton) and Babylon (Sargon the Great). Until they became settled city builders about 900 B.C., the nomadic Hebrew tribes must have been go-betweens, transmitting legends throughout the riverworlds and eventually hybridizing them into their own song stories, which, between 950 B.C. and 750 B.C., became the first written fragments of the Old Testament; and by the time of the Babylonian captivity of the Hebrews about 550 B.C.: "It's final then," says Mazar.

And by the time of the Dead Sea Scrolls, written over a four-hundred-year period commencing about 300 B.C.: "It's final," Mazar says again. "That's the final development."

〰〰〰〰

The Semitic tribespeople—Canaanites, Hebrews, and Phoenicians—had started their national histories as polytheists. Even as the Old Testament reached its final form, in Genesis 1:1, the Hebrew word translated as God was *Elohim,* a plural form of the Canaanite god El, meaning "all the gods" (with the singular form, *Elyon,* translating as "the most high," or *El-Shaddai,* "the God almighty"). It is possible that in the earliest song stories from which the Book of Genesis was derived, the Creation was the work of several gods and that the monotheistic compilers of approximately 500 B.C. had to contend with an oral tradition which by then had become so widely sung and so firmly ingrained, if not sacred, that to change it would have been sacrilegious.* Further traces of polytheism, which seem to have survived the editing, include a passage in Genesis 11:7, wherein God, angered by man's arrogance, announces: "Come, let us go down, and there confound their language."

〰〰〰〰

Somewhere between 1650 B.C. and 750 B.C. the plurality of gods had been transformed by the Hebrews into a single entity with infinite power. In time

---

*A similar situation has been with us ever since the Hebrew Bible was translated into the Greek Septuagint. "Moses" was originally "Mosheh"; Adam and Eve were "the man" and "Havva." In the original New Testament texts, Christ is not Jesus' (more correctly Joshua's) last name. "Christ" is a title translated as "the anointed one," the Greek equivalent of the Hebrew word *Messiah.* The proper name for Christianity's founding prophet is *Joshua, the Messiah,* but few today would tolerate the change.

the founding prophets of Christianity and Islam would come to worship this same Hebrew god.

Up there in the walled city, the Intifada, a rebellion in which young Palestinians throw rocks at assault vehicles driven by equally young Israeli soldiers, is now gaining worldwide news coverage. It is not news, really. People have been throwing rocks at one another in these parts for nearly four thousand years. The only real news would be if it ended. No one can predict where this ancient conflict will eventually lead, but there are, in texts sacred to both sides, accounts of its beginnings.

According to the compilers of Genesis, the unsettled Hebrew tribesmen migrated twelve hundred miles up the Euphrates River, leaving behind what would one day become the oilfields of southern Iraq for the fertile plains of the Jordan Valley. They pitched camp on the West Bank, and it is here that God promised the tribal leader Abraham, "It is to your descendants that I will give this land."

At age eighty-six Abraham had no heirs. At the urging of his wife, Sarah, he took Hagar, her Egyptian slave girl, as his concubine. In Genesis 16:15 we read that Hagar bore Abraham a son, Ishmael. Fourteen years later Sarah delivered a second son, Isaac. From the viewpoint of the biblical scribes, though Ishmael was the firstborn, his descent through a concubine diminished his standing, and both Hagar and her son were promptly expelled from Abraham's home. They settled "east of Egypt on the way to Assyria," in Saudi Arabia on today's map.

The Arabians, even after the establishment of Islam in the seventh century A.D., accepted the Old Testament writings and actually incorporated the stories into the Koran, calling Abraham a great prophet and common ancestor of the Arabian and Hebrew tribes. The Koran also tells of Allah ordering Abraham and Ishmael to journey to Mecca, where they built the first temple enclosing the sacred black rock from Paradise. According to Islamic tradition, Ishmael and his mother are buried in Mecca.*

The Old Testament tends to lose track of Ishmael's clan. Most of the attention is focused on Isaac's lineage, particularly Jacob, who after a time changed his name to Israel, and thus, we are told, subsequent generations began calling themselves Israelites. Viewed from a biblical perspective, this

---

*According to the Bible and the Koran, Ismael and Isaac came together at the Cave of the Patriarchs, in Hebron, for the burial of Abraham. The cave is now the site of the Ibrahim Mosque, and contains tombs sacred to Moslems and Jews as the burial places of Abraham, Sarah, and Isaac. As it turns out, the rock from heaven is a large nickel-iron meteorite. Its connection to heaven is not likely rooted in mere coincidence. Someone must actually have seen it fall from the sky.

blood-spattered river valley that defies resolution can be regarded as the site of the world's longest-running inheritance dispute. Viewed biologically and historically, the picture is not very much different.

When the Hebrew tribes began pasturing their herds in this fertile land, the Canaanites they found here not only spoke a Semitic language, many of them spoke actual Hebrew. As the tribesmen began living between and occasionally within the Canaanite cities, they often adopted the customs of the locals, including a long-standing aversion to pig meat—which tended, in these regions in those times, to be so laden with parasites as to make people immediately and lethally ill. Ultimately they adopted even the language of the Canaanites, yet they managed, even as they evolved from nomadic tribesmen to conquering armies and city builders, to preserve their own religious traditions and values.

The ancient Hebrews, like the desert sands, had a tendency to shift and scatter unexpectedly. One day they would found a settlement, or settle in among other founders and form a dune. The next day a wind would lift them and the dune would be gone.

The relative wealth of settled farming communities was often a temptation for nomadic tribespeople either to infiltrate peacefully or to invade, especially as the settled people had a tendency for expansionism, claiming more and more of the best pastureland around the Tigris, the Euphrates, the Jordan, and the Nile for themselves, forcing the nomads to travel deeper into the desert in search of hidden pastures (oases) for their herds.

With but one major exception, history has always been written by members of settled agricultural societies. From their vantage point, settled scribes depicted the nomads, the people outside the gates, as a cruel barbarian tide without culture, guided by instinct and not civilization. The Hebrew wanderers, however, became a literate people, and the Old Testament is one of the very few glimpses we have of the nomad's side of the story. In Genesis 4:2, for example, Abel is described as a keeper of sheep (a nomadic herdsman), "but Cain was a tiller of the ground" (a settled farmer). Cain, the farmer (and villain in the tale), is frowned upon by God. Abel, the shepherd (and tragic hero), is favored by God over Cain, who becomes so filled with jealousy that he rises (in Genesis 4–8) and kills his brother.

Farmers, like snakes, have come down through early Hebrew tradition with a bad reputation. As with the serpent and Eve's children, there is, at the beginning of time, enmity between the people of the cities and the people of the desert, and thus does the story of Cain and Abel provide mankind with a rare and valuable window on how those who lived outside the city walls viewed their world. As punishment for his deed, God (in Genesis 11–15) forces Cain—the builder of the first city, the founder of civilization, and the proverbial first murderer—to take up his brother's nomadic existence, to experience firsthand, as he travels from village to village, the hatred directed by the people behind the gates toward the wanderers. But God puts a mark upon Cain's head, so the other people will not kill him.*

Around 1628 B.C. one or more groups of Hebrew nomads (identified in the Bible as descended from a line leading from Cain through Abraham and Isaac), having taken up a sedentary life in Egypt, left the fertile pastures of the Nile and rediscovered their nomadic roots in the desert. After many years (four decades in Sinai, according to the Hebrew account in the Book of Exodus, or a general upsurge of migratory activity among *all* people lasting two centuries, according to archaeological accounts) they arrived like a sheet of wind-driven sand in the Jordan Valley, became farmers again, and, dunelike, formed new settlements among Canaanite and Philistine city builders. As the Hebrew populations grew, and as the walls around their towns thickened, other bands of nomadic people, driven in part by an interglacial warming that had been shrinking the fertile corridor and any outlying oases for more than six thousand years, swept in from the Arabian desert to ravage the farms outside the walls. These were the Midianites, whom Gideon chased across the Jordan. After the Babylonian king Nebuchadnezzar exiled the Hebrews to Iraq in 586 B.C., other people, including nomads, swarmed in to take the suddenly deedless land on the banks of the

---

*The mark of Cain is often interpreted as God's gift to the first murderer, that he may live to find redemption for his evil act. But equally important: Just who were these *other people*? The early chapters of Genesis give us Adam and Eve and their two sons, Cain and Abel. If we are to read the Bible literally, the total population of the world at the time Cain sets forth with God's mark upon his head (eventually to find a wife who bears him Enoch) was three. The confusion probably arises from Babylonian scholars, who, about 550 B.C., while assembling the first copies of the modern Bible, brought together the cultural prehistories of many tribes and, in the process, made Cain and Abel from Tribe A the sons of a legendary Adam and Eve from Tribe B. Other changes included deletion of almost all ancient references to Lilith, except in Isaiah 34:14: "There shall Lilith alight." The name is derived from the Babylonian Lilitu, a name given to "the monster of the night," which is itself derived from the Semitic word for "night." According to Hebrew tradition, Lilith, a woman of striking beauty, was Adam's first wife and was, for one offense or another, cast out of Eden before Eve was created. Since this was before the fall of Adam and Eve, she presumably lacked the knowledge of good and evil and therefore retained her immortality, but she could live only by night and subsisted on the blood of children and wild beasts. She was, as Babylonian tradition would have it, the first vampire.

Jordan. Later the Greeks took over; then the Romans; then the Arabs; then the Mongols and the Turks; then Christian Crusaders; then the Arabs again; then Hebrews—each bringing materials for new buildings, each leaving behind impressive traces of their existence, each raising higher and higher the mounds of debris upon which cities like Jerusalem continued to thrive.

Even during the first years of the last decade of the second millennium A.D., nomadic herdsmen still live outside the city gates. They have little comprehension of, and no respect for, passports and transport papers. Because they follow the watering holes and the grasses, their way of living has remained unchanged since antiquity, except for such essentials as gasoline to run their portable electric generators (which they purchase by bringing coins and other artifacts mined in the wilderness to Jerusalem, where they fuel the antiquities trade; a four-pound sack of Roman coins may bring the equivalent of fifteen American dollars from an Israeli dealer). In the deserts, far from the Israeli-occupied and well-irrigated West Bank, I have seen TV antennae sticking out the tops of tents. One herdsman told me that the most popular program among his people is an American soap opera called *Falcon Crest*.

It is a fascinating turn of events, when we take pause to think that the once-nomadic Hebrews are the people who now rule behind the walls of the cities. As history tends to be written by those in control of city walls (which in modern times translates to control of the press), recent history looks down upon the increasingly landless Palestinians as the world's newest barbarian tide. The atrocities committed on both sides of the wall are appalling, and in this land of "an eye for an eye," with virtually every Palestinian and every Israelite knowing a brother or a cousin killed by the opposing side, can there be any end to the downward spiral? On one point the oral traditions that became the Bible agree heartily with the evolutionary history written on our mitochondrial genes: In terms of geologic time, the ancestral parting of Palestinians (Cain→Abraham→Ishmael) and Israelites (Cain→Abraham→Isaac) occurred only a nanosecond ago. Out there in the night, somewhere beyond the helicopter search lamps; and above, in that false galaxy of city lights, the drama is brother killing brother.

As of late, anthropologists have been trying to portray man as a kinder, gentler species. They tend to cast us in a light of true cleverness, having come a long way from our remotest forebears.

An older tradition held that we must be even more brutish than the beasts who preceded us because man was the only species that murdered its own kind. It became a major theme of Rod Serling and Pierre Boulle's original screenplay for *Planet of the Apes* ("Man destroys Man. Apes do not destroy apes"), and once elevated to cultural icon, it seemed destined to endure forever, at least until explorers like Jane Goodall began living among the apes.

We now know that murder is both blatant and widespread among our closest mitochondrial cousins, the chimpanzees, including such clearly inhuman behaviors as cannibalism committed against children. By most accounts, any individual chimpanzee's odds of death by violence before reaching adulthood are more than twice that of a child growing up in Israel's occupied territories, or in almost any other human war zone. And so the anthropologists are reversing the old tradition, patting humanity on the back and saying many good things about us. The argument goes something like this: "Murder is nothing new to the world, and in fact, man is less skilled at it than the animals, so we must therefore be better than the animals."

I'm not so sure about that. When a chimpanzee bites another chimpanzee to death in the wilderness, it is very obviously murder. When a man stabs another man in a robbery attempt, it is very obviously murder. But I suspect there is another, more insidious form of murder that we fail to see, perhaps because it is so commonplace as to be socially and even legally acceptable. The anthropologists have used, as a basis for comparing man with the beasts, only the most blatant acts of killing, which usually involve the spilling of blood, followed by arrests (which become statistics). Among humans such acts generally take place during moments of recklessness in which logic has suddenly dropped dead. It's the more logical, bloodless, and officially uncounted murders that worry me. I've known only one person, during the past ten years, to die by what would officially be called murder (a victim of terrorism in Sri Lanka). During this same decade I've known at least two cases in which an employer tried to get out of paying benefits to near-retirement-age workers by deliberately, and with calculated precision, setting out to harass them into heart attacks. In one instance the victim had just returned to work after a triple bypass operation (in the business world, this is called "just business"); I've known professors who ganged up on and mentally tortured a student for three years, then stood in the tearoom the day after he leaped from the eighth floor of a university building, joking about how the "poor fool" couldn't even kill himself in the right way, for if he had jumped from a higher place, he would not have ended up screaming on the sidewalk for twenty minutes before he died (in the academic world, this behavior, though rare, is sufficiently commonplace to have a name: "playing snap with your students"); I've known an exceptionally bright high school student similarly driven to suicide by his peers (in the adolescent world, this is known as "having fun with the kid who doesn't fit in"); I've known a woman who, upon being told by doctors that if her alcoholic mother continued drinking, she would be dead within six months, immediately began delivering cases of vodka to her mother's house (in the world of estate inheritance, this is known as "if you can't teach an old dog new tricks, why bother?"); I know of a second case in which this same "old dog" strategy was used against a husband.

Legal murder—it's all around us. In America, health insurance companies are notorious for arbitrarily refusing to pay for a wide and ever-widening range of medical procedures. Knowing that a cancer patient awaiting an expensive bone marrow transplant is living in a time-critical situation, it is no accident or oversight that insurance company lawyers consistently draw litigation out for months or years, until the patient either is dead or has deteriorated to a state that ultimately precludes the procedure. No district attorney has ever called this murder, and I'm sure even the insurance company lawyers don't think of it that way. "It's just maximizing the profit potential."

Here, in Israel, when Saddam Hussein's rocket attacks were imminent, every Israeli citizen was given a gas mask, every citizen, that is, except the Palestinians, who on government census records "did not count as Israeli citizens." A woman in Bethlehem, a Palestinian, sued the Israeli government in the Supreme Court, urging that "Arab life is as dear as Jewish life and we should get gas masks, too." As the first rockets fell on Tel Aviv, the Supreme Court overruled the government's hard-liners, decreeing that gas masks would be made available to the entire population. It was a moot point. Under a logic that dictated letting Arab-launched missiles gas Arabs at their targets, the necessary number of gas masks had never been produced in the first place. Of the 1.75 million Palestinians living in Israel, only one tenth were issued masks before the government "ran out."*

Looking around, I see no evidence that the frequency of killing necessarily goes down with increasing intelligence. If anything, our species has become far more skilled at it. The same swollen brains that gave us a knowledge of death and the ability to distinguish between good and evil simply increased the complexity of our reasons for killing. As American

*During his Rosh Hashanah sermon of September 28, 1992, in celebration of the Jewish New Year 5753, the senior rabbi of the world's largest Jewish temple offered a glimmer of hope. "Every Jew has to weep when anyone is killed," Rabbi Ronald Sobel told his congregation, "because the cycle of hatred has not been broken. There is a better possibility of peace now than ever since the creation of Israel. . . . This is central to Jewish religious life: The saving of a human life means all human life, irrespective of religion, race, or gender "

insurance lawyers and "old dog" strategists demonstrate so vividly, we stand apart from the beasts in that we can think our intentions out more clearly and with a little planning carry them out more subtly.

We are *Homo sapiens*—"the wise primate"—planet Earth's compulsively clever species, makers of helicopter gunships, rockets, false galaxies, legal and ethical systems that all too often are not. If this is where we have come to, we have not come very far.

II

BLOOD UPON THE WATERS

6

THE
RIVERS
OF
BABYLON

In the sky came a black cloud from the foundation of heaven. Inside it Adad thundered . . . turning all light to darkness . . . when the seventh day came, the cyclone died away . . . the flood ceased. And all mankind had turned to clay. The ground was flat like a roof.
—EPIC OF GILGAMESH

And He will stretch out his hand against the north, and destroy Assyria [the Tigris-Euphrates region]; and will make Nineveh a desolation. A dry waste like the desert.
—ZEPHANIAH 2:13

THE LOST CITY OF UR, SOUTHERN IRAQ, A.D. 1929

It had to be the flattest, most desolate place on Earth. Freezing in winter-time; 114°F at midnight during the summer. Mud everywhere. And when the dry season came, you wished for the mud to return. And when the wet season came, you wished for the dust. Anyone taking in air and in his right mind hated this land. Leonard Woolley loved it.

There was never any question about where to dig. Every dune, every protrusion on the horizon, was without exception man-made. Six hundred fifty miles north, the Arabian microcontinent was in collision with Turkey, as it had been for seventeen million years. The impact was pushing up the Taurus and Pontine mountains, which, as they increased in stature, were slowly being hewn down by the erosive forces of wind, rain, and ice. Thousands of tiny streams had coalesced to form the headwaters of the Tigris and Euphrates, and as they carried bits of pulverized mountain southward, they were building, an inch at a time, a great outwash plain between the rivers, known to Woolley as Mesopotamia and to the ancients by such names as Babylonia, Sumer, and the Plain of Eden.

At 7:49 A.M., on the three-month anniversary of the Black Monday now ending the Industrial Revolution's Age of Optimism, Leonard Woolley and his chief excavator, Hamoudi, stood on the eastern flank of an eroded mud-brick mound. The sun was still low in the sky, and the air was so cold that both men knew the pools of water on the mound would not even begin to thaw by day's end; yet even the faintest warming of the sun's rays was more than welcome. From where they stood, they could look across the great pit, where two hundred workers had removed protective tarps and were now sifting through an unfrozen bed of bone-dry silt whose upper-most reaches began about ten feet below the plain. For a long time the two men said nothing, but stood on the ruined temple, taking in the morning air, feeling the stiff wind from the south tugging at their coats and raising the chill factor. Finally Woolley said, "I think I understand, now, why ancient people worshiped the Sun."

"Maybe you would rather be somewhere else," Hamoudi urged.

"Not on your life."

"Nor I," said the excavator, who had begun his career as a bandit and whose four sons, also outlaws, had recently joined him at the site. "Understand, Professor, that there have been two passions in my life: archaeology and violence."

"You wicked old murderer."

"Murderer?" Hamoudi cried, either expressing true indignation or feigning it very well (Woolley could not tell which). "I have never in all my life killed a man for money, only for fun!"

~~~~~

Like most archaeologists of his generation, Leonard Woolley started out studying theology. During the spring of 1904, at New College, Oxford, he discovered to his dismay that theology really did not interest him. He was, during that last semester, moving through life like a rudderless ship, without a decided course. Fortunately his anxiety did not go unnoticed, and so it was that he received a summons to the office of Warden Spooner.

"Ah, Mr. Woolley," began the warden, "am I correct in saying that when you came here to Oxford, you had every intention of taking the holy orders?"

"That is correct," Woolley said.

"And am I correct in saying that you are now having second thoughts?"

Woolley weighed truth against caution.

"Perhaps."

"Then what do you propose to do with the rest of your life?"

"I might become a schoolmaster," the student murmured, almost at a whisper.

"Oh, yes, a schoolmaster, really," Spooner said flatly. "Well, Mr. Woolley, I have decided that you shall become an archaeologist."

At the time Woolley was not entirely certain what an archaeologist was. The only thing he *was* sure of was that one did not argue with Spooner; that was the long and short of it, plain and simple. So he shifted his attention toward vanished civilizations, and never regretted it.

He began his apprenticeship under a man named Arthur Evans, who had just found splendid palaces, warehouses, and multistory apartment buildings in four buried cities on the island of Crete. Beneath the ancient streets ran a sewage system of a sophistication to rival the one then being planned for San Francisco. The apartments were equipped with bathtubs and flush toilets, built more than three thousand years before the British Empire "invented" them. And strangest of all was the fact that Evans and Woolley had to dig down through layers of Classical Greek debris before they could reach the palaces, warehouses, and apartments. This meant that the more advanced and hitherto unsuspected civilization was also the older civilization. Evans quickly named the lost people Minoans, after the mythical King Minos. His colleague K. T. Frost proposed that the Greeks had in fact inherited or copied the earlier civilization's art forms and technological know-how—just as Plato said they had done, though he'd called the mother culture by a different and more immediately recognizable name: Atlanteans.

In 1907 Woolley followed a trail of Minoan shipping routes to the Nile, whose inhabitants had called them by still another name, the Keftiu. They were the only people the bold and daring Egyptians considered civilized enough to merit a name of their own (all other outsiders were simply "barbarians who are an abomination against Ra"). As way led unto way, the young archaeologist wandered for three years in the Nile Valley, became an expert Egyptologist, then began following the campaigns of Tuthmosis III out of Karnak into the Jordan Valley, up the rivers that fed the Galilee, and finally to the headwaters of the Euphrates, where, in 1912, at age thirty-one, he was given command of his own archaeological site.

His two best-known assistants were Max Mallowan, destined to become a famed archaeologist in his own right, and a young radical named Thomas Edward Lawrence, destined not merely to explore history but to make his mark upon it. Both were veterans of Warden Spooner at Oxford, who had likewise ordered them into archaeology.

Woolley found Lawrence a competent and enthusiastic, if not sometimes an *overly* enthusiastic archaeologist who darted with lightning speed from the work immediately at hand to a half dozen other and at best tangential interests. To begin, there was his obsession with the mythical city of Ubar—"the Atlantis of the Sands" he called it—said to have been swallowed whole by the desert in a single day and night, after more than two thousand years of prosperity. If it had existed at all, it was located all the way downriver, somewhere near the Persian Gulf. And then there was his excessive interest in the railroad the Germans were building to Baghdad. Lawrence and Hamoudi were always stealing away to vantage points on the mounds, taking with them five cameras and a telephoto lens. The intelligence work for which Woolley's digs provided cover focused largely on vulnerable bridges and culverts. Britain was concerned that if the growing tensions with Germany should escalate to war, and if Turkey should ally itself with the adversary, a new overland supply route to the increasingly important oilfields near Basra (sidestepping a British-controlled choke point at the Suez Canal) would give the Germans an intolerable advantage. Lawrence was under orders to fit in with and gain the trust of the local people, for among them he would recruit the saboteurs of the German line. As with everything he did in life, he approached this task with fanatical vigor, not merely voicing an admiration for and an understanding of the Arab and Kurdish people but genuinely developing these qualities, until at last he became indistinguishable from the tribesmen he was sent to infiltrate, even to the extent of turning with them against British rule.

Leonard Woolley had decided from the start that he liked the man. They shared three mutually agreeable traits: an impish sense of humor, a penchant for luxurious living, and an approach to archaeology that was, by the standards of any time, curiously erratic. In the wilderness of Iraq they built an archaeological villa such as none the world had ever seen before or would ever see again. There were ankle-deep sheepskin rugs, a huge fireplace, coffee tables topped with ancient Babylonian statues, and a large bathtub with beaten copper trim. "Woolley fancies himself in the bathroom: a gleaming mosaic floor reflecting his shining body against the contrast of the red-stuccoed walls," wrote Lawrence to a friend.

"There is gold scattered under our beds," added Mallowan. One of the royal tombs had looked to him like a golden carpet, and the treasures were overflowing the explorers' capacity for storage and shipping back to Oxford. The villa was constantly receiving visitors, ranging from Arab chieftains and magistrates to the British mystery writer Agatha Christie, who seemed to have developed a crush on Max Mallowan and was becoming inseparable from him. One afternoon, when all four were together, Mallowan served up dates in a gold cup from the tomb of Shub-ad and tea in Hittite clay goblets.

Christie's nose wrinkled. "Isn't that dangerous? Won't these things break?"

Mallowan let out a loud laugh, then motioned toward Lawrence, who wore a robe of gold and silver thread weighing sixty pounds.

"If I drop this goblet," Lawrence said, "the British Museum will be glad to have the pieces."

"That's crazy," Christie said.

Mallowan downed the last of his tea. "That's archaeology, my dear."

Leonard Woolley and his flamboyant team were, if anything, committed (committed to *what* was not always clear, and from time to time Agatha Christie suspected commitment to an asylum would not have been out of order). The excavations began every morning at daybreak and rarely ended before midnight. Woolley himself often stayed awake repairing or trying to draw meaning from some artifact until two or three in the morning. He belonged to the exotic and enviable minority who required only about three hours' sleep out of every twenty-four.

"He had a genius," Mallowan recalled in his memoirs, "for finding what he set out to look for, and the patience to know when to wait." When his trial trenches probed the edge of a tomb crammed with fossil impressions of wooden harps and fragile, gold-inlaid objects, he saw at once that at least four years more of technical experience would be required before he was up to the task of excavating the treasures without reaping more havoc than preservation. So he sealed the tomb and waited. "My aim," he wrote, "is to get history, not to fill museum cases unless both me and my men are duly trained." When Woolley and Mallowan led Agatha Christie down a winding alleyway into the foundations of an abandoned Bronze Age house, they amazed her by actually mentioning the man who had lived there by name. From a pile of baked-clay tablets heaped in a corner, they began reading off how many urns of grain the master of the house had purchased one season and how much he had paid for them. To these men the Euphrates River was not a graveyard of forgotten cities but a thriving civilization with merchant ships in its canals, children playing in the streets, and households purchasing wheat. To them, as they picked through its ruins, the past was still alive. They were, in a very real sense, unstuck in time; and time travel, any committed paleontologist or archaeologist could have told Christie, is addictive. That very first fossil or pottery shard given to the wrong child may begin a quest no less obsessive than that resulting from the first sip of whiskey given to an alcoholic. Thanks to Warden Spooner, Woolley, Mallowan, and Lawrence were mainlining ancient Babylon, meaning that anyone who tried to stand in their way would be taking a terrible risk.

Although they had built camps and villas amid the ruins, it could not be thought that they lived there. The archaeologists, like their Kurdish workers, were nomads who went wherever their work called them, breaking down and carrying their tents and villas from place to place, and it did not matter to them what kind of land they lived on. Their Kurdish excavators were a fiercely proud people, coming from diverse tribes that existed in a constant state of suppressed warfare. They tolerated the Arabs, loathed the Turks, and hated the Germans. The Germans, in turn, hated Woolley. The animosities came to an inevitable boil one spring morning when a party of German engineers asked if they could run the Baghdad railway through the walls of a city Woolley had been exploring.

"You start digging away that wall," Woolley announced, "and you had better dig yourselves a few extra holes, because I intend to shoot you."

Without any counterthreats or fuss, the Germans left. They waited patiently for Woolley's team to move downriver in early summer, then swept in and began demolishing the site. Lawrence and Hamoudi soon discovered the destruction. They sent a small contingent to lie down on the threatened walls and hold the railroad construction team off at gunpoint while Lawrence telegraphed the authorities. The Arab and Kurdish workers employed by the Germans allied themselves with Woolley, Hamoudi, and Lawrence rather than with their own employers, who had always regarded them as inferior nomadic races and had let them know it. The Germans had three choices. First, they could fight Woolley and their own traitorous workers, with certain loss of life, ill will among surviving workers, and intolerable work slowdowns resulting. Second, they could reroute the line around Woolley's precious mounds of rubble, again with intolerable delays resulting. Third, they could use their influence to bring government pressure on Woolley, bog him down in bureaucratic procedural games, and stop his excavations dead.

A local governor summoned Lawrence to trial on a technicality involving "the unauthorized transport of rocks." When Woolley accompanied him to court, the governor refused to show up, and a magistrate ordered all the excavation notebooks confiscated by court officials. Woolley loudly challenged the case's legal validity, informed the magistrate that he refused to accept any judgment rendered against his team, and demanded the immediate return of his papers.

"I appreciate your opinions," the magistrate said. "One of the sad things about this case is the total breakdown in communication between yourself and the governor. It seems as if you are determined to dispute our decision even if I find that your excavations were correctly closed down."

"This case is over," Woolley said. "And I now demand that you bring the governor here and return my papers."

"The governor will not see you."

"I'm not going to leave until I get the papers!"

"In that case, you won't leave till we are both very old men." The magistrate then broke into laughter, and at least twenty curious onlookers who had filled the courtroom joined him. Woolley was unmoved. Cold and cheerless, he ascended the bench, drew a revolver, and held it against the magistrate's left ear. Hamoudi drew two revolvers, and Lawrence, also brandishing firearms, left to waylay the governor. As the laughter died away, Woolley proceeded to give the magistrate a short but intensive course on field archaeology in general and his excavations in particular: "Do you see that table? Upon it, during the next five minutes, one of two things shall be produced. All of my papers. Or your brains."

Two minutes later the notes and excavation permits were back in Woolley's possession. Hours later he arrived at the excavation to much fanfare from the workmen, who, in addition to tolerating each other, loathing the Turks, and hating the Germans, were coming to adore Woolley's direct dealings. Lawrence, too, was deeply impressed, "We are all very amused," he wrote home. No one ever learned what the Germans thought. Without any further words, they rerouted their tracks around the disputed walls and then replanned the entire Baghdad line, carefully avoiding any mound or irregularity on the plains that even looked like a place Woolley might want to dig up.

SUMMER A.D. 1991

Looking across Arabia from a height of five miles, I see a low wall of black haze spanning the northeast horizon. It is Kuwait afire, a dead end. Like a thirty-six-year-old Huck Finn, I started out down the Nile on a boat, probing gradually forward into archaeological time as I tracked the fertile corridor north, then east, then north again. Now, aboard a jet, I am skipping effortlessly over the desert between the Jordan River and the Tigris-Euphrates Delta, crossing in minutes over lands traveled only with the greatest difficulty, by wayfarers of another kind, in those millennia before men began pulling bauxite from the rocks and fashioning it into wings and engine casings. I leave the Jordan behind, flying east toward Ur and Ut-Napishtim's flood. But I will be able to see, from the air, only the general direction of Leonard Woolley's campsites. No one is permitted to land at Ur, or to visit Babylon, or Mashkan-shapir; not this season, at least. Too few weeks have passed since the Scuds, the Patriots, and the Tomahawks scratched fire across the sky. Even from a height of five miles, even without binoculars, civilization's cradle now looks more like its tomb. Today the wall of darkness completely obscures the Persian Gulf, into which the Tigris and Euphrates rivers drain. According to the latest issue of *Nature,* the soot rains acid upon India's rice fields and is detectable as far away as Japan.

By the end of 1929, Thomas Edward Lawrence had struck out on his own, moving south into the land over which I now fly. He was searching, among other things, for his "Atlantis of the Sands." Seen from the air, the ghostly imprint of an ancient road points like an arrow in the right direction. Lawrence had searched in vain for that road, and must have crossed over it many times without even knowing, because from the ground it was completely invisible. According to the oral traditions of the Arabs, not until men climbed a rope to heaven would the roads and the lost city be seen again. And so it came to pass that a half century after Lawrence's death a machine he could never have anticipated deposited behind itself a column of exhaust that stood away from the Earth like a mighty length of rope. Riding the space shuttle *Challenger,* six men climbed two hundred miles into the far sky and photographed, for the first time, the city and its roads.

According to the latest archaeological evidence, the city appears to have been inhabited from 2800 B.C. until about A.D. 100, when, if one believes in this sort of thing, "God commanded the earth to swallow it up." The Christians knew the city as Ubar, the shipping center for frankincense (a fragrant form of amber) and traditionally the source of the frankincense offered to the infant Jesus by three wise men from the East. Known as Iram, "the city of towers" in Islam's sacred Koran, it came to have a fate like that of Atlantis in Plato's *Critias* and Sodom in the Bible. As with most destroyed cities, even in relatively recent times (including the 1902 volcanic destruction of Saint-Pierre, said afterward to have been deserving of death because its inhabitants practiced voodoo), when an explanation is sought for evil events in a presumably just universe, history brands the people villains because they became victims, victims because they became villains. In the Koran the Iramites grew too proud of their city. Thinking themselves more powerful than nature itself, they began living sinful, unrepentant lives. Predestruction Iram is described as a "many-columned city, whose like has not been built in the entire land." Erected as an "imitation of Paradise," the settlement was renowned for its porcelain towers, irrigation channels, vast groves of fruit trees, and incomparable wealth. It is not difficult to understand how, rising out of the desert, six days from the nearest drinking water, it took on a mythic quality.

One oral tradition holds that between 722 and 586 B.C., when their wars with Babylon seemed lost, the Hebrews moved their most sacred possession from the Temple of Solomon in Jerusalem. According to legend, Iram is one of the possible resting places for the lost Ark of the Covenant. Arab tradition holds that the Ark is hidden in a secret chamber called the Cave of Blue Light.

Out there in the desert, the land under and around Iram is now known to be honeycombed with watery underground passageways. I've seen pictures of their entrances, and to judge from these, decades will pass before any of them are fully explored. Thousands of centuries old, the caves run for miles and miles in every direction.

A hole in the Earth marks the center of Iram. It is easy to see what happened, easy to see how an Atlantis-like legend got started. The Iramites built their city over a huge limestone cavern. Pools on the cavern floor provided a limitless water supply, with which the people irrigated the desert into luxuriant bloom. For generations the palaces, warehouses, towers, and apartment complexes grew in stature, until one day they collapsed the roof of the cavern under their own weight. The very heart of the metropolis went down into the Earth, as if summoned to hell; and what remained, according to legend, was consumed by the desert in a sandstorm that lasted three full years.

∿∿∿∿∿

Although at least one city appears to have fallen into a cave, and although it is true that archaeologists tend to dig things out of the ground, this is not to say that the structures and implements of ancient civilizations have somehow sunk below the surface. Typically, as in the case of Karnak, Jerusalem, or Ur in southern Iraq, old cities in locations favorable to succeeding generations (that is, prime real estate) are covered over with newer buildings, and over the course of four or eight thousand years, the most ancient and consistently inhabited settlements raise tier upon tier to become mounds covering several square miles and standing more than eighty feet above the original ground level.

A look at lower Manhattan provides a convenient analogy to how archaeological sites evolve and what items are most likely to remain behind for the archaeologists.

Construction workers digging new foundations sometimes come across the remains of Indian campfires.* Other times they expose hundred-year-old ships where colonial ports have silted in and been paved over. Yet even among today's glass and steel towers, Trinity Church stands much as it appeared in A.D. 1732. If you could travel back to Ur about 1628 B.C., you

---

*The word *Indian* is actually an erroneous name. Columbus, when he sailed east, miscalculated the diameter of the Earth. Everyone else told him that the Earth, then known to have been round, was much larger than he believed and that one could not circle halfway around it by sailing across an Atlantic Ocean only two thousand miles wide, thus forging a shorter route to India. When Columbus found America on the other side of the Atlantic, he believed he had indeed landed in India and proved his small-Earth hypothesis correct. He called the islands he found the West Indies, and he called the people Indians, and we have been calling them that ever since.

Archaeologists are always finding cities underground, but only rarely is this the result of a city sinking into the Earth. If, for example, the present blacktopping of American streets (which has, in only three decades, put once raised sidewalks and medians below street level) should continue as a trend, then in a thousand years most first-floor dwellings will become basements. A similar street-building process is illustrated near the gates of Karnak, where a woman stands on the original Egyptian street level. More than twenty feet overhead, the back doors of a mosque show the height to which street level had climbed by the Islamic Era, after which the archaeologists came along and excavated the rear exit down to the old street level, producing the world's most harrowing first step.

would see the same phenomenon: older structures still maintained and preserved among the new.

Now, if we permitted the East River to dry up and added a massive economic collapse, and a war, leading to the virtual abandonment of New York City, anything of value would likely be removed by the departing masses, and by scavenging nomads. Even a common Coca-Cola bottle might be valued as a vessel for carrying water and would likely be left behind for future archaeologists only if it were rendered useless by breakage— which explains why most of the pottery one finds beneath Jerusalem and Ur is broken and discarded as either road gravel or the contents of rubbish heaps. If fire, invasion, and neglect further reduce Manhattan to a twenty-foot layer of rubble, the foundations of the World Trade Center's twin towers and Trinity Church will still be there for the archaeologists, along with the sidewalks, sewage systems, the layout of the streets, broken bottles, chips of chinaware, perhaps even a barely functioning 1927 Remington typewriter thought too heavy and too useless to carry off, found amid plastic containers that might cause our hypothetical explorers to wonder who Dannon Yogurt might have been. Beneath the churchyard of Trinity, future construction workers, excavating a new foundation upon the twenty-foot mound, may find the graves of priests; and the archaeologists summoned to the site, as they examine the contents of rapidly buried, well-preserved, and unplundered caskets, will learn much about the people who once inhabited old New York. From the silver crucifix, or the Saint Christopher medal clasped in two bony hands, they will learn something about ancient religion. From the teeth of individuals ranging over three hundred years they will follow the development of dental technology and shifting eating habits.

The mounds of Jerusalem and Ur are not much different. If you dig down deep enough, you will find nomadic campsites from tens of thousands of years ago, and everything in between. In Iraq, one Babylonian mound whose most recent inhabitants, about three hundred years ago, left a row of mud-brick houses crumbling in silence just as New York City began to take shape, is capped by plaster columns protruding through rubble. The old, pre-seventeenth-century ground surface, on which the columns stand, is more than a yard below the mud-brick town. They were built around A.D. 1220 by Genghis Khan, the nomadic Mongol conqueror of most of Asia and of Europe to the Dneper River. Passing through the land between the Tigris and Euphrates, he paused to lay a group of captives upon their backs, chained together by their ankles, with their feet and a centerpost forming the axis of a circle. He then laid another circle of ankle-bound prisoners on top of the first, and another circle upon the second—another, another, another—until at last the upper layer of prisoners crushed the life out of the layers beneath. When the column of human bodies had grown as tall as flesh and blood engineering would allow, Khan's men plastered it over, even as the upper layers still writhed and strained, and began construction of another column a few yards east. Much of the plaster has long since fractured and tumbled away in slivers. A visitor to Iraq's pillars—some of them standing twenty feet high—will find the tops of prisoners' skulls pushing through into sunlight.*

One vanished civilization after another had flourished beneath Genghis Khan's pillars—successful ones had lasted four hundred years, or even a thousand; unsuccessful ones, only a few generations—but each had deposited a distinctive layer of broken pottery, midden heaps, and the rubble of its buildings as they were ravaged (often with their foundations reused by the next civilization) and their owners were led away to slavery or buried in mass graves. Downward and downward, the picture is one of ruins growing on top of ruins until at last, near 2800 B.C., the spade slices through thick brown clay, nearly twelve feet of it. It looks at first glance like the original ground surface upon which the settlers of the Tigris-Euphrates riverworld built the first cities, but closer inspection reveals the ground surface to be an enormous accumulation of river sediments left behind by a flood that must surely have spanned much of the world between the two rivers, from horizon to horizon. The water, tens of millions of tons, tore great chunks out of the Earth and dispersed them impartially. Far below Babylon, Ur, and the City of the Dead lay thick foundations of clay, and when you probe beneath the clay, you find the smashed remains of dwellings belonging to an earlier, unsuspected civilization. East of Karnak and Jerusalem, in the land identified in the Bible as the location of Eden—the birthplace of Eve, Cain, Shem, Noah, and Abraham—a whole world came to an end.

~~~~~~

*Genghis Khan left a trail of human pillars across Iran and Iraq. Along with "the bloody eagle" (an invention whose details history has not recorded but which was apparently a means of hanging a man in such a manner as to draw the lungs out, winglike, through his back), the pillars became a sort of calling card. Like the first-known scalping of a *Homo erectus* skull, it is impossible to say precisely why such things were done. What can be said is that the past often provides signposts to the future, and as some of us begin to consider the colonization of distant worlds, knowing that behaviors no less violent than Khan's were recently practiced in the country that produced Beethoven and Brahms (and as late as the 1990s were being copied by the Serbs), we'd better take pause to question what we may be propagating out there.

It looks like the most unlikely place on Earth for the birth of a civilization, the broad, flat plain where the Euphrates meets the Tigris and they drain together into the Persian Gulf. At the far end of the corridor of riverworlds, in Egypt, the *adj mer* had flint, copper, and tin. There were limestone and granite, gold and silver. "Everything the Egyptians needed existed within Egypt," explains Stony Brook University archaeologist Elizabeth Stone.

"The inhabitants of ancient Iraq had neither stone nor timber from which to build their cities. The only tree you will find growing there is the date palm, which does not provide you with very good construction material. Any pebble that you find in southern Iraq is imported; it doesn't matter how small it is. There is nothing larger than a fleck of silt that occurs naturally in that part of the world. The people who first arrived on the plain found no limestone to be cut into blocks, no copper to be shaped into tools with which to cut the blocks. There were no gems or deposits of gold that could be traded for imports; there was nothing, in fact, besides mud and sunshine."

Evidently, mud and sunshine were enough. The alluvial plains, built from pulverized bits of mountain, were a nourishing mineral broth awaiting only the introduction of agriculture. A favorable combination of fertile topsoil and virtually year-round sunshine must have produced a grain surplus almost from the moment people began settling along the riverbanks, and the rivers that sustained the surplus also provided a convenient trade route along which grain could be exported to the Iramites and other city builders in exchange for copper tools. With the arrival of copper tools, deeper, longer lines were drawn in the mud. Irrigation channels expanded at a furious rate. So, too, did the grain surplus.

As the channel diggers became canal builders, the mud they plowed aside began increasingly to shape their civilization. Cast in wooden molds to produce uniform blocks, it was used to construct mud-brick houses, or fired to rock hardness, it formed the foundations of palaces, temples, and the estates of wealthy merchants. Mud, shaped on potter's wheels, became storage jars of all shapes and sizes. Mud, flattened into tablets, etched with a stylus, and baked, would allow the Babylonian canal builders, long after they were dead, to speak.

To a civilization that raised itself from a mud plain, a Creation legend in which the first man was shaped from mud by a divine being, much as a potter would form a clay figure, does not seem at all out of place. The biblical Creation story, which took its present form among Hebrews exiled to Iraq in 586 B.C., was in fact the Babylonian Creation tale with most of its polytheistic references removed. Indeed, in Genesis 2:8–14, the Old Testament actually places man's first residence squarely on the Euphrates.

By 2500 B.C. pictographic symbols, in which a wavy line (water) next to a mouth meant "to drink," had evolved into phonetic symbols in which patterns of triangular shapes pressed into slabs of mud represented sounds much like the combinations of letters on this page (archaeologists have named these triangular indentations cuneiform writing, after the Latin word for "wedge-shaped"). Once baked, the slabs, like all pottery, became, in the archaeological sense, virtually indestructible. On such slabs Babylonian schoolchildren began recording lessons in Euclidean geometry more than two thousand years before the birth of Euclid. Merchants recorded transactions, kings recorded laws, and lawyers recorded charges, countercharges, and summations. And the poets began etching into clay their outlooks on the world, including cherished legends and cautionary tales that had been passed from generation to generation through fireside epics sung to the accompaniment of harp or lyre.

On more than one occasion Woolley and Lawrence found whole libraries of cuneiform tablets buried near city gates. One deposit ran three feet deep, and at the bottom someone had conveniently left a dictionary. The ruins were no longer mute. A linear mound, its features softened by erosion, ran more than a mile over the plain of the Euphrates, revealing where a city wall had once stood. A clay tablet brought that same featureless wall immediately alive, for according to the epic of Gilgamesh, it once "glistened in the harsh sun with the brilliance of copper." The archaeologists found remnants of a canal cutting through the heart of Ur's houses, shops, and bazaars, and the houses themselves, as reconstructed from their foundations and whatever forgotten debris had fallen into them, were sophisticated two-story villas whose rooms were tiled and plastered. Shards of pottery near a front door were parts of a basin whose purpose could only be guessed at until a piece of contemporary writing revealed that visitors, when they entered a house, were obliged to wash the dust from their hands and feet before proceeding to the inner courtyard and reception room. "The houses themselves reveal comfort and even luxury," wrote Woolley. "We found copies of the hymns which were used in the services of the temples and together with them mathematical tables. On these tables were anything from plain addition sums to formulae for the extraction of square and cube roots."

The purpose of the city gates, and the miles of fortresslike walls, was obvious to the archaeologists; but the walls themselves could not identify the enemies of Ur. The triangular etchings, however, revealed quarrels arising from the expanding boundaries of the city-state encroaching upon the borders of other cities. While the geography of settlements deposited linearly along a corridor of riverworlds made it virtually impossible for any one kingdom to spread a universal empire over the entire length of the corridor and, by precluding this, sustained in perpetuity the sort of competition that stimulated technological development, this same favorable condition made the inhabitants of the Euphrates, about 2500 B.C., the first people to learn how easily one could overdose on a stimulant. Around the cities the wheat fields expanded to rival the American Midwest today. But every new canal, every new irrigation grid lessened the amount of water available to the cities downstream. According to the cuneiform annals, intercity hostilities were already progressing to open warfare when the very technological advances that had made the deserts bloom began transforming them back again into deserts. Every gallon of water carried a microscopic quantity of salt, and as the water evaporated from the soil, the heavier salts remained behind. From this same process the fresh water of the Jordan River had, molecule by molecule, built the Dead Sea into the saltiest body of water on Earth. Here and there on the Euphrates plain the ground surface began to resemble a winter frost. The land could not recover until wind blew the salt away and river flooding deposited new layers of topsoil. This required, at best, decades, during which no new stresses could be put upon the land. Yet if they were to prosper, the city-states had to expand. From a strictly economic perspective, they could not afford to wait; and so, as irrigation networks continued to spread like a mighty web, warehouse scribes recorded diminishing returns and wildly inflated food prices. Property values declined. There followed economic collapse, the charging of city gates, the pulling down of monuments. "Evil is descending on our land," wrote one inhabitant. "Ur is destroyed. Bitter is its lament. Bodies dissolve like fat in the sun. The gods have abandoned us like migrating birds. Smoke lies on our city like a shroud."

Populations crashed, and as the land slowly began to recover, Semitic tribespeople who had long been drifting in from the deserts of the Arabian Peninsula found the city gates more poorly defended than they had been in previous centuries. Indeed, many had abandoned city life, following the caravan routes into Arabia, becoming refugees among nomadic shepherds and traders who had never known a city. Among the Semitic insurgents from the western deserts was the man referred to in cuneiform texts as Sargon the Great. His tribe settled along the Euphrates, in a land that came to be called Akkad, and the Semitic tribespeople, Akkadians. Sargon was born about 2485 B.C., and in his search for the past Leonard Woolley came across tablets linking a curiously familiar legend to his name. Nearly a thousand years before the story of Moses and the plagues of Egypt, the infant Sargon was said to have been placed by his mother in a pitch-covered basket and set adrift on the Euphrates. A canal builder found him, and from this beginning he rose to become a king. For fifty years he ruled from the Persian Gulf to the headwaters of the Euphrates. About midway on the Euphrates he built the capital city of Agade (Akkad), from which he appointed his fellow Akkadians to administrative posts in other, slowly recovering city-states.

Of himself, Sargon wrote: "Sargon the mighty king of Akkad, am I. My mother was lowly, my father I knew not. The brother of my father dwelt in the mountains. My city is Agade, which lieth on the bank of the Euphrates."

Agade itself is referred to in several Ur tablets as one of the most magnificent cities ever built by the hand of man. Rising from virgin, unsalted ground, it boasted the widest canals, the largest gate, the most people, and a pyramidlike temple two hundred feet wide at its base. Yet of this city, not one brick stands upon another to mark Sargon's achievements. Today archaeologists cannot guess within ten miles where the king's palace once stood. Not quite a century had passed before the next subsidence into famine and chaos ended the Sargon dynasty.

About 2100 B.C., as the land slowly began a new cycle of agricultural recovery and collapse, another wave of Semitic nomads, the Amorites, swept in from the west. The Amorites were known to the scribes of Ur. For decades they had been moving peacefully into the riverworld, pasturing their sheep in salt-encroached fields that could no longer support wheat and could barely support scrub grass. Some had even become mercenaries in the armies of city-states caught up in a new outburst of infighting. But during the final century of the third millennium B.C. the migrants entered the Euphrates plain in unprecedented numbers, overrunning battle-weary cities and establishing a new capital in the hitherto-undistinguished town of Babel, known to us by its Greek name, Babylon.

Little is known of the Amorites, save that they are mentioned in Genesis 15:21 synonymously and seemingly interchangeably with the Canaanites, whom God promises Abraham he will displace in favor of his descendants. Evidently they had established considerable power in both Israel and Iraq.

The sixth Amorite king of Babylon, a man named Hammurabi, ascended the throne about 1950 B.C. He referred to himself as "the one who makes wealth and plenty abound; the one . . . who revived [the city of] Uruk; who supplied water in abundance to its people . . . who stores up grain." Like Sargon and the Akkadian insurgents before him, as Hammurabi and his people took command of the Tigris-Euphrates civilization, it became diffi- cult for outside observers (including present-day archaeologists) to distinguish whose culture was taking over whose. The Amorites absorbed the writing, art, pottery styles, lifestyles, construction techniques, educational systems, and, with an occasional addition or deletion, even the religion of the vanquished.

Hammurabi's most lasting legacy is his collection of laws (some appear to have been borrowed from the writings of Ur-Nammu, king of Ur, who preceded Hammurabi by some two hundred years). They were a code of life that included among its edicts penalties for criminal offenses ranging from damaging a neighbor's wall and bearing false witness to committing murder and adultery (the penalty, in all four cases, was death).

Like the story of the baby Sargon being sent downriver in a reed basket, Hammurabi's laws foreshadow the biblical Moses. As with Moses, Hammurabi receives the laws by divine revelation (they are communicated to him in a covenant with the Sun-god Shamash). As with the Mosaic laws, they are engraved on a sacred stone tablet, and although the penalties for crimes may sometimes differ, there are instances in which Moses echoes Hammurabi with such spine-chilling fidelity that it is easy to believe the Hebrew tribes heartily absorbed Amorite/Canaanite culture, even as they strove to displace it.*

During the twentieth century B.C., Hammurabi wrote, "If a seignior's ox was a gorer and his city council made it known to him that it was a gorer, but he did not pad its horns or tie up his ox, and that ox gored to death a member of the aristocracy, he shall give half a mina of silver. . . ." More than three hundred years later Exodus 21:29 echoed, "But if the ox has been accustomed to gore in the past, and its owner has been warned but has not kept it in, and it kills a man or a woman, the ox shall be stoned and its owner also shall be put to death."

"If a seignior has destroyed the eye of a member of the aristocracy, they shall destroy his eye," proclaims the Hammurabi stone, and, "if he had broken another seignior's bone, they shall break his bone." In Exodus 21:23–25 we read, "You shall give life for life, eye for eye, tooth for tooth, hand for hand, foot for foot, burn for burn, wound for wound, stripe for stripe."

On January 16, 1991, about one hour into the Gulf War, the persistence

---

*Unlike the Mosaic tablets, housed in the lost Ark of the Covenant, the whereabouts of the Hammurabi tablet is not unknown. About 1450 B.C. it was taken from Babylon as a war trophy by King Shurruk-Nahhunte and set up in Susa (in the mountains east of the Tigris-Euphrates junction). In 1902 one of Leonard Woolley's associates found it lying in three pieces beneath the ruins of Susa, and it can be seen today in the Paris Louvre

of the Hammurabi and Mosaic codes became astonishingly apparent to NBC News correspondent Tom Brokaw. He could not avoid thinking of the Iraqi citizen with whom he'd struck up a brief friendship. He was a university-educated man who said he admired Americans and liked them very much. The Iraqi was very concerned about the coming war and hoped it could somehow be avoided. And then, as Brokaw walked to his airplane on the last day of peace, he turned to the man and asked, "Do you have a family?"

"I have three small children," he said.

"I hope that you will be able to get them out of harm's way," said Brokaw.

"I think I will be able to" came the reply. The man was silent for a moment. Then for another. The correspondent watched a mask come down, and very gravely the Iraqi said, "If something happens to my children, any one of them, I will fight you forever, and my children will fight you as well."

It came to Brokaw then, and it stayed with him through the night of January 16, that his friend's statement was probably representative of the entire population of the city, that he had, perhaps, underestimated the complexity of the Mideast, marching to its own, very different drummer from that of the Western cultures.

Later that night, when he related the incident and his thoughts about it to a U.S. Embassy official, the official nodded in agreement: "These are the sorts of people whom it is not wise to cross, except when you have to. So that the man you just mentioned, who said he would never forgive us if anything happened to his children, evokes the idea that where we, in the West, tend to think more of our New Testament heritage—where you turn the other cheek and you let bygones be bygones and forgive and forget— the people of the Middle East are the people of the Old Testament. They believe much more in an eye for an eye, a tooth for a tooth; and you don't forget, and you don't forgive, and you carry on the vendetta and the struggle long after people in the West would be prepared to say, 'All right. It's over. Let's not worry about it any longer.' "

To the historian and Jesuit priest John MacQuitty, the Hammurabi and Mosaic codes, brutal as they may sometimes appear, were actually a step upward in mercy. He sees in this first small step the working of two different concepts of God: the anthropomorphic God, which is the God created by man and culture, and the divine or spiritual God, which is trying to sift through and re-create man through culture. "Man's concept of God evolved as man evolved," he says, as if to embrace the notion that man evolved God in his own image. "And so, eventually the vengeance type of survival of the first civilizations gave way to concepts no one had ever encountered before. One way of interpreting the Old Testament message behind the first laws is that, while radical to people of Hammurabi's or Moses' time, the new covenant was not so alien to human experience as to be met with immediate rejection. In other words, man's spiritual evolution had to filter down through civilization's patterns of dysfunction.

"We are talking here of ancient brutality. From a modern Western perspective, 'an eye for an eye, a tooth for a tooth' may seem like cruel and unusual punishment, but it was at that time a giant step toward justice. 'If you kill my brother, I will kill you' was more civilized than what had gone before, whereby if someone killed another's brother, his whole family or his whole tribe would be killed in retribution. In the beginning, there was perfect vindictiveness and dysfunction.

" 'An eye for an eye' was a necessary step leading toward the even more radical steps taken by Jesus and Buddha, Gandhi and Sadat. In the time of Hammurabi and Moses, no one would have accepted 'Love thy enemy.' Yet when Tiberius ascended to power in Rome, the time was evidently ripe to convince a small number of people to embrace a higher standard of civilization. Even so, Jesus' ideas about being kind to each other and 'giving unto Caesar what is Caesar's' were too radical for most of his followers, who really wanted a revolutionary general who would lead conquering armies to Rome, as Moses had led the Israelites to Canaan. Ten years later both Romans and Judeo-Christian reformers—hundreds of them—would come to wish they had not nailed him so hastily to a tree (for he was just starting to get interesting). At the time of his crucifixion, Jesus' core following numbered only about a dozen men, but it was enough."

WINTER A.D. 1929

Once Woolley, Lawrence, and a handful of other explorers on the Tigris-Euphrates plain had learned to read cuneiform, they could enter into the lives and even the feelings of people who had dwelled in the ruins when the temple walls were still gleaming white and smoothly plastered. They could read the confidential dispatches of Hammurabi to his generals or peer over the shoulder of a Babylonian accountant as he made entries in his ledger and scrawled in a corner, "You can have a lord, you can have a king, but the man to fear is the tax collector."

And when they began to read epic poems, the archaeologists became aware that the people of Ur had been speaking to them all along, even as they were children, long before they ever heard of cuneiform. The Chaldean flood tablets, as they were sometimes called, told of a man named Ut-Napishtim who survived a universal deluge sent by the gods to punish mankind. The tale was copied hundreds, perhaps thousands of times, for fragments of it, inscribed in cuneiform and fire-hardened, had been found as far away as Megiddo in Canaan and in the archives of Karnak. As near as Woolley could tell, the Babylonian deluge must have been the world's first international best-seller. According to the legend, Gilgamesh, in his search for immortality, sought out Ut-Napishtim, who was said to be an immortal born before the flood. When finally they met, Ut-Napishtim told Gilgamesh how the world had become evil and how he was commanded by the Babylonian god Ea to build an ark and to load the animals upon it in preparation for "the end of all flesh." The ship eventually ran aground on what had, before the flood, been a mountaintop, from which Ut-Napishtim sent forth a dove to find land. Thus was Ut-Napishtim frequently referred to (by Woolley's financers) as the Babylonian Noah, though Woolley suspected that it might be more accurate to call Noah the biblical Ut-Napishtim.

There was fascination enough for Woolley in the idea that Noah's flood was not a Hebrew story but a Babylonian one, taken over by Hebrews. What he could not have anticipated, as he stood with Hamoudi in the wind and the ice on the dawn of a nuclear age, was that his two hundred workers were about to expose even greater fascinations. And—oh, the stories he would have to tell Lawrence, if ever they met again.

His onetime site mate had moved south, where he was taking up new tools and finding a new calling in life. Already he was making quite a name for himself. To Hamoudi and all the others at Ur, he was no longer simply Thomas or Lawrence but Lawrence of Arabia.

To Woolley he had always seemed one of the world's greatest storytellers. He was at his best when standing over a campfire in his robe of gold and silver thread, weaving a tale of ancient life for young Arabs and Kurds. Woolley remembered a story, entirely made up, about a young Babylonian priestess whose spirit still roamed the plain at night. He was hypnotic, with a grand master's sense of timing, and with dramatic swoops of his hands, Lawrence explained that the priestess could still be heard shifting uneasily in the wind, but he'd held back this detail until precisely the moment a desert breeze came up from the Persian Gulf and riffled everyone's hair. Woolley himself had to admit to being transfixed and to feeling a shiver steal into his spine.

"It is not the tale that counts," Lawrence had said, "so much as he who tells it."

Woolley was inclined to agree, but an apt and by analogy opposite argument was about to come out of the ground, revealing to the archaeologist that it was not whether the tale of the flood was first told by Hebrews or Babylonians that counted so much as the evidence, buried deep under the Euphrates plain, telling him that the story of Ut-Napishtim and Noah might have germinated from a very real kernel of truth.

One of the workmen came running toward him. "Come see! We have struck bottom. Sterile river mud." When Woolley joined the men in the pit, he agreed that indeed, they did seem to have located the bottom of Ur. They had gone down through the floor of a tomb, down through a layer of scattered clay tablets recording early attempts at cuneiform writing. Now, on what appeared to be the original prehabitation ground surface, somewhere near 2800 B.C., all traces of human artifacts had broken off. But something about the ground seemed out of place to Woolley. On a hunch he began digging into the thick clay and instructed the workers to join him. Twelve feet deeper their spades broke through into a horizon of collapsed walls and broken pottery. Difficult to believe, but there it was: a deep, uniform, water-deposited stratum of Euphrates clay, poised between two levels of civilization . . . Woolley was quite convinced of where it was all leading, but that way lay madness. "We all agree that your theory is mad," one colleague inevitably commented. "The problem which divides us in this: Is it sufficiently crazy to be right?"

Madness . . . he was prepared to swear himself never to mention his suspicions about the meaning of the clay stratum until his wife, Catherine, climbed into the pit with him, glanced at the slice through alternating layers of ruins, river deposits, and ruins, and casually remarked, "Well, of course, it's the flood."*

Could it really be the origin of the Deluge legend? Could it really be? Woolley thought so, but he knew that he could scarcely argue a theory of such epic proportions on the strength of a single pit a few yards square. He began another pit seventy-five feet by sixty that ended up plunging down sixty-five feet through a Babylonian potter's workshop, then through an eleven-foot deep stratum of Euphrates clay and again through a horizon of human occupation beneath the clay. Other pits more than thirty miles from

---

*Catherine Woolley had joined Leonard as an apprentice around the time Lawrence departed. She became an accomplished field archaeologist in her own right but was to die tragically young.

Ur brought increased resolution, revealing multiple destruction layers—multiple floods—separated, in some cases, by centuries. From the depth of clay, Woolley estimated that at least one of the disasters had engulfed an area four hundred miles long and a hundred miles wide. Looking at a modern map of the world, he would have been inclined to regard the inundations as local occurrences, but when he put himself in the place of a citizen of Ur, looking from horizon to horizon and seeing only water, he could understand how a flood could be regarded as all-encompassing and go down into oral and written traditions which, to a later civilization, with a much broader knowledge of geography, could come to be read as a global event, though originally it engulfed not even a notable fraction of 1 percent of the Earth's surface.*

"It was not a universal deluge," wrote Woolley; "it was a vast flood (or series of floods) in the valley of the Tigris and Euphrates which drowned the whole of the habitable land between the mountains and the desert; for the people who lived there that was indeed all the world."

From the cuneiform annals, Woolley knew that Babylonian records spoke of events as happening either before or after "the flood." The Babylonian Deluge of approximately 2800 B.C. appeared to have been capable of washing away all or most records, especially as writing was then coming only sparsely into use. To those who built upon the destruction layers, who set new foundations down on top of the lost cities, events "before the flood" must have taken on a legendary, larger-than-life nature. They began writing about individuals who, in the time before the end of the world, could expect to live for many centuries. Ut-Napishtim was immortal; Noah, the closest thing to it. He lived into the time of Abraham and the postflood zenith of Ur and Sodom, when, according to the Old Testament, he died at the age of nine hundred fifty.

A final piece of the Deluge puzzle shifted into place when workers reported that one of Ur's mounds was neither a fallen temple nor the residua of multiple levels of civilization but a large hill that had towered over the land thousands of years before the arrival of man. Though people had built palaces on top of it, it occurred to Woolley that the hill must have been the only natural outcrop on the whole plain standing above water when the

---

*"A well-known example of this," observed scientist-author Isaac Asimov, "is the statement frequently met with among the ancient historians that Alexander the Great conquered the world and then wept for other worlds to conquer. What was meant was merely that Alexander had conquered a large part of those sections of the world which were well known to the Greeks of the time. Actually, Alexander conquered only 4 or 5% of the earth's land surface and had plenty of room in which to extend those conquests."

floods came. So he sank a new shaft into the mound itself, descending, as he expected, through the same pottery styles he was used to finding above the clay layer. Also as he expected, at the place where the intermediate clay layer should have been (but was not, for he was digging into an island), he moved directly, without interruption, into the artifacts of the people who had lived below the clay. "About sixteen feet below a brick pavement," Woolley noted in his journal, "which we could with reasonable certainty date to about 2800 B.C., we were among the ruins of Ur which had existed before the Flood."

Late at night, when he stood upon the hill, he imagined it as a lone island on a turbulent sea. For miles and miles in every direction mud-brick walls were slowly dissolving or crashing down under the weight of running water. Barges and rafts would be approaching, trying to make landfall on his hilltop-turned-island. Some ferried caged geese and probably a few heads of prime breeding stock hastily loaded aboard; the origin, perhaps, of the story of the Ark. The people of preflood times had not been obliterated by the disaster, Woolley guessed. A few had survived to lay down the foundations of the next cycle of empire building and decline. And even as they rebuilt upon the clay layer, they tried to make sense of the destruction, to find a reason for it. At night, around cooking hearths, you heard stories of the gods' vengeance against an evil humanity, the beginning of legend.

A silver meteor ran down the sky, breaking up into fading red sparks over the Arabian Peninsula, where Lawrence had gone. "You were wrong!" Woolley called out across the desert. "It is the tale, not he who tells it!"

7

SOUTH
OF
SODOM

*On December 2, 1942, man achieved here the first
self-sustaining chain reaction and thereby initiated the
controlled release of nuclear energy.*
—FROM A BRONZE PLAQUE AT THE WEST STANDS
   SQUASH COURT, CHICAGO, ILLINOIS

*History is on our side. I cannot get those words out of
my head. Land of Lincoln and Franklin and Melville, I
love you and I wish you well. But into my heart blows a
cold wind from the past; for I remember Babylon.*
—ARTHUR C. CLARKE, *The Nine Billion Names of God*

T HE TENTH CHAPTER of Genesis describes a mighty hunter and king named Nimrod, and while it is a fact that the farther back in time we go (that is, the closer we get to the beginning of the Old Testament chronology), the more likely events are to have accumulated embellishments and distortions before finally being committed to biblical texts between 900 and 500 B.C., Nimrod is the first description we have of an individual who, to one degree or another, becomes familiar to us in historical writings independent of the Bible.

According to Genesis 10:10, "The beginning of his kingdom was Babel (Babylon), Erech, Accad, and Caleh, all of them in the land of Shinar." The city of Babel was on the Euphrates River. Erech (known in cuneiform inscriptions as Uruk, the city founded by Gilgamesh) was also located on the Euphrates, about one hundred miles downstream of Babel; and Accad, or Akkad, is in cuneiform Agade, the lost city of the Akkadian king Sargon.

"The land of Shinar" is clearly the Tigris-Euphrates region, evidently at a time ("in the beginning of his kingdom") when Sargon ruled over it. In Genesis 10:11 we read, "Out of that land went forth Asshur." The land of Asshur (better known by its Greek name, Assyria) encompasses the upper reaches of the Tigris River and was passed from Akkadian rule (Sargon's kingdom) to the Amorite kingdom under Hammurabi (who reigned about 1950 B.C.). Genesis 10:11 further tells us that in Asshur Nimrod built the cities of Nineveh and Caleh. One of Hammurabi's successors, Shalmaneser I, did indeed build Caleh, nearly five hundred years after Hammurabi's death. Another monarch, Tukulti-Ninurta, built Nineveh.

Some biblical scholars (among them scientist-author Isaac Asimov) have suggested that verses 10:10–11 are a brief history of nearly a thousand years of kingship on the Tigris-Euphrates plain, ranging from Sargon the Great (born about 2485 B.C.) to Shalmaneser I (who ruled about 1450 B.C.). Compressed in this way, Nimrod can be seen as a composite king, telescoping the deeds of Sargon of Agade, the Amorite king Hammurabi, and the Assyrian city builders Shalmaneser I and Ninurta (with even a hint of Gilgamesh thrown in).

To the compilers of Genesis, living in the first millennium B.C., Assyria was the newest, most powerful kingdom in the Babylonian world. Among its first kings was Ninurta. He took the throne about 1520 B.C., and during the six hundred years between his ascendancy and the time in which the earliest chapters of the Bible began to take shape, his name had grown to legend. To the Greeks he was Ninus, founder of Nineveh, conqueror of all Babylonia. To the editors of Genesis, Ninurta, or Ninus, apparently became Nimrod, founder of Nineveh, conqueror of all Babylonia, and the biblical picture of him clearly points to an Assyrian monarch. Genesis 10:9 tells us, "He was a mighty hunter," and cuneiform accounts reveal that the Assyrian kings sought to legitimize their power in the public eye by sustaining the image of the victorious hunter. One clay tablet tells of 920 lions, 10 elephants, and 4 bulls imported and released into the king's game park for him to kill with spears and arrows.*

Into the story of this first and (from a first millennium B.C. perspective) most memorable of Assyrian kings, the deeds of earlier kings seem to have been woven. By analogy, the story of Nimrod is not much different from what we might be reading today about the founding of America if only fragmentary and sometimes misplaced or misunderstood information had survived from that period. While we would correctly learn that George Washington was the first president of the United States, his biographical sketch might also identify him as the man who crossed the Atlantic Ocean to discover the New World, who conquered the American Indians, built Washington, D.C., and was so strong in character that as a child, when he chopped down his father's cherry tree, he immediately admitted the deed.†

Viewed in a strictly archaeological and historical perspective, Nimrod and other larger-than-life biblical figures could easily have acquired similar telescoping biographies—which brings to mind Abram, a patriarch born, according to the eleventh chapter of Genesis, at some point after the reign of Nimrod, yet (somehow) only 292 years after Noah's flood.

---

*The Persian word for the king's game park was adopted by the Greeks as *paradeisos*, defining an enclosed park. In the Greek version of the Old Testament it was applied to the Garden of Eden and came down to the English language as *paradise*.

†The story of the cherry tree is simply an embellishment arising from America's own oral traditions. And far from building what most modern people regard as America's always-existent capital city, George Washington did not even spend his presidency in the city that bears his name. America's first White House was in New York City, at 3 Cherry Street, now buried under the foundations of the Brooklyn Bridge.

Abram (known as Abraham in later years, as Ibrahim to the writers of Babylonian cuneiform) was, according to Genesis, the first Hebrew patriarch to visit Canaan and the first to whom modern Jews universally trace both their physical and spiritual origins. His birthplace, we are told in Genesis 11:27–28, was the city beneath which Leonard Woolley found the mysterious clay layers: Ur of the Chaldees. High above the youngest clay stratum, in a zone marking the time in which Genesis was evolving into its final, written form, Ur was an obscure village of no more than a few hundred people whose mud-brick homes were heaped between already ancient mounds. Erosion had long ago softened the features of ziggurats and palaces, had winnowed their former glory down beyond recall. But the compilers of Genesis, in identifying Ur as the birthplace of a founding prophet, were thus required to define it by more than name alone. They called it *Ur kasdim,* which is translated as "Ur of the Chaldeans."

The Chaldeans were a tribe of people who migrated from Arabia into the Euphrates region about 1150 B.C., more than four hundred years after the time of Ninurta and Nineveh, nearly a thousand years after Sargon, Hammurabi, and (by inference) Abraham. During the Assyrian period the Chaldeans were the most important tribal element of the empire, and "Ur of the Chaldeans" became a quick and easily understood way of identifying Abraham's birthplace in the minds of first millennium B.C. readers, despite the fact that the biblical Abraham and the historical Chaldeans were separated by a rift in time as great as any continental division.

When Leonard Woolley arrived in Ur and plunged shafts through the remnants of a small Chaldean town, he found a mighty Babylonian city beneath. There were courthouses and terraced, multistory apartment dwellings. It was a city of paved streets and libraries whose tablets urged young Babylonians to "marry for pleasure; get divorced when you've thought it over."

"We must radically alter our view of the Hebrew patriarch," wrote Woolley, "when we see that his earlier years were passed in such sophisticated surroundings. He was the citizen of a great city and inherited the traditions of an old and highly organized civilization."

If in fact a patriarch named Abraham did exist and did live in Ur, the idea of a city-dweller-turned-nomad is not necessarily so contradictory as Woolley once supposed, especially in the Tigris-Euphrates region. Semitic tribespeople were always known to settle down for a while in and around the cities, and a few, among them Sargon the Great and Hammurabi, even rose to high office. They were also known occasionally to abandon the cities, sometimes without any warning or fuss. It is not at all inconceivable that Abraham, or someone very much like him, left Ur and rediscovered his nomadic roots, perhaps following one of the regions' all too frequent cycles of environmental recovery and collapse, rise to prosperity, and return to ruin. If one is allowed to speculate on this possibility, the speculation receives some small support from Genesis 12:10, which identifies famine as a driving force behind the wanderings of Abraham's clan, sending him as far afield as Egypt.

When Max Mallowan and a half dozen other archaeologists began unearthing cuneiform tablets bearing the name Ibrahim (Abraham), they could not help jumping to the conclusion that the Tigris-Euphrates plain was offering up clues that the patriarch of Genesis was in fact a historical reality. But the theory soon began to suffer from too much of a good thing. As the excavations continued, more tablets—scores of them—turned up with the same name inscribed. The tablets ranged from the time of Shalmaneser I through Sargon the Great, spanning an entire millennium and, instead of drawing Abraham closer to the archaeologists, confounded them. Because they found their discovery so "unusual and disturbing," they failed to ask the obvious next questions: (1) Might the multiple Abrahams explain the peculiar telescoping of times and places as perceived by later biblical compilers, and (2) does it not seem possible, if not likely, that after a patriarchal figure named Abraham became famous, subsequent generations passed the name to their children in his honor? If so, it should not seem strange or contradictory to us that Babylonian tablets list even an Egyptian and a Cypriot among those named Abraham. In fact, the name is a tradition that continues to this day, and by analogy, we can begin to appreciate the task confronting the compilers of the Old Testament if we try to imagine a scholar ten thousand years from now attempting to untangle the mystery of the "historical" Abraham, given only a few random scraps of biblical text and a partial history of the past four thousand years. While fairly certain that there had lived an important man named Abraham, she will not be able to determine whether he was the nomadic founder of a mighty tribe, a mayor of New York City, an Egyptian noble, an American president, the co-founder of a department store chain, or a telescoped composite of all five.

Because there was a tendency among biblical scribes to compress events from diverse time frames into a single image, the first millennium B.C. image of Abraham, while capturing certain kernels of truth, may at the same time describe a social background very different from what actually existed in the second and third millennia B.C. For example, while Genesis 16:2 tells us that Sarah presented her slave Hagar to Abraham, and while Babylonian cuneiform texts tell us that this is an accurate portrayal of second and third millennium legal traditions, in which a wife was usually urged to provide a "substitute wife" if a marriage produced no heirs, another section of that same chapter tells us that an Egyptian pharaoh rewarded Abraham with camels. Cuneiform texts and fossil remains reveal that camels were not domesticated until the fourteenth century B.C., nearly a thousand years too late for Abraham to have owned one.

The Old Testament further explains that after fleeing a famine and living for a time in Egypt, Abraham and his nephew Lot settled in Canaan. Their herds were so multiplied that it became necessary (according to Genesis 13:10–12) for them to seek separate pastures:

> And Lot lifted up his eyes, and beheld all the plain of the Yarden, that it was well-watered everywhere, before the Lord destroyed Sedom and 'Amora, like the garden of the Lord, like the land of Mizayim, as thou comest to Zo'ar. Then Lot chose for him all the plain of the Yarden; and Lot journeyed east; and they separated themselves one from the other. Avram [Abraham] dwelt in the land of Kena'an, and Lot dwelt in the cities of the plain, and pitched his tent toward Sedom.

The Yarden is the Jordan River, the fertile corridor of an otherwise mostly arid Kena'an (Canaan), described in Genesis 13:10 as being the central focus of two equal and opposite riverworlds: at one end "the garden of the Lord" (that is, the Plain of Eden, or the Tigris-Euphrates plain) and at the other end "Mizayim" (Egypt). After traveling east from Egypt, Abraham settles in Canaan, and we encounter him later in the south, on the banks of the Dead Sea. Lot's destination is less clear. If he settles in Canaan, then the Jordan Valley is the logical place for him; but Genesis 13:10–12 distinguishes Abraham, not Lot, as the settler of Canaan. Lot separated from him and went somewhere else. Unless his journey east refers to a mere stroll downhill from a western slope on which he and his uncle surveyed the Jordan Valley, then "east" suggests a return to Babylon, and the "cities of the plain" (the Jordan Valley is a rugged land, lacking smooth plains) would thus be the cities on the plain of the Tigris and Euphrates rivers. Among Lot's cities is mentioned Sedom, better known by the Greek version of its name, Sodom.

〰〰〰

"The first thing you must understand," says Father John MacQuitty, "is that much of the Bible is analogy. It's *the story of* Abraham, or his nephew Lot, or his grandson Dedan, who gave comfort to Job. There may not necessarily have been a Lot, a Dedan, or a Job, and events may not necessarily have happened as the Bible describes them. It was all put to writing as a moral and spiritual lesson.

"The Book of Job, for example, is the story of a good man who endured great misfortune without ever losing faith or blaspheming his God. The legend must go back to the very edge of prehistory, for there is a version of it existing in the most ancient literature of the Tigris-Euphrates region.

"Satan said to God—do you remember?—'Naturally Job is upright: seven sons, three daughters, seven thousand sheep, three thousand camels, scores of servants, the wealthiest man in the land of Uz. Why not be upright, see how well it pays? But put forth your hand and take away all that he has, and see how he curses you to your face.' So God allows the enemy to put Job through every trial except death. A great wind kills his children. Bandits take his camels and slay his servants. A fire falls from heaven and burns up his sheep. A mysterious wasting disease strikes him. And so, reduced by a whim of God to a man whose days are spent without hope, whose flesh, blackened and clothed with worms, hangs down from bones burning with internal heat, Job looks up and says, 'I shall not put away my integrity. Naked I came from the womb. Naked will I return. God has given, God has taken away. Blessed be God's name.'

"Job has made his stand against the enemy, the oldest enigma of human existence: senseless evil. But now comes the worst horror of all. Job's comforters arrive, Dedan and the others. 'In a just universe ruled by a just God, evil does not arise without cause,' they say. 'Since God is just, Job must have become too proud of his accomplishments, kissed his own hand, or committed some other terrible sin. Search your deeds, Job. Repent.'

"But Job fights back: 'I have not sinned!' He will submit to God's divine will, but not to the pious judgment of his comforters. And so, from the depths of unexplainable iniquity, Job rises to face his God and to demand an answer to the timeless mystery of senseless evil."

Job gets his answer, which on first hearing answers nothing, and yet at the same time answers all. God himself speaks to Job out of a roaring storm:

> Who is this that darkens council by words without
> knowledge? Gird up now thy loins like a man; for I will
> demand of thee, and answer thou me:
>
> Where was thou when I laid the foundations of the earth?
>   Declare, if thou hast understanding.
> Who determined its measurements, if thou knowst?
> Or who has stretched the line upon it?
> Where upon are its foundations fastened?

Or who laid its cornerstone, when the morning stars sang
    together, and all the sons of God shouted for joy?
Or who shut up the sea with doors?
Hast thou entered into the springs of the sea?
Or has thou walked in the recesses of the depth?
Or has thou seen the doors of the deepest darkness?
Canst thou bind the chains of the Pleiades or loosen the
    cords of Orion?
Canst thou send lightnings?
Hast thou comprehended the expanse of the earth?
Declare if thou knowst it all.

Has thou commanded the morning?

To Father MacQuitty, the raft of questions set before Job define the wonders of creation and dare man to comprehend them. According to the Book of Job, man is incompetent to cope with the universe; man, a poor insect that, scaled against the mind of God and the time frames of the Earth, lives for an instant and then dies. The roaring storm has declared only that God's reasons are hidden, beyond Job, beyond his comforters, beyond all mankind. The spiritual lesson behind the story is simply this: The universe does not necessarily make sense.

Before each expedition into the deserts or onto the high seas and under them, Father MacQuitty always gives me a special blessing. He knows me as an agnostic who questions virtually everything, and on our first meeting I'd automatically assumed that he, being a religious man, would be offended even to be in the same room with a scientist who questioned God's very existence (after all, I'd seen fellow scientists respond brutally to lesser heresies surrounding such irrelevant oddities as amber, fossil crabs, and the ash of Thera). But I'd underestimated the Jesuit. "By all means keep questioning," he urged. "Your questions are God's gift to you."*

I hope Father MacQuitty will not think me to have gone too far this time, but if the questions penned in the Book of Job more than three thousand years ago were put to a scientist today, she could answer, "Yes, thou hast understanding," and follow through with a few questions of her own.

Have not the archaeologists been probing back ten thousand years to prebiblical Babylon and Jericho? Have not the paleontologists, in becoming accustomed to thinking in hundreds of millions of years, begun to compre-

---

*I once told him that the day I go from agnosticism to faith or atheism is the day I cease to be a scientist, for that will be the day I have stopped asking questions. "So, go," he said. "Go in the grand old tradition of Saint [Doubting] Thomas."

hend the expanse of the Earth? Have not the physicists, tooling with deep-space telescopes and antiprotons, pushed back the frontiers of time more than nine billion years to the first three seconds, in which the physical foundations of matter were fastened and the cornerstone of the universe laid?

Yes, I guess you could say we have done that.

Can we not send lightning? At Indian Point the cleaving of uranium nuclei feeds the electronic gridwork that runs through Manhattan. Seen at night, from on high, the grid thins in the direction of Philadelphia, dims from dazzling rectilinear lines to phosphorescent cobwebs draped over the land, telling any who may chance to look down upon this Earth that we city builders evolved from daytime hunters and foragers, for we try, at the expense of extraordinary energies, to banish the darkness, to create artificial daylight wherever we go (and we cannot help asking, if there be other sentient, city-building beings amid the stars, and if, unlike us, they evolved from nocturnal creatures, would their cities display lights after sunset?).

Send the lightning? Yes. We've done that, too.

Can we not shut up the sea with doors? Yes, we can. There are plans on the drawing boards to dam the Strait of Gibraltar, lower the level of the Mediterranean, and provide electricity for all of Europe, all of Africa, and all of the Mideast. And yes, with a coming generation of relativistic rockets, we can bind the chains of the Pleiades or, if we so wish, explore the shoulder of Orion.

With engineering miracles named *Argo* and *Alvin* we have explored volcanic springs more than a mile and a half down on the beds of the Pacific and Atlantic oceans.

Yes, we have entered the springs of the sea and walked in the recesses of the depth. Indeed, I've done these things myself.

And have we not determined the Earth's circumference and stretched the line upon it? Yuri Gagarin and Scott Carpenter did precisely that when they hurtled beyond the mountains and the clouds, beyond atmosphere into the darkness of space; and as they traced the circumference of the Earth once every ninety minutes, flying into sunrise seventeen times daily, one might say that they really did command the morning.

There was a time, not very long ago, when such successes would have filled me with pride and a wild optimism for this civilization's future. But I remember the fall of the rocket ship *Challenger,* and I have seen, from Karnak to Babylon, the repeated falls of earlier civilizations. I remember an oceanographic expedition that carried me more than two thousand miles from the nearest city, yet one morning there was a lens of reddish-brown smog on the horizon, and lightbulbs floating. Oh, yes, we can answer for Job, now, but it is the question that could not be asked in the time of Job that becomes most relevant: What will civilization do with its successes?

We live in a solar system provided with everything necessary for the further development of humankind, if we are wiser than the Babylonians and the Egyptians, and pay attention. The Sun, our parent star, is tugging us five miles closer to Sagittarius every second. We have traveled more than a light-month since the building of the Pyramids, tens of light-years since *Homo erectus* first fashioned stone hand axes near Thebes. Even if we choose to go nowhere, we shall go far. In a figurative sense, our system may be viewed as an eternal spaceship that grew its own passengers while en route. The wanderings of Abraham and Lot over a relatively small corner of the ship have come down to us as legend, but if we follow the lives of archaeologists attempting to uncover what it must have been like to live in the time of Abraham and to walk among his clan, we see that the ancient patriarchs were not the only ones with a touch of the wanderlust.

Benjamin Mazar was nine years old when the 1917 Russian Revolution erupted around him. He was living in the Crimea, best remembered for the Crimean War, in which British cavalrymen charged madly and headlong against Russian cannons . . . and won. The charge was led by Lord Cardigan, who passed his off-hours knitting sweaters in a style that became as famous as his charge—which went down into history as the Charge of the Light Brigade. A decade later a czar had been compelled to sell Alaska and the Aleutian Islands to the United States to recoup his losses.

For Benjamin, one or even a half dozen visits to the local museum had not satisfied his sense of wonder, though to every other child in town the place had become a bore after the first tour. And outside the museum were ancient buildings predating the Crimean War by centuries. Some even went back to the Roman period. Of particular interest was a synagogue dating to the eleventh century and still in use. A thousand years ago its builders had conceived the temple as a direct descendant of Solomon's Temple in Jerusalem, and Benjamin's community still regarded it, and all synagogues, as such. The Torah Shrine was the receptacle of the scrolls, God's word preserved in endlessly copied and recopied Scripture. It was, his parents explained, one of an ever-multiplying lineage of spiritual successors to the biblical Ark of the Covenant, which had been carried away from the Temple of Jerusalem never to be seen again, sometime around 700 B.C. The long habit of multiplication had guaranteed that, if not the Ark itself, then at least a few of its successors and, above all else, the words they contained would endure any natural or man-made catastrophe.

In 1924 Vladimir Lenin died, triggering a brutal power struggle in which Joseph Stalin ultimately eliminated more than two million real and imagined enemies. Benjamin, now age sixteen, had joined several Zionist youth organizations and had begun writing historical dramas of the Second Temple period. His plays derived from two constantly growing interests: On one side he saw the history of his people, on the other side he saw the ancient buildings, and both drew him more and more toward vanished civilizations. By 1926 he decided that he would travel abroad and study archaeology. As Stalin, having seized complete control of the Communist party, began to list Russia's Jews among his enemies and prepared to focus his bloodlust on Zionist factions in particular, Benjamin Mazar left the Crimea and took up residence in Berlin, where, just a few doors away, there lived a young, obscure artist who had recently completed a nine-month prison term in Bavaria for disorderly conduct. The artist had behaved well and even written a book during his imprisonment, so the authorities let him out early, citing him as a model prisoner who had made productive use of his time and rehabilitated himself. The book, then in the press, was titled *Mein Kampf.*

Around the time Woolley and Hamoudi found the Babylonian flood layers, just a few years ahead of Germany's *Kristallnacht,* Benjamin traveled to Israel, where local Arabs were already responding to an influx of Jewish immigrants with suspicion and animosity. When Benjamin first arrived, he wondered what ruins might lie hidden beneath the streets and shops of modern-day Jerusalem. By the time the state of Israel formed, bomb craters would expose Roman walls, and the old ruins would be indistinguishable from the new.

En route to Israel Benjamin crossed paths with a Greek archaeologist named Spyridon Marinatos, who, on the island of Crete, had been excavating under the directorship of Leonard Woolley's mentor, Sir Arthur Evans. Probing beneath the residua of Classical Greece, Marinatos was exploring the lost Minoan culture. Unlike their contemporaries in Egypt and Babylon, the Minoan kings did not erect giant monuments to themselves. There was nothing on Crete to compare with the Pyramids, which reigned (along with the Eiffel Tower) as the largest, most spectacular useless objects ever built by the hand of man. In striking contrast with the Egyptians, no one really knew who the Minoan kings were. As near as the archaeologists could tell, they had reined in their egos in favor of civil engineering projects. All or most of the nation's economic surplus was reinvested in giant warehouses, aqueducts, sewage systems, the world's first navy, the world's first dams, and apartment complexes equipped with sophisticated plumbing, including communal showers with hot and cold running water. But the mighty civilization had disintegrated, seemingly at the height of its power and with astonishing rapidity. Marinatos knew from the remnants of palaces in strata immediately overlying the collapse that at least a few Minoan architects had survived and were being employed by people who later came down from the north and took over. They were the predecessors of the Classical Greeks. Arthur Evans believed it was these northern conquerors who had destroyed the Minoan cities. Marinatos was not so sure. On the northeastern shore of Crete the young apprentice found the bottom half of a foundation stone weighing several tons. The top half, also weighing several tons, had been shorn off and carried nearly three hundred feet away. He could imagine invading people looting a town and enslaving its people or even setting the buildings afire. But what, he wondered, would compel them to break worthless foundation stones in half and drag the pieces hundreds of feet?

Nothing, he decided, and he began turning his imagination away from a human cause, more and more toward a natural upheaval, possibly a mighty surge rearing up from the sea. There were echoes of a great Cretan disaster in ancient texts. They read to him like descriptions of a volcanic explosion or tsunamis, or both—eyewitness accounts, perhaps, recorded from vantage points very far away. Apollonius' *Argonautica* told of a "deadly darkness" rising from within the Earth and covering Crete. The prophet Balaam referred to "the dust cloud of Israel" shutting out the skies with darkness and provoking terror. Homer's *Odyssey* described a city with distinctly Minoan traditions being shut in by a mountain. In the Old Testament Crete was identified by the name Kaphtor. According to the Book of Amos (as described in 9:5–7), the land had melted, and risen up "wholly like a flood," whereupon three mass migrations were triggered: Israel out of the plague-ravaged land of Egypt (an unmistakable reference to the Exodus), the Syrians from war-ravaged Kir, and the Philistines from Crete.

The Philistines settled the area around what is now Tel Aviv and called the land Philista. In later years the name would be pronounced Palestine, and the apparent descendants of a displaced Minoan remnant, who would stand in direct conflict with Benjamin Mazar and other displaced peoples, came to be called Palestinians.*

---

*During his six-and-a-half-year captivity in Lebanon (1985–1991), journalist Terry Anderson was given only one book to read: the Bible. He was known to be a religious man, and on April 17, 1992, interviewer Barbara Walters asked him if the Bible had provided him with some measure of comfort. "No," he said, and then explained to her that outside his window he could hear constant gunfire and screaming, people doing terrible things to other people. And as he read, it was all the same. The technology and the names of the combatants

Seventy miles north of Crete, in a place called Thera, Marinatos began exploring a hole in the Earth, eight miles wide and thousands of years old. Could this crater be the source of the Minoan decline? he wondered. Was this the reason for the broken foundation stones? Was this the trigger for all those migrations? Could this be true?

Thus did Spyridon Marinatos become one of the most widely traveled archaeologists of his time, moving from Crete to Israel to all the eastern Mediterranean islands, across the Atlantic to meetings with volcanologists in the United States, and back again. During one of his visits to America, in 1960, he stayed in the home of a young scholar whose son had become somewhat of a Jacques Cousteau addict ever since the release of a movie called *The Silent World*. The boy's name was Paul Zimansky, and it seemed to Marinatos that he just couldn't tell the child enough about exploring the Mediterranean islands. Their mutual enthusiasm reached a flash point, at which Marinatos began pulling apart his slide trays and holding up to the light every picture he could find showing the steaming black mound of volcanic rock that was slowly rising in the center of Thera Lagoon. He never did manage to rearrange the slides in their proper sequence in time for his lecture. More than three decades later Zimansky would still feel badly about the mix-up, which actually turned parts of Marinatos' presentation upside down.

About that same time, half a world away in England, six-year-old Elizabeth Stone had already decided she wanted to be an archaeologist. Digging in her backyard, she had found bits of broken china and rusted implements representative of a Victorian farmyard in Oxford. More than ten times her own age, anything from the turn of the century truly qualified, from a child's vantage point, as ancient. Her father, a historian, identified the objects and explained them to her. They amazed her, and when he told her that there existed remnants of civilizations a hundred times older than the Victorian shards, she was hooked. At age ten she joined a team of university archaeologists, working through an entire summer on her first "real" season of excavation. For her, this was bigger and better than any summer camp, and until her family moved to America when she was fourteen, she joined the archaeologists every summer, burrowing with them into abandoned churchyards and medieval foundations. Upon entering Pennsylvania State University, she promptly explained to her supervisor, a specialist in

---

had changed, but nothing else. This sort of thing had been going on in the Mideast almost forever. And so it was that Mr. Anderson wished he had been given anything, *anything* else except the Bible. He would have much preferred the *Oxford English Dictionary*.

European archaeology, that she had grown tired of European archaeology and wanted to travel farther east, to look at some of the early civilizations that were, relative to European civilizations, inaccessible and largely overlooked. He happened to have colleagues working in Iran and Iraq, so he passed her over to them. Simultaneously Paul Zimansky, the man who was to become her husband, landed in the same place by accident.

"I was sort of following my early interest of doing things underwater," recalls Zimansky. "So all my interest focused on Greece and the eastern Mediterranean. The problem was, there just weren't that many ancient shipwrecks being excavated. I even went to the island of Thera in August 1967, when *Time* magazine announced that Spyridon Marinatos had just found something important there, a Minoan city shut in under two hundred feet of volcanic ash. But nobody on the island would admit that there was a site there at the time. I remember wandering around the village of Akrotiri—the lost city was right under my feet—and I was looking around and asking people about Marinatos, asking where the ruins were, and everybody just shied away. They had obviously been told to keep very quiet about this. I never did find the excavation.

"When I began my archaeological training, I was studying classical European archaeology, but at that time I also started taking courses about the Near Eastern civilizations, just because those were the only other archaeology courses being offered at the university. As it turned out, there seemed to be so much more to learn about the Tigris-Euphrates region. In Greece everything seemed to have been pretty much nailed down, whereas there were whole empires missing in Mesopotamia. I began to see that the chance of actually finding something new and exciting, the chance for a budding archaeologist to actually get a toehold, was far greater in Iraq than in Greece. So I started learning cuneiform and reading everything Woolley and Lawrence had written about desert archaeology, and it was by happy circumstance that I ended up digging where, save for the occasional spring flood, there wasn't any water at all."

SPRING A.D. 1990

Nergal, the ancient Babylonian god of death and king of the underworld, had resurfaced in the desert of southern Iraq. He was the original grim reaper. Baked clay reliefs showed him carrying a sickle, with which he presumably harvested human lives. Elizabeth Stone and Paul Zimansky were slowly resurrecting Nergal, in the form of a city built in his honor more than four thousand years ago. The city itself had grown more than a half mile in diameter before being mysteriously abandoned, but four millennia of springtime rains and desert wind had flattened it so utterly that to untrained eyes surveying from ground level it would seem that nothing, not even a brick upon another brick, had ever stood here. The temples, the palaces, and the houses all had disintegrated, but the implements they once held were still present in the dust. Wandering across the plain and scuffing at the soil, Stone and Zimansky found the ground studded with copper blades, clay figurines, the keys to a house, and the contents of a courthouse.

Two hundred miles above their heads a satellite had traced the bed of the ancient Tigris. Though now flowing some eighteen miles east of the City of the Dead, the river was once only a mile from it. The satellite also revealed long, straight lines—the fossil impressions of canals—running like arrows into the very heart of what could only have been a network of streets. There

Seen here as if from a helicopter, the city of Mashkan-shapir, a probable kernel of truth from which the legend of Sodom sprang, is reconstructed from archaeological survey maps and satellite data (notably, the exact shapes of harbors and walls are constantly shifting under the accumulation of new evidence). A system of canals linked the Euphrates River with Mashkan-shapir, and also sustained an extensive irrigation network outside the city's defensive walls. The city behind the walls was a jumble of mud-brick buildings and narrow alleys. A huge temple complex enclosed a multi-tiered ziggurat, where the people worshiped the underworld god Nergal, a prototype of Satan.

was truth to the old saying that nature abhors straight lines, while humans revel in them. The trail from the river meant that ships were able to sail directly from the Tigris, through the city gates and into Nergal's canal system. The harbor, huge and rectilinear, was located in the center of the city, where a baked commemorative seal inscribed with cuneiform identified the place as Mashkan-shapir. Texts from another city described a military campaign in which more than two hundred ships were moored in Mashkan-shapir Harbor, ready to attack the city of Kish. The enemy city stood thirty-seven miles across the western desert, on the banks of a completely different river, the Euphrates. Zimansky often wondered how, given the geographic barriers, anyone could even have dreamed of carrying out a naval assault across the plain, and it occurred to him that the city builders, during periods of trade, must have connected the two rivers with a system of canals, more elaborate, more ingenious, and probably more beautiful than anything he had dreamed possible.

~~~~~~

"At the time of the Nergal cult, there was much more interlinking between the Tigris and Euphrates systems than there is today," said Elizabeth Stone. She was looking at a satellite image, trying to trace a faint depression running westward from the Tigris.

Paul Zimansky shook his head. "That's why I'm very worried, Charles, by what you're saying about linearity and the river civilizations."

He was referring to my view of the riverworlds as a vast isolating mechanism, wherein life was confined to a narrow, three-thousand-mile ribbon of fertility that force-fed competition, rendered universal takeover by any single competitor difficult, if not impossible, to maintain (at least for very long), and thereby fueled technological advance. According to this view, life along the riverworlds, with deserts on either side, became much like Spyridon Marinatos' Minoan islands, surrounded by natural moats that hindered empire building yet at the same time sustaining interisland competition in trade and even warfare. Had not Marinatos' volcanic catastrophe befallen the Minoan world, and had not the Roman Empire expanded into the riverworlds (subsequently collapsing and dragging them with Rome into a great Dark Age*), it is possible to believe that the competition would have

---

*Rome, during its more than five hundred years of prosperity, had conquered all horizons—which may explain why, in spite of the fact that the empire possessed steam engines and multiple gearshift devices, they were prized as little more than elaborate toys. There were no competitors in neighboring, unoccupied territories threatening to transport trade goods

continued indefinitely and that by the time of Columbus the first transatlantic airline service and moon landings might already have been ancient history. A comparison of Eastern and Western Europe during the past five centuries provided a timely and relevant analogy. In the West, small, fiercely independent nations learned how to exploit a wealth of peninsulas, mountain ranges, and islands, to the frustration of all would-be empire builders. Even in the twentieth century A.D. such barriers defined national borders. In the East, where geographic barriers were less clearly defined, nation-states had been less stable against takeovers, there had been less political freedom, scientific progress had been slower, and multinational empires had prevailed. Peace also prevailed, a peace in which Eastern versions of Leonardo da Vinci and Isaac Newton, when they came into existence, were less needed and less likely to be listened to. It was a devil's bargain, by which empire building ultimately thwarted competition, and peace became the kiss of death for technological advance. Conversely, the Crescent of Fire resisted resolution, forestalled peace, and emerged as a powerful vector of change.

Zimansky disagreed with this view of history. "That linear quality of civilization is very true of Egypt," he said. "But it is not really true of Mesopotamia, of the Iraqi side of the story, which is more a tale of competing networks growing up along two separate yet intertwined river systems, rather than little kingdoms lined up in a row, as if they were markings on a ruler. When we read cuneiform and learn of ancient battles (including Mashkan-shapir's preparation for a naval assault on Kish), a lot of the conflicts seem to arise from something akin to network alliances. Historic events are not always moving upriver or downriver but will sometimes cross sideways between the rivers."

Zimansky raised a valid point, insofar as the easternmost end of the system was concerned, but even when all the Tigris-Euphrates networks were unified under the rule of men like Sargon the Great and Hammurabi, if one could step back from Mesopotamia and view the entire riverworld at a glance, from end to end, the most powerful Babylonian kingdom added up only to a localized unification. There were still river kingdoms in Lebanon, Israel, and Egypt to contend with (and each of these, even in the most

---

or military supplies by rail and steamship, so sails, horses, and an ample supply of cheap slave labor would do just as well. Except for such events as the Germanic wars, there were few periods of technological advance in Rome. Had someone else built a rail line, Rome, with the necessary tools at hand, might have responded quickly to the challenge. But there were no challengers, so the empire stagnated under prolonged success, enfeebled from within, until it was all so clearly too late.

unifying of times, was always fragmenting along its length). The Tigris-Euphrates system did not necessarily invalidate the linearity theory. Much like portions of Israel, Syria, and Egypt, wherein fertile valleys may vary from a few inches to more than twelve miles in diameter, the Tigris-Euphrates plain was just one of several broadenings of the corridor, a simple knot in the ribbon.

Stone noted that today, when we look at a map of Iraq, it is easy to think of the Tigris and Euphrates as two separate rivers. But satellite photos tell her that in antiquity there were at least four or five different branches cutting through the plain (and Genesis 2:10–14 seems to agree with the satellite view, describing a river parting into four heads, their names Pishon, Gihon, Hiddeqel, and Perat*).

"The branches were all much closer to each other," Stone explained. "It was in fact a more linear system than you will find on a modern map, though this by no means eliminates the argument that *our* rivers—the Tigris and Euphrates—made very different demands on civilization from, say, the Nile. Essentially, the Tigris and Euphrates are being pushed apart by accumulating silts, a consequence of the first two or three millennia of agriculture. Originally the system could more accurately be described as a braided stream that just flowed through the center of the alluvial plain."

"It may be for that reason," Zimansky added, "that the Babylonians took a worldview, in contrast with the Egyptians, in which chaos was a normal state."

Elizabeth Stone got the giggles. We all did. There was something about the image of whole rivers shifting course under a civilization's feet that provoked laughter when one tried to imagine how this might shape an emerging culture's belief systems.

"For the Egyptians," Zimansky continued, "chaos was an abnormal and terrifying thing. Their river was stable, unshifting. During the flood season the waters rose only so high and then stopped, and the annual flooding could be predicted almost to the day."

"Unpredictable overflows and shifting watercourses were a standard feature of ancient Iraq," said Stone. "So the people came to know and expect chaos. That's why, for instance, the biblical flood story, I think, is a Mesopotamian story. It is perfectly at home on a Mesopotamian land-

---

* *Perat*, meaning "great river," was changed from its Hebrew name (which preserved the ancient Assyrian syllables) by Greek translators of the Bible, who came to call it Euphratés. The name Hiddeqel refers to the most turbulent, undisciplined of the four branches, which the Greeks came to associate with the powerful tiger, hence the modern name Tigris.

scape, where, when you come right down to it, there is no landscape at all. I'm telling you, nobody can really appreciate how flat it is. In a flood everything will vanish underwater, all the way out to the horizon. Even if the water is only four inches deep, it can look like the end of the world."

Stone did not believe that, archaeologically and geologically, any of Leonard Woolley's clay layers (some of which she suspected were actually layers of wind-deposited silt, blown in during periods of abandonment) could ever be identified as the one and only origin of the biblical flood story. What she did believe was that Leonard Woolley had been a very clever promoter who, financed mostly by private donations, built up and perhaps deliberately overblew all possible biblical connections (she knew, for example, that he had once handed out little vials of silt, telling wealthy contributors, "This is from Noah's flood"). She also believed that flooding was at least as familiar to the ancient Babylonians as to the inhabitants of Jamestown on the Mississippi. Everyone was bound to see at least one great flood during the course of a lifetime, and in both Jamestown and Babylon, when the floods came, people would tell of a legendary flood that was even worse.

"Can it be," I said, "that the Gilgamesh epic of the flood is just a recollection of many floods—perhaps *all* of Woolley's deposits—combined into one story?"

"Yes, I think so," said Stone. "It's really a matter of taking something they're already familiar with and increasing it for its mythological power. There are certain places where floods are part of the natural landscape. And whereas people living in Jamestown today, with modern American architecture, tend to clean out all the mud and start again, what the people of the Tigris-Euphrates plain did, with their mud-brick architecture, was just knock the house down, save the wooden roof beams, and build on top of the ruins."

And here arose another difference between the people of the plain and every other civilization. The ancient civilizations of the Nile, Crete, and the Indus Valley had impressed me with examples of indoor plumbing and city-spanning sewage systems. In fact, I'd come to believe that the complexity of a culture's sewers was directly proportional to its level of technological prowess. At most sites, a city's sewers were all that remained unmolested (on account of their bearing nothing of value to subsequent invaders). When I approached any lost civilization, it was becoming my habit to look first beneath its streets and ask, "What did its sewers look like?" until I met Elizabeth Stone and Paul Zimansky. I remember my disappointment when they told me that the people of the plain had built no sewage systems, and for a moment there was an inclination to look down my nose at the

Babylonians, to regard them as the most primitive of all the riverworld civilizations. At first glance it would have been difficult to believe they had built schools and libraries, invented Euclidean geometry, and designed the world's first electric battery. But when I looked closer, I noticed immediately that it was not the people of the plain who were in any way lacking but the yardstick by which I was so quick to judge them.

For their household drinking water, as near as any of us could tell, people went down to the canal, filled pottery jars, and carried them home. The toilets were simple holes in the ground with seats built over them and were enclosed in a shed or "outhouse." Pigs and other farm animals were evidently kept in the courtyards. Paul Zimansky believed that it was a mistake to impose models from modern (or even Roman and Egyptian) cities on Babylonian houses. While Iraq's ancient courtyards might have contained gardens, they were not simply after-dinner resting places but were intended for more general use, including garbage disposal. Zimansky thought of the house he and his wife had rented near the dig. The walls were mud-brick, and while the courtyard provided, among other things, a cool place to sit in the evenings, the landlord buried garbage there and kept chickens. In one corner stood a pile of foul-smelling debris that the archaeologists did not want to get terribly close to. Homelife in ancient Mashkan-shapir looked much the same: a rapid buildup of debris in the houses and courtyards. The floor levels, as they rose higher and higher on piles of refuse, were covered over with reed matting.* Inevitably the roof would have to be lifted to accommodate the rising floors, but this did not seem to be a cause for worry.

This was a civilization that had grown from mud, and mud-brick buildings simply did not last very long, meaning that there was an ongoing

---

*The refuse includes broken pottery and other discarded implements of everyday life. Because they were garbage, they were left undisturbed by later civilizations, and because they were left undisturbed, layer upon layer, garbage dumps are, and always will be, highly prized time portals to archaeologists. During the twentieth century A.D. two time capsules were buried in New York City's world's fair grounds, with artifacts and messages to future generations. During a recent checkup, researchers were unable to determine where the first of these, dating back to the 1930s, was buried. Probably no one will ever see it again. In the future New Yorkers should place their time capsules in the Oceanside landfill. Like all landfills, it is an anthology of our civilization, complete with newspapers decaying so slowly, under alternating layers of clay and plastic sheet, that some of them may still be readable four thousand years from now. Archaeologists, if any exist, will definitely look beneath the mound, finding, among other items, the half-built and discarded moon ship *Apollo 20*. It sounds odd, I know, but to someone like me, landfills are beautiful—the pyramids of our time.

process of renewal, in which the truly expensive, durable items, such as roof beams and wooden supports, were periodically pulled down, cleaned up, and remodeled. One did not have to work very hard to knock down a house and manufacture a few more bricks. All that was necessary was straw, dirt, and water, which were available everywhere. Nor was any great effort required to block up an old door with new bricks and plaster it over or to open up a new door in an old wall. The Babylonian home was curiously malleable. Even a preliminary mapping of Mashkan-shapir's foundations revealed an urban landscape that must have been continually under reconstruction. Not even the courses of streets were immutable. A road might suddenly be cut off by an expanding home, only to open up somewhere else. This was a totally alien situation to the Egyptians, or to anyone else building from limestone, granite, and other more permanent materials.

It occurred to me that a permanent, inflexible network of underground sewers could not possibly accommodate houses and administrative buildings whose floor plans and dimensions were constantly shifting. One could no more belittle the Babylonians for their latrines than one could dismiss the pyramid builders of Palenque and Teotihuacán as technologically deficient because, though they attached wheels to children's toys, they evidently never built a horse-drawn cart (horses and oxen were long extinct in the Americas until European invaders reintroduced them. Indeed, the only animal available for domestication was the jaguar, and history does not record anyone who succeeded in teaching a jaguar to pull a cart or who lived very long trying).

Paul Zimansky believed that if he could travel back four thousand years, he would have seen already legendary towers and ziggurats on the horizon, some of them eroded into unrecognizable shapes.* The Mesopotamian monuments were not, like the pillars and obelisks of Karnak, built for the eons. If not rebuilt and maintained, the mightiest of them would crumble and fall in a single human lifetime.

"And they knew this," said Zimansky. "They knew all of their buildings were going to fall down. They were very familiar with ruins themselves. So inevitably, from even an ancient Mesopotamian perspective, the past was going to be in ruins, and this could not help but shape their concept of permanence."

"They were always writing inscriptions to the future," said Stone, "to whoever was going to dig them up"—she broke into laughter—"messages to me."

---

*Some of these legendary, crumbling ziggurats are believed to have formed the basis for the story of Babel's tower.

"Right," said Zimansky. "They just kind of assumed—they *knew*—that these things were going to be studied by the equivalent of archaeologists."

Knowing that inscriptions on baked clay tablets would outlast mud-brick temples, the ancients often inserted them into the walls of their buildings. A cylindrical text near the vanished gate of Mashkan-shapir declared to future generations that the city wall was a contribution of King Sin-iddinam, ruler of Larsa. It contained some fourteen lines of what could only be described as official propaganda for the king, including mention of lavish wages for construction teams, which Stone suspected were greatly exaggerated to heighten Sin-iddinam's glory in her eyes. Another line of text read: "Anybody who pretends this inscription was written by somebody else will pay for all time." Evidently the people of the plain, while having a keen understanding of chaos and interruption, also tried to retain, to some small degree, a sense of continuity. When a temple had to be added to or restored, the builders often dug into walls and supporting structures of earlier building periods, and what the earlier builders did not want to have happen was for someone else to take credit for their work. So they placed inscribed cylindrical tablets in the walls, cursing anyone who removed an inscription and replaced it with his own.

About 550 B.C. the neo-Babylonian king Nabonidus* described on his own cylinders the experience of digging down through different construction levels of a temple and finding an inscribed cylinder from a king who had ruled five hundred years before. Eventually he dug all the way back to Sargon the Great, prototype of the biblical Nimrod, and discovered the king's statue lying in the ruins of his temple. He restored both the statue and the temple to their original forms. As Nabonidus described it, digging up buildings and artifacts and putting them back together was, to him, an act of piety, part of an already ancient tradition of rebuilding the past.

"About five miles southwest of Ur, near the ancient gulf coastline,"† Zimansky said, "is Eridu, where the ancients believed kingship first came down from heaven. It dates back to around 5000 B.C., perhaps even a little earlier. It's one of the earliest focal points of settled life."

"There's a real continuity," said Stone, "between what was going on at the beginning up through the higher strata of more recent civilizations."

"Yes," said Zimansky. His eyes were bright, and he leaned toward me,

---

*He is the same Nabonidus known from the Dead Sea Scrolls for the Prayer of Nabonidus, referring to a period of exile in which he is said to have been afflicted with malignant boils.

†The Iraqi coast is nearly 130 miles south of its 3000 B.C. position as the result of silt deposition by the Tigris and Euphrates, which are continually creating land where once there was water.

as if pleased to be revealing one of the Earth's oldest secrets. "You can actually watch the evolution, as you excavate down through a fully developed temple already dating back past 2000 B.C. You descend through seventeen or eighteen layers of civilization, until, near 4000 B.C., you get down to a tiny one-room shrine. And then you realize that you have just witnessed about two thousand years of the same site being sacred. I mean, that's nearly as long a span as the Temple Mount in Jerusalem!"

Elizabeth Stone nodded. "Well, it ran longer than that, actually. Because in Eridu they built a ziggurat over the temple, so that the continuity continued for many centuries beyond 2000 B.C. What we have here is a one-room shrine erected by a simple agricultural population and a continuity in which that shrine grew with one of the first civilizations to become a vast temple and pyramid complex."

I could not keep a grin from breaking across my face. Not that there was any cause for restraint. It was one of science's great "Wow!"s in which something can seem highly improbable, maybe even impossible, yet at the same time be true. "Imagine that: rivers and whole coastlines shifting, cities changing shape so fast that they become unrecognizable in the life span of an ordinary house cat, and yet in the middle of all this chaos, one site that remains sacred for thousands of years. It's so contradictory."

"Well, maybe not," said Stone. "That's still consistent with the terrain. I think it is possible, perhaps even necessary to have this long continuity of temple development in a region dominated by chaos."

"After all, Charles, having a single, sacred temple in Jerusalem hasn't done much to produce stability, has it?"

"True," I said, after a pause. "We can't argue with that."

"I have a theory," Zimansky said, "that if you could really follow the Eridu temple from the late sixth to the third millennium B.C., you would see profound changes in religion and—who knows? Perhaps you would find entirely different religions worshiping in that same place at different periods. We can't really document it because we don't have preserved texts going back that far; nevertheless, the idea of the same piece of ground being sacred for thousands of years—"

"Right!" I said. "Atop the Temple Mount in Jerusalem, the sacred rock under Abd al-Malik's Dome [built A.D. 687] is where Abraham was said to have offered his son Isaac to God and where, sometime during the seventh century A.D., Muhammad is said to have made a night journey into heaven and to have conversed with Jesus. It's one rock touched by three religions. And—"

"And the black rock, the meteorite at Mecca," added Stone. "It was worshiped in pre-Islamic times. So we do have this sense of mystic places."

Her husband smiled in agreement. "Maybe the more chaos there is, the more you need that."

"So," I said, "if your civilization is caught in turbulent waters, that's when you most need an anchor."

"Exactly."

The contrast between the Tigris-Euphrates world and the Nile was clearer now. As a matter of fact, their worldviews could not have been more in contrast. Egyptian civilization was remarkably static, and relative to the Babylonian end of the riverworlds, the Egyptians were a traditional, conservative society. "What's remarkable about it," said Stone, "is these real flashes of genius that they had at the very beginning of Egyptian civilization, but they did not do anything about it."

She was impressed by the fact that within the first five hundred years of pyramid and canal building, the Egyptians had written advanced medical texts, with approaches not only to the diagnosis of disease but to its treatment with surgery and plant-derived drugs. "And you would have thought that from that point they would have continued to move forward."

She leaned back, shook her head, and, as if speaking about a favorite student who had become lazy and thrown away a promising career, said, "But you look one thousand years or fifteen hundred years later, and they're still doing exactly the same thing they were doing at the very beginning. And yet what you see in Babylon, in terms of development, is a civilization that is always changing, always being reworked. The art styles change. The technology changes. The language changes. Political organization changes. And I think this may have something to do with the environment always catching them left-footed."

As Stone saw it, nature handed the Egyptians everything they needed on a golden platter, and they were quite happy, once they had reached a certain level of development, just to sit on the banks of the Nile and regard Cheops (Khufu) and Karnak as the loftiest of civilization's achievements. With no parallel rivers to worry about, with nothing except open desert on both sides of the Nile, and the Mediterranean in the north, Egypt's boundaries were more secure than Babylon's. And its river was stable, flooding and receding in predictable ways at specific times of the year. Though dynasties came and went, and though Upper and Lower Egypt occasionally fractured into rivals, civilization did not periodically collapse and have to be rebuilt. The past was not always in ruins. Rivalry there was, but no anarchy. Although it was impossible to extend a universal empire over the entire riverworld system (and difficult even along the length of the Nile), most generations of Egyptians grew and died through a lasting peace, and I think history is teaching us, time and time again, that peace can be the kiss of death for innovation.

In the eastern riverworld there were six thousand years of chaos, "and yet," observed Stone, "we find this tremendously innovative civilization." I was reminded of something Orson Welles said in his film *The Third Man:* "In Italy they had five hundred years of bloodshed and they produced Michelangelo, Leonardo da Vinci, and the Renaissance. In Switzerland they had brotherly love. They had five hundred years of democracy and peace and what did they produce? The cuckoo clock!"

By the time Julius Caesar met Cleopatra at Alexandria in 48 B.C., the Great Pyramid of Cheops was twenty-six centuries old. Impoverished citizens had long ago stripped away the gold plate and silver hieroglyphics that once capped the monument. Even its casing of polished white limestone had been scavenged to build new dwellings. The history of Egypt is one of impressive advancement up to about 2000 B.C., possibly spurred on by fierce competition with Evans' and Marinatos' emerging Minoans (the only people whom the Egyptians, by their own writings, considered worthy of being treated as a competing civilization). After the Minoan decline of 1628 B.C. there were few other challengers, and it might be said that Egypt became much like America after the Apollo moon landings: a victim of its own success. The lunar expeditions, of course, would never have happened if not for a new species of competition—the Cold War—between two giant, fiercely independent nation-states. After the Russians orbited the first satellites and the first test pilot, American technology became more difficult to sell overseas. Removing the challenge by force was precluded by a chancy "balance of terror," in which both sides ascribed to the thermonuclear equivalent of "an eye for an eye." So the only response remaining was for America to outperform its rival, and in so doing, it performed a deed for which there was no response.

The world watched Neil Armstrong and Buzz Aldrin walk on the moon, and nothing happened. Truly nothing. No other nation answered the challenge, and a deadly complacency set in: "If we can go to the moon, we can do anything." America after Apollo became like Egypt about 1500 B.C., master of all it surveyed. The policies developed by the builders of Apollo— among them General Electric and Chrysler—in which any worker could pull a "panic cord" and stop the assembly line if he saw a defect, were quickly abandoned. Why bother? Anyone, anywhere in the world, could look up at the moon and see that American technology was superior. And as the market for American products overseas became analogous to a universal empire and as, over the next fifteen years, American cars became characterized by stagnation and even declining quality, executives in Japan, learning everything they could from NASA and Grumman, applied quality control methods originally developed for Apollo (including panic cords), to automobile and VCR assembly lines.*

By 1990 there had emerged in the East a powerful new competitor without whom the quality of American technology might have continued to spiral down. In 1993 American executives, making a study of Japanese production techniques, began adopting the panic cord to their own assembly lines, never guessing that it was an idea inspired by Cold War necessity more than twenty-five years before that had by economic necessity come back full circle to General Electric and Chrysler.

Empire building, leading to a successful empire, leading to stagnation and collapse, is also a cyclical force, and while empire maintenance was virtually impossible along the three-thousand-mile length of the riverworlds, the Nile offered the fewest obstacles to local control. If this view of civilization were to survive the test of time, then the story of Egypt's final takeover by Greeks and Romans was a tale not of displacement by newer, more competent empire builders but of replacement by opportunists, made possible by Egypt's own internal weaknesses.

It was plain that Stone and Zimansky did not share this view, at least not entirely. Stone attributed the arthritic nature of Egyptian civilization primarily to the relative stability of the Nile, to an absence of periodic agricultural collapses, unpredictable floods, and shifting river courses. I had seen the relative absence of small, isolated, yet competing city-states as the source of the stability. In reality, both stabilizing forces might have been simultaneously at work, although I suspected Egypt's greater success at empire maintenance to be the more dominant one. That the last surges of Egyptian creativity ran parallel to the conflict between Queen Hatshepsut and her stepson Tuthmosis III, and with Ramses II's wars, could hardly be ascribed to coincidence.

Archaeologically the issue was (and still is) a long way from being put to

---

*In 1976, as America's lunar landers, abandoned in mid-construction, were rebuilt into everything from Christmas tree ornaments to birdbaths, Japan commissioned Grumman Aerospace to build one last lunar module for a Tokyo space exhibition and sent their executives to observe NASA-Grumman quality control and construction methods. They were most impressed by an Apollo veteran named Joel Taft, who told of the day a quality control man had pulled a panic cord and held up construction of three lunar modules for a whole afternoon because he found one small defect. "That must have been a very bad day for you," said one Japanese executive. "No," said Taft. "It was a good day. We located the source of the defect, corrected it, and were satisfied that it would never happen again."

rest. We did not have all the science, all the evidence. The jury would probably be out for many years. Yet I was intrigued by what Stone had said about the environment always catching the Mesopotamians left-footed. She believed the Tigris-Euphrates in the east, and the Nile in the west, had shaped their respective civilizations' expectations of the afterlife, shaped their religions. Sooner or later, whatever influence the rivers exerted on eastern and western thought was bound to filter down to the nomadic and ultimately literate Hebrew tribespeople who moved between the Nile and the Tigris and who, in time, were going to write the Old Testament.

Some of the evidence was as obvious as the desert sands. Mashkan-shapir had been built within reach of the Tigris, somewhere near 2000 B.C., yet the river was now eighteen miles away. Karnak had risen about the same time, yet the Nile still flowed before its temple gate. Stone said that the Egyptians, perhaps because they had this blatant regularity in their world, perhaps because they had the unshifting and predictable Nile, developed a worldview in which the afterlife was a field of rushes growing on the banks of a placid, infinite river. "And life for them, in many ways, became focused on the afterlife. The Egyptians embraced an optimistic view of life, whereas the people of the plain were always pessimistic. They saw the gods as being fickle. Their myths—preserved for all to read on cuneiform tablets—spoke of the gods getting drunk and doing irrational things, including destroying all humanity with a flood. They saw the universe as being created from chaos, to which it would eventually return. They saw death as being a thoroughly unpleasant place to be, in which you wound up in the underworld hung up on a hook like a lump of meat, or where crabs—creatures you had cracked open and eaten during life—surrounded you and cracked open your skull and pecked out your brains, and no matter what happened to you, you were always conscious, and whether you were virtuous or cruel during life, you could still end up in eternal damnation. The gods and the universe were like that: very unstable."

And so it was that at two extreme ends of the riverworlds, two concepts, equal and opposite, began moving through the oral and written histories of a dozen cultures, and came down to us as heaven and hell.

〰〰〰〰

The City of the Dead, Mashkan-shapir. The remains of its occupation, most of them lying within inches of the ground surface, were scattered over an area nearly a mile wide. The city was comparable in size to Ur and was of sufficient stature to be mentioned in the prologue to the Code of Hammurabi, but unlike most major centers, Nergal's city was occupied but briefly. That it had been built, and then mysteriously abandoned before, growing layer upon sequential layer, it could rise seventeen strata above the plain, was precisely what attracted Stone and Zimansky to the site. Unencumbered by the mass of prior and subsequent settlement debris characteristic of every other Iraqi site, they could actually hope to find, discarded among the foundations, cuneiform contracts and letters still lying in their true historic context of marketplaces, courthouses, and houses.

"The people of the plain wrote everything down," Stone had said. "We find their laundry lists, their bills, documents complaining about their wives, every detail of what their world was about."

Located eighty-four miles southeast of Baghdad, the lost city had been identified only as site 639 before Elizabeth Stone, shielding her eyes against a dust storm, looked down and saw a newly exposed nub of baked clay jutting up. She pulled a fist-size cylinder out of the ground, brushed the silt-encrusted cuneiform letters with her fingers, and whispered six syllables to the blowing sand: "Mashkan-shapir . . . Nergal."

According to the cuneiform annals, the monarch Sin-iddinam's army had cast mud brick for thirty days and built a wall around Mashkan-shapir. Sin-iddinam was said to have ordered the wall under the direct instructions of Nergal, who had the pointed ears of a wild bull and adorned his scythe with a lion's head. Orbiting satellites had revealed long, barely discernible shadows in the soil, suggesting that the foundations of the wall were still intact. A gap marked the place where the city gate had once stood, and when Elizabeth Stone went to the spot and began to uncover the entrance, a snake reared up before her, hissing and spitting and flexing wide the wings of its hood. Respecting the threat display, she gently backed away and decided to avoid the place—which, ever since, has been called the Cobra's Gate.

〰〰〰〰

During the spring of 1990 the archaeologists were the only foreigners who lived in Iraq like Iraqis. Paul Zimansky observed: "There is a real divide between the world of Baghdad, in which the diplomats and journalists are all living in the fancy hotels, and the world we are living in, out in the countryside, where we are renting a little house and we have to deal with the problems of water and electricity that [in peacetime, at least] are just unknown to the citizens of Baghdad. And we have had the privilege of seeing both worlds. We can travel through Baghdad, where we know the urban, educated [political] elite, and we also know the people out in the countryside, including the men who had just been discharged from the army, who are delighted to have survived the Iran-Iraq War and do not want to have anything further to do with guns and rockets."

Most of the diplomats were restricted from venturing outside Baghdad. Under Saddam Hussein (who fancied himself as the rebirth of the Babylonian king Nebuchadnezzar, and who was reproducing the monarch's murals on the walls of the capital, with his own head replacing Nebuchadnezzar's), foreigners had to apply for permission a week in advance if they hoped to leave the city. April Glaspie, the American ambassador, had once lamented to Zimansky that her people were required to specify exactly where they were going and to provide maps detailing which route they would take to get there, from which any deviation was strictly forbidden. And yet, as the emulator of the cruel Babylonian king who had burned Jerusalem plotted to make war against Kuwaitis, Saudis, Israelis, and Americans, two American archaeologists drove freely on the Iraqi international highway, from Baghdad to the Jordanian border, down to Kuwait and back again, without ever being stopped at checkpoints and asked to show their papers. The desert was paved with a six-lane, beautifully engineered road. All the markings were in Arabic and English. There was almost no traffic and no speed limit. So freely did they move that it came to Zimansky as a shock when he drove out of Iraq to discover that the Jordanians were much more concerned with checkpoints and looking at transport papers. Glaspie envied the archaeologists their freedom.

Mashkan-shapir was barely more than an hour's drive out of Baghdad. The archaeologists lived in a mud-brick village that had grown up nearby, "and I'd say," Elizabeth Stone later recalled, "that we were the only foreigners who really came to know the Iraqis." Those who worked for foreign corporations tended to live in little enclaves that imported their own national foods. In Baghdad the American diplomats received frozen White Castle hamburgers flown in daily by jet, and whenever travel permits could be obtained, embassy personnel drove full speed into Kuwait, seeking the nearest Safeway shopping centers.

Kuwait: That was where everything exotic in Iraq—car parts and Levi's jeans—came from. Zimansky would long remember the morning he stepped out of the excavation house to find that someone had broken into his rented car. The thief had tried without success to pry a cassette player from the dashboard, but Zimansky was surprised to learn that what he really wanted most, and had been unwilling to give up on, was the rear window. The American soon found out why. There were any number of front windows for sale in Iraq, but the only way to get a rear window was (A) to order it from Kuwait at a cost that, after the importer's travel permits were added in, could equal a substantial fraction of the car's value or (B) to steal it from another car.

Iraqui law required that the car be returned to the renter intact. Two days of excavation were lost to the trip to Kuwait, where a border dispute over an Iraqi oil pool (one of the world's largest), whose southernmost boundary edged about two thousand feet under Kuwait, was beginning to heat up.

Zimansky and Stone were living in a village of ten thousand people, two mosques, one police station. Most of the men they hired to assist with the exploration and excavation of Mashkan-shapir had just gotten out of the military and were just beginning to rebuild their lives after the long, bloody trench war with Iran. But though the war was over, it seemed to the archaeologists that as the season wore on, some sort of draft was still in effect. Men were simply being pulled off the streets. There had been a kindly old guard, old enough to remember Woolley and Mallowan. Most of his teeth were missing. He was so skinny that to Zimansky he looked more like a bird than a man; and he was on his way to work one morning when a car from the Popular Army drove up and two men grabbed him. The car sped south, toward the Kuwait border, and no one in the village ever saw the old man again.

When the archaeologists looked up from the site, there seemed to be more air traffic than usual, and heavy vehicles were practicing military maneuvers near the village. There were tank tracks running over the grave of Mashkan-shapir.

The main problem with our site, Zimansky thought, is that it is submerged under a nice, empty stretch of desert where there are no farms, no oil rigs, no railroads, or anything else of importance to the national economy. It was a condition that made the place attractive to archaeologists (no angry farmers or German railmen to worry about), but it was also, for the same reasons, attractive to the army. This problem was compounded by the fact that the city had only existed for a flicker in time, on account of which there were no telltale hills rising above the plain. Put simply, it did not look like much. At the Cobra's Gate, the tallest ruin jutted a foot and a half out of the earth, and if you bent down and took it apart, and knew what you were looking for, you might notice that it was part of a mud-brick wall. But there was nothing else for miles around; no stone columns, no porcelain towers, no Greek Acropolis, and it was easy for a traveler to tear up a lot of historical information without ever realizing that he had driven over a city. As if to drive home the message, three oil explorers came through one morning, digging holes and setting off dynamite charges under the ground.

When Paul Zimansky ran to the middle of the city and asked them what in the world they thought they were doing, the oilmen laughed.

"Seismic waves," one of them said. "We're listening for oil."

"We've been finding it everywhere we go," said another. "But finding oil isn't really the problem. It's all a question of how much oil."

"Everywhere you go?" said Zimansky. "How far have you come?"

"From almost as far north as the Turkish border. Kurd country."

Zimansky did some quick mental calculations. The men had been threading their way between the rivers, down through the very center of Mesopotamia. They could easily have passed near or through a half million archaeological sites.

"What's that woman doing over there?" one of them said pointing. "Is that a kite I see?"

"She's an archaeologist," said Zimansky. "There's a whole city under these sands, and we're mapping it. We've got a camera with an automatic timer mounted on the kite."

"A city?"

"Come. I'll show you."

Zimansky led them to a series of trenches, about four feet deep, and showed them the foundations of ancient mud-brick walls. They were amazed and could not believe that the archaeologists knew where to dig in order to find those walls. They assumed that Zimansky and Stone must be doing something far more sophisticated than looking at subtle color variations from the air, that the archaeologists must somehow be seeing through the ground, as they were now "seeing" with seismic charges. They were convinced that the kite cradled a tremendously powerful camera when in fact it was an ordinary 35 mm that could be bought in any camera shop for two hundred dollars. There was nothing extraordinarily sophisticated about what Stone and Zimansky were doing, and the fact that the oil explorers refused to believe this gave Zimansky some idea of how subtle archaeological sites really were.

The kite was simply providing more detailed coverage of what the satellite had already photographed, and offered the advantage of being able to take photographs at whim, within minutes, whenever the humidity of the soil and the angle of the sun were just right. What they were looking for was simply this: Mud-brick was slightly better at retaining water than the silt and sand that surrounded it, and while the eroded tops of walls were normally indistinguishable from the eroded ground surface, if a morning dew had settled or if a light rain had fallen and then begun to evaporate, there was a ten- to twenty-minute period, after the surface sediment had bleached dry, when the walls would still be stained dark. The color difference could even be seen from the ground, but from six hundred feet in the air, for a few brief moments before the walls, too, bleached out, the streets and floor plans of the entire city stood forth as clearly as hand-drawn blueprints.

Living as they were in "the outback," the archaeologists had no way of knowing that Baghdad had just placed a strict ban on aerial photography or that should the wrong Iraqi soldier notice the kite, they might be summarily executed as spies.

Forty-five miles west of Mashkan-shapir, on the Euphrates, Saddam Hussein was rebuilding Babylon. He had arbitrarily created a giant earthen monument, sculpted it into a pyramidlike ziggurat, and planted terraced gardens down its sides. These he called the Hanging Gardens of Babylon and declared them the second wonder of the ancient world, along with the Pyramids at Giza, that could still be seen today. When he could not figure out how to water the plants, he offered a million dollars to any archaeologist who could tell him how the Hanging Gardens of Babylon were irrigated. Paul Zimansky did not know very much about the original Hanging Gardens—no one did, for no detailed descriptions had survived—but what he did know for certain was that they were never located on top of a great mound of earth. They were probably down low, near the river, and did not look anything at all like the scar upon the landscape (covering at least a half dozen *real* archaeological sites) that Saddam had built more or less as a monument to himself, to show that he too could build a ziggurat.

Nearby he had erected baked-brick walls on the original neo-Babylonian foundations, with every fiftieth brick proclaiming "built during the reign of Saddam the Mighty." This was not the Babylon of Abraham and Nimrod, the time period Stone and Zimansky were probing. This was the Babylon of more than a thousand years later, the Babylon of Nebuchadnezzar, who sacked Jerusalem and carried many of the surviving Jews off into captivity. Turning his attention toward the possession of Kuwaiti and Saudi oilfields, and toward a modern-day assault on Israel, Saddam saw himself as conqueror of the Jews just as the biblical Nebuchadnezzar had.

Before A-6 bombers and Scud missile launchers put an end to exploration, Stone and Zimansky had found the terra-cotta shields of model chariots, decorated with serpents, horned figures, lion's heads, and sickles—all symbols of the lord of the underworld, Nergal. They had seen model chariots at other sites; but at Mashkan-shapir they were finding more than had ever been seen by anyone, anywhere else, and the symbols—the sickles and the serpents—meant that the chariots were somehow related to the ideology of the main god of the city.

They ranged from six to ten inches in length, about the right size for children's toys, but it seemed unlikely that they were toys. The chariots were covered with religious motifs (some dedicated to the lord of light, Shamash, but most to Nergal), and they were concentrated around what, if the surrounding scatter of cuneiform debris was anything to guess from, appeared to be the site of an administrative building. The one administrative function Stone and Zimansky had read about in texts discovered elsewhere in Iraq was that Mashkan-shapir was the location of something akin to the Babylonian supreme court. It was not uncommon to find tablets in other cities describing cases that had to be taken to Mashkan-shapir for trial. Most of them involved disputes over property. One man had sued his neighbor because they shared a mud-brick dividing wall between their courtyards, but the defendant had decided to add an extension to his house and in so doing stuck a peg in the wall and put a heavy roof on top of it, which weakened the entire structure and threatened collapse. According to the final court decision, the defendant had to buy the damaged wall from his neighbor, who was obliged to build a new wall with the money, creating a double wall between the two homes, at the cost of approximately two feet of property to the plaintiff. The ruling only served to move the level of animosity between the neighbors up another notch.

Mud-brick walls were taken very seriously in Mesopotamia. According to the Code of Hammurabi, the judge acted with uncommon leniency toward the defendant. The penalty for putting a hole through a neighbor's wall could just as easily have been death by impalement on a stake at the gates of the city.

Elizabeth Stone's favorite case stemmed from a man's marriage to an heiress. Six years after taking the vows, he had spent all or most of her money, which was presumably the basis for his attraction to her in the first place. And as the Babylonian equivalent of a stenographer etched claim and counterclaim on a clay tablet, the husband charged that his wife had failed to treat him like a man.

"He is *not* a man," said the wife. "My husband has never slept with me."

"Oh, she's just a scold," answered the husband. "She's been nagging me, and she's been saying bad things about me."

The court came down on the wife's side: "Listen, husband, you have taken her money. Now, you have to treat her right, like a wife."

The cuneiform dispute escalates to the husband yelling at the judge: "I don't care. You can put me in prison. Do what you like. I will not live with her!"—at which point the record trails off into missing tablet pieces.

Exactly where the model chariots entered this process is anybody's guess.

Was evidence somehow stacked on them and wheeled ceremonially across the judge's bench, as if to be weighed? I don't know. And we may never know, unless some legal manual turns up describing the procedure in detail. My personal guess is that the chariot was a forerunner of our more figurative scales of justice.

Paul Zimansky once said that what attracted him and his wife most to the people of the plain was their peculiarity of keeping a record of private life on nonperishable materials, allowing everyday men and women who have long turned to dust, turned to less than dust, to speak clearly after four thousand years.

SPRING A.D. 1992

"The record is curiously mute on the last days of Mashkan-shapir," I told Father John MacQuitty. "And I think you'll be very interested in what I'll be showing my editor in a few weeks."

I had spread open a folder of notes, chapters under construction, maps of ancient riverbeds, penned reconstructions of artifacts, satellite photos, and, of course, Dr. Stone's "blueprints via kite."

"Some of this gets rather technical, and before we come to it, let me give you an idea where this city stood in time and whence the difficulty arises. During the Late Bronze Age the volcanic island of Thera blew up, sending out an ash cloud that shaded the Earth, plunged temperatures all over the Northern Hemisphere, and left summertime freeze scars in trees ranging from Turkey to California. Since tree rings thicken and thin in yearly cycles, they actually provide us with fingerprints through time, with which we have traced the explosion to 1628 B.C. . . . Autumn, presumably."

"What? You can actually come that close?"

"Give me a few more years, and I imagine we'll be able to come even closer than that. Today we find ash from the volcano deposited thickly across Turkey, eastern Crete, and the island of Rhodes; yet the cloud seems never to have touched Greece or western Crete—which is consistent with one of the Mediterranean's powerful autumn squalls. We also have evidence of sudden and extreme cooling in an alpine Chinese bog during that same year—we can read it in the sediment layers—and the Chinese histories of the emperors, which are easy to date because the scribes kept a record of eclipses in the margins, tell us there were false winters in summer at that time, and a famine so terrible that in some provinces man became a man-eater."

"You're talking about a global catastrophe, then."

"Yes," I said. "Now here's the important part. The paleontologist Daniel Stanley, of the Smithsonian Institution, has discovered remnants of the ash cloud buried under the Nile Delta. Carbon dates on the mud above and below the ash layer, though subject to errors of thirty years or more, bracket tree ring dates from Turkey and California, also the Chinese bog. Moreover, we know that the government on Crete shifted from Minoan to Greek control when Thera exploded, and a foreign minister to the pharaoh Tuthmosis the Third recorded the changeover in his tomb, meaning that if we take the Thera date literally, then Tuthmosis the Third ruled Egypt in 1628 B.C., at least a hundred twenty years before most Egyptologists say he did."

That last statement had caused me some discomfort in recent years. Most archaeologists, working from dates derived by assigning every new pottery style an arbitrary life span of fifty years, would not even discuss the Theran ash layer. One had even threatened suicide if I persisted, and I could not begin to understand what everyone was so upset about. My own training was in paleontology. When I found a fragment of bone or amber dating from the time of the last dinosaurs, I took it as a part of everyday life that, looking across a span of nearly 70 million years, I would never know within 2 million years when my fossil had actually lived.* We bone peddlers are used to being off by at least 4 percent. How, then, could there be cause for despair in looking across nearly 4,000 years and learning that you were off the mark by only 120 years? To me, they should have been amazed that their pottery clock was accurate to better than 3 percent. Someone had obviously done something right. I began to realize that insofar as concepts of time were concerned, archaeologists and geologists lived and breathed in different worlds. And thus was it inevitable, if not entirely understandable, that we few paleontologists who were now surfacing into the shallows of archae-

---

*The 1992 discovery of pieces of impact glass (called tektites) splattered across the Caribbean by an asteroid that evidently struck the Yucatán Peninsula about the time the last dinosaurs died out has changed our experience of paleontological time. We can easily date the glass by knowing rates at which certain rare isotopes of the element argon have decayed. We now know that at least one of the (possibly two or three nebula-derived) mountains of ice and rock that struck the Earth at this time, depositing a global "ash" layer, did so in 64.4 million B.C. (give or take a hundred thousand years). For years we paleontologists have been rewriting the last days of the dinosaurs on the basis of what fossils occur above and below the ash layer and how far they are found from it. Now, as I approach the Theran ash layer, I bring with me all the experiences (and no doubt a few cumbersome prejudices as well) gained from my encounter with the dinosaur puzzle.

ological time were sometimes viewed as intruders who should have remained content with our dinosaurs and ammonities, instead of sticking our noses where they did not belong.

"That 1628 B.C. ash layer is the first real time probe we have," I continued. "With it, we can push Tuthmosis the Third back to the time of Thera, and because records of the pharaoh's contacts with foreign kings still survive, we must push back kingship all over the Middle East."

"You'd better watch what you say, Charles. There are some who would probably hang you for less."

"There's one who tried," I said quickly. "Almost destroyed a couple of years' work. But not to worry: He won't be troubling me anymore."

"How's that?" said MacQuitty, looking worried.

"Oh, he's still plenty alive, Father. But forgive me, I had to 'Leonard Woolley' the bastard."

"Good God . . ."

"Yes, that's just what he thought of it, and if my guess is right, that's what they thought at Mashkan-shapir in the end. As you'll see from this chart, if we use the 1628 B.C. time line as a calibration, the sixth Amorite king, Hammurabi, who ruled about three hundred twenty years before Tuthmosis the Third, must have come to power in approximately 1950 B.C.

"Now, the great wall of Mashkan-shapir was dedicated during the six-year reign of a man named Sin-iddinam, who ruled approximately four hundred years before Tuthmosis the Third and eighty years before Hammurabi, meaning that the wall went up about 2028 B.C. The city still stood during the time of Hammurabi, and Elizabeth Stone has tentatively placed its final abandonment about thirty years after the end of Hammurabi's reign, which brings us to roughly 1896 B.C., some two hundred sixty-eight years before Thera.

"Stone was schooled on the traditional chronology, which would place the end of Hammurabi's reign plus thirty years about 1746, not 1896 B.C., but she has expressed puzzlement over the fact that from what little wood is available for carbon dating, it has always looked to her as if everything ought to be older than traditionally believed. If we follow the Theran time line, in which Hammurabi's reign ends about 1926 B.C. and Mashkan-shapir is abandoned thirty years later, then the Hittite raids occurred (putting an end to the dynasty of Amorite kings about one hundred fifty-five years after the death of Hammurabi), about 1771 B.C., some one hundred forty-three years before Thera.

"There are descriptions from this time of fierce mountain tribes raiding the cities of the plain. My bet is that the invaders were made bold by the cities' own internal weaknesses, for about this time another long cycle of salted farmland, failed crops, economic collapse, and anarchy had begun. When Thera exploded in 1628 B.C., Mesopotamian civilization had been in eclipse for almost one hundred fifty years and would not begin recovering for almost fifty or a hundred more. When the recovery began, under what became known as the Hurrian Dynasty, the major construction effort was centered on digging channels to get water back into the old canals, to bring back the abandoned cities, rebuild the ancient temples, and get all the old dwelling places of the gods functioning again—all, that is, except Mashkan-shapir."

"How unusual is it for a city that large never to be inhabited again?"

"Quite unusual. Mashkan-shapir is the only one that we know of."

"As if the land were somehow poisoned," MacQuitty suggested.

"Or in the very least rendered useless," I said. "Stone's feeling is that the Tigris River had already shifted several miles west by the time civilization returned, and because the city had not been around long enough to become sacred, it was not worth the effort of dragging the Tigris back by canal. She also quotes something from a cuneiform text about bitter water just to the south of the city. Perhaps the local water table became too salty."

"You're not satisfied with that explanation?"

"No. When you read my file, you will be able to tell that we are all a little frustrated. If there's anything nature abhors more than straight lines, it's a vacuum. While total abandonment can be a blessing to archaeologists trying to reconstruct everyday life in a lost city, it creates a very stubborn 'Why?' As I see it, Mashkan-shapir puts the *Mary Celeste* and the mystery of the Roanoke colony in the Little League. The disappearance of the population is not the whole problem, however. These people were building out of mud brick. All their wood had to be imported, so there was not very much of it around. According to Stone, what little fuel they had for cooking came mostly from sun-dried cakes of sheep and cow dung, so there was really very little in that city available for burning."

"Then what have your friends found, Charles? What's the problem?"

"I can answer that in one word, Father John: *firestorms.*"

SPRING A.D. 1990

They had wanted to make sure that the kite-revealed blueprints they were seeing on the surface went down to a point where there would be something worth digging up. If in fact, they were just seeing the bottom bricks of a foundation, then there really wasn't an archaeological site at Mashkan-shapir after all. As it turned out, the trial trenches that Stone and Zimansky sliced in the alluvial plain followed the architecture down at least four feet. Only the tops of the mud-brick walls had completely eroded away. There were more than floor plans in the earth. The floors themselves were strewn with debris.

The first room they found was badly burned, so badly that with the application of a little more heat, the bricks themselves would have baked rock-hard and the upper walls would still be standing above the plain. The chamber was filled with ash and soot, and Paul Zimansky noticed that when he climbed out of the trench and examined the ground surface, virtually everywhere he went the soil was peppered with specks of ash and solidified shards of black, tarry matter. "All right," he recorded. "This whole thing looks like it's burned."

Stone pointed out to me that a large part of the site appeared to be covered with a scatter of burned bone: "Animal bone, presumably. And you do not normally find burned bone without there having been some kind of widespread conflagration."

Raiders, I thought. But that answer did not quite fit the picture. Invaders tend to loot as they burn. Animals are valuable. Livestock would be virtually the first thing taken from the city. Why burn the animals? It made no sense. It was like throwing gold down the sewer. Stone added that the bone field did not cover *all* of Mashkan-shapir, "and so we may still have some large burned areas without having the whole city burn. Just large parts of it."

But even a quarter of it burned to a crisp was difficult to explain. It is not very easy to char livestock to the bone, especially in a city built of mud, with nothing to sustain a fire except a few imported wooden beams, reed mats, wood-and-reed roofs, reed baskets, and, in some of the wealthier homes, wood-fired stoves. The mud itself was fireproof. The worst imaginable conflagration was a burning roof setting a whole neighborhood of roofs aflame: pretty easy to escape, not likely to trap either men or cattle. Even today, in the mud-brick village that stood within sight of the City of the Dead, there had been no need to create a fire department.

"A fire that consumes a home and then leaps from house to house just does not happen," said Zimansky. "There just isn't enough material available for burning. It wasn't like the city of Chicago, built almost entirely from timber and waiting to make its acquaintance with Mrs. O'Leary's cow. There was essentially nothing in or around those buildings that could have given rise to a firestorm."

And the permanent abandonment? Maybe, in the aftermath, the people came to believe there was a curse on this place, as certain oral traditions were known to suggest. I had a number of weaknesses, but jumping to superstition was not one of them. It was true, however, that there was nothing in or around the city that could explain a firestorm. But what about *under* it? One detail had been overlooked, perhaps because it was so large no one had taken serious note of it before.

Underneath the plain there were whole lakes of flammable hydrocarbons. From Mashkan-shapir to Kuwait's southernmost border, in an area barely larger than the state of Massachusetts, lay nearly a quarter of all the known petroleum reserves in the world. Unlike the basement rock, oil had a tendency to wander about. One never found it where nature first put it. Pile a few thousand feet of alluvial sand on top of it, and slowly, like an amoeba grown out of all rational proportion, it collected itself and started to move. Sometimes it bubbled up into daylight, dried there, and became pools of tar. Other times the fluids and gases, emerging through cracks and propelled by the weight of the desert sands, erupted geyserlike onto the surface. I could not stop thinking about the three oil explorers Stone and Zimansky had encountered at the site. They had found oil everywhere, even under Mashkan-shapir. There were a half million archaeological sites on the plain. Sooner or later it was bound to happen.

Where the rivers flowed, the edges of continents were squashing together like sheet metal in a car crash played at such extreme slow motion that only one frame passed every few centuries. This was a zone of earthquakes—the microscopic creaks and snaps within the crash—and whenever a new fissure brought a gaseous fountain to the surface, it was usually a harmless event, producing just another tar pool and perhaps polluting the local water supply. But if a gusher emerged into a domestic setting, dotted with ovens for baking pottery and bread, the gases would encounter ignition sources in all directions, every hundred feet or so, and the fountain would become a pillar of fire and soot jetting out of the ground.

If my theory bore currency, the fire could never be put out. It would have burned until the pressures from below stopped pushing petroleum to the surface and until all the erupting fuel had been consumed by the flames. It would also have left traces: little speckles of combusted petroleum products raining down over the city and long since solidified to rock hardness. The speckles may be difficult to differentiate from the bitumen, or tar, that seems to have been used as waterproofing in virtually every house throughout the city. But if indeed Mashkan-shapir was the victim of an oil fire, then closer to the actual source of the fire, some residue of natural gas or masses of tar from a flaming oil pool must still exist, awaiting only the excavator's spade.

I had to admit that, like most hot speculations, the one I was heading toward was probably wrong. But if not an oil fire, then some unknown incendiary agent had killed the animals and caused the city (or a significant portion of it) to go up like the smoke of a furnace.

Sodom and Gomorrah (Sedom and Amora, in the original Hebrew) were said to have met that same fate, along with at least two other "cities of the plain." Like Abraham and Nimrod, Sodom, if it existed at all, must have entered the Old Testament as a telescoping of oral traditions from diverse places. The City of the Dead fitted many requirements of the biblical Sodom and as such might have brought some very real grains of truth to the heart of the legend. It died by fire and was never inhabited again. Its last citizens are said to have served abominable gods, for which (in Deuteronomy 29:15–27) the "anger of the Lord burned against this land." It was fascinating to know that here they worshiped an underworld deity who had all the characteristics of a satanic prototype. And what was I to make of the justice chariots? They could only bring an ironic smile. If, as the Bible would have it, we were to believe that all of history had God's hand in it, then what was the point of burning a city full of lawyers and judges? Was God trying to show us that he had a sense of humor?*

Then Abraham fell upon his face and laughed.†

Or was he just trying to get the world off to a good start?

And what the hell was Sodom doing in Iraq? Weren't the biblical cities of the plain supposed to be near the southern shore of the Dead Sea?

Not necessarily.

Benjamin Mazar had cautioned me to bear in mind that many of the old Israelite tales were also well known in Egypt and Babylon. Quite naturally, later biblical writers could have displaced them to Canaan, which became the focal point of the story of the Hebrew tribes. Because of this, Mazar had lamented, it would be difficult to distinguish which Bible stories were original Israelite legends and which ones were borrowed from other lands. Much as the Classical Greeks exhibited a tendency to displace very old and mythical events westward into the mysterious and seemingly infinite Atlantic, the Hebrews exhibited a similar tendency to relocate Babylonian traditions (to judge from those traditions that have survived in cuneiform tablets

---

*I could not help being reminded of the old joke in which someone breaks the wall separating heaven and hell. God complains to Satan that the damned are escaping to heaven through the break. "So what?" the devil says, and refuses to mend the wall. "Then I'll sue," God answers. Satan smiles indulgently and asks, "Now where are *you* going to find a lawyer?"

†Genesis 17:17.

predating the Old Testament scribes) westward, toward Israel. I reminded myself that the Sabbath in its holiest form, had been derived from the first chapters of Genesis, which were, in turn, derived from a Babylonian Creation story, which had in turn adopted a Babylonian "week" (concluding with a seventh day of rest) based upon the seven heavenly bodies visible to unaided eyes. The laws of Hammurabi and the story of the infant Sargon's passage down a river in a reed basket were also distinctly Babylonian traditions that appeared to have been knitted into the books of Moses and displaced west. In like manner, the Babylonian flood, originally a telescoped view of terrible events on the plain between the Tigris and Euphrates, had, by the time of the Genesis compilations (during the Babylonian captivity of Hebrews, about 550 B.C.), spread out of Mesopotamia to engulf Turkey, Israel, and Egypt; and in more modern traditions, the flood has grown to encompass the entire world.

When I looked more closely at the Bible, it was easy to see that cities tended to move around a lot. According to Job 1:1–3, the city-state Uz was located somewhere east of Canaan, possibly in the Tigris-Euphrates region. The genealogical lists of Genesis, which named major population centers after legendary founders, identified Uz (in Genesis 10:22–23) as a district in Aram, known on a modern map as Syria (support for this identification could be found in cuneiform tablets referring to a Syrian district called Ussai). Later in the Bible, the prophet Jeremiah (25:19–20) listed kingdoms that had angered God and were overdue for punishment. He worked his way up the Mediterranean coast from Egypt to "the land of Uz" and then continued northeast to include "all the kings of the land of the Philistines." The land of the Philistines was Canaan itself, therefore placing Uz somewhere southwest of Israel. If the Old Testament was to be taken literally, then a single city-state was to be found east, north, and south of Jerusalem.

The city of Sodom also seemed to wander. One reads in Genesis 14:1–3 that Shinar (a district between the Tigris and Euphrates rivers) and four other districts made war with the kings of Sodom, Gomorrah, and three other cities. These were the cities of the plain, which Genesis 13:10–12 said Lot traveled *east* to after arriving with his uncle Abraham on the banks of the Jordan. "East" would suggest the Tigris-Euphrates region, and indeed, cuneiform tablets ranging from 2000 to 1700 B.C. referred to the land between the Tigris and Euphrates rivers as the cities of the plain. Genesis 14:3, however, clearly stated that the warred-upon cities were "joined together in the vale of Siddim, which is the salt sea."

Though the cities of the plain referred to in cuneiform texts were near the Persian Gulf (a salty body of water), the biblical text had always been thought to be identifying the cities of the plain with the most distinctively salty of all seas: Israel's Dead Sea.

This appeared to me as just another erroneous placement on the map. In the first place, the names Siddim, Sedom, and Amora (as originally transcribed in Hebrew) had a distinctly Mesopotamian flavor. In the second place, the Dead Sea seemed a long way for the king of Shinar (located, according to Genesis 11:1–31, on the plain between the Tigris and Euphrates rivers) to be sending his armies, especially as the cities of ancient Babylon were generally too busy warring against each other (except, as in the case of Nebuchadnezzar, under brief and infrequent unifying dynasties) to have spare legions available for foreign wars. And in the third place, the Jordan River and the Dead Sea were situated on the bottom of a giant crack in the Earth. When I stood in the Jordan Valley, surveying rugged hills in every direction, I knew that the Atlantic Ocean must have looked much like Israel nearly seventy million years ago. Bet-She'an, Masada, Tell Quasile—from end to end, it was the most fascinatingly violent and beautiful terrain I had ever seen. And if anything was certain, it was that no tract of land anywhere in the valley qualified as a vast plain, much less a plain large enough to accommodate multiple cities.

Genesis 14:10 said, "the Vale of Siddim was full of slime pits." This aroused my curiosity. In ancient Babylon, tar, or bitumen, was also referred to as slime, and in order for it to have been used so extensively for the waterproofing of houses, pools of tar must have existed relatively near Mashkan-shapir and the other cities.

"The biological evidence that we are getting suggests that there may have been a lot of water around Nergal's city," said Stone. "And one of the cuneiform descriptions of Mashkan-shapir notes that it was just to the north of—quote: 'bitter water.' "

Bitter with what? was the question. "Fouled water?" I suggested. "Oil-rich water?"

"There might have been a large marsh somewhere south of the city," said Stone. "What we don't know is how saline that marsh would have been because the fish bones we are finding [in courtyard rubbish heaps] represent freshwater species, not saltwater."

To her it was a question of whether, over the centuries, there had occurred an accumulation of salts that shifted an initially brackish marsh barely able to sustain freshwater species over the line to one that could sustain only saltwater species, hence the reference to "bitter water." If Stone's salt intrusion ever did occur, it was limited to the city's last days, or to some point after the fall, because there was, according to the story told by bones in rubbish heaps, no interruption of the freshwater fish supply during the life of the city.

The same rule applied to my idea of an oil extrusion turning the waters bitter. The event, if it occurred, would have been limited to the end of Mashkan-shapir's brief existence, for if the water supply had been fouled at the start, no one would have settled there in the first place.

In Deuteronomy 29:20–27, Moses told the tribes of Israel that many generations after Sodom's destruction, one could still "see the plagues of that land, and the sickness which the Lord has laid upon it; and that the whole land is brimstone, and salt, and burning, that it is not sown, nor bears, nor does any grass grow on it." To the tribes it was a warning of what could happen to them if they forgot the words of the Torah. To me it was, at first reading, an apt description of the cyclic saltings and collapses of Mesopotamia's agricultural infrastructure. More telescoping, I guessed. If there were any truth behind biblical accounts of brimstone (sulfur), slime (oil tar), salt, and fire assailing the cities of the plain, then Nergal's city, if we assumed it was to be part of the tale, was simply one among several cities that, over the course of a millennium or two, fell victim to disaster and gave rise to stories of God's vengeance against a sinning humanity.

But from every indication Nergal's city had inexplicably burned to a cinder, just like Sodom, and had been forever abandoned, just like Sodom. Was it really more than a mere part of the tale, perhaps even the actual prototype for the biblical Sodom? There were times when I was tempted to think so, and in some of my least disciplined speculations I kept coming back to the similarity between "the Vale of *Siddim*" and the name of the king who had built the great wall of Mashkan-shapir . . . *Sin-iddinam* . . . *Siddim* . . . *Sedom*. If I'd ventured a wrong guess, the guess itself was not without some precedent. The eighth century B.C. capital of Assyria, on the left bank of the river Tigris, had been named after King Nimrod. Ussai, too, was said to have been named after a legendary king. Sin-iddinam . . . Vale of Siddim . . . Was the connection legitimate? I doubted it. And I regretted that none of us would probably ever know one way or the other.

Of this much I could be reasonably certain. The city of Sodom, if it existed at all, existed in or near Iraq. In 1976 Giovanni Pettinato, the original epigrapher of the Italian expedition to the archive of Ebla (in what is today northern Syria), had said otherwise, announcing that the five cities of the plain were mentioned on one of the tablets and were probably located in Israel. Five years later his successor, Alfonso Archi, came up with a different interpretation of the mid-third millennium B.C. texts. Though locked in what was then mushrooming into one of archaeology's most bitter rivalries, both translators were forced to agree that some early inter-pretations of Eblaite references to Sargon, Agade, and the cities of the plain should be abandoned. Archi pointed out that Sidamu, mentioned in a text concerning the delivery of textiles and proposed by Pettinato as the city from which the Hebrew name Sedom was derived,* was also mentioned in two other tablets, which listed one of Sidamu's neighbors as the town of Lubaan, in northern Syria. If Sidamu was in any way connected to Sodom, then it was never anywhere near the shores of the Dead Sea. As for the location of the biblical Gomorrah, the cuneiform name of the city Amorah would read *imar-at*, and indeed, an almost identical name did appear on two Eblaite tablets; but it was associated with the delivery of six oxen to the king of Kakmium, a city near the Tigris, in Mesopotamia.

"This may come as a surprise to you," said Zimansky, "but the one thing that had absolutely nothing to do with our interest in Mesopotamia was the Bible."

"Absolutely nothing," Stone agreed.

"Nobody I know," Zimansky continued, "nobody working as an ar-chaeologist in Mesopotamia has any interest in the Bible whatsoever. If anything, it's a kind of antagonism. We sort of feel we're stigmatized in having to deal with it because a lot of the popular interest in this region comes from people who are interested in the Bible."

Zimansky was right. It did come as a surprise to me. In my journey through the riverworlds I had encountered a broad range of belief systems among archaeologists working on the Nile and the Jordan, and along the thin bands of date palm oases lining the northern coast of Sinai. "On the Nile," I blurted, "I even knew one Egyptologist who, when asked to indicate her religion for the national census, wrote: 'ancient Egyptian.' A lot of them seem to be into New Age cultism."

"Well," said Zimansky, "Egyptology is a funny field. Very isolated, very self-contained. Again, it's the same reason Egypt is a funny ancient civiliza-tion."

"And Israel . . ." I trailed off. I'd never met an atheist Egyptologist or

*The derivative, Sedom from Sidamu, may be of even less substance than Sedom from Siddim and Sin-iddinam. The builder of Mashkan-shapir's wall (Sin-iddinam) was at least a contemporary of his city. As an example of how faulty such derivations can be, one might, at first glance, be tempted to suggest that the name Sidamu is also descended from Sin-iddinam, but this derivation is precluded by the fact that the Sidamu reference was written some six hundred years before Sin-iddinam was born. (For leads to literature on the original Ebla tablets and early interpretations, the reader is referred to the December 1981 issue of the *Biblical Archaeology Review*.)

a particularly devout one, but the Israeli archaeologists I knew ranged the entire spectrum from atheist to devout (while never touching New Age cultism). And in the third major riverworld, the Tigris-Euphrates, spirituality was at best an irritant. Zimansky and Stone had presented a fascinating new puzzle, whose first piece, it seemed to me, was brought to light when a journalist for *The Philadelphia Inquirer,* reporting on the AAAS astrobiology symposium of 1986, noticed that every one of us on the panel was either an only child or a firstborn son. Subsequently I noticed that a high percentage of the world's dinosaur experts were dyslexic. And now came this odd convergence of religious belief and the specific riverworld toward which an archaeologist was likely to gravitate.

"Granted there are a lot of people who go off and dig in Israel because they are interested in the Bible," Zimansky finished for me. "But that's a completely different crowd from the Mesopotamian crowd. Completely different. With a completely different agenda and completely different priorities."

It seemed incredible that one could look at a field of endeavor and immediately know something about a scientist's birth order or level of spirituality. It was a completely new perspective on the riverworlds, strange and unexpected, crazy even, but there it was.

Late in that first spring season of the last decade of the second millennium, as Stone and Zimansky broke camp, said good-bye to their workers, and prepared for the flight home, they took a long last look toward Ur. Something was changing. During the past few days and nights, new columns of dust had begun rising from the southern plain. Military vehicles were converging, Stone guessed, organizing for some sort of practice maneuvers. The expression on her face was one of puzzlement creased with worry. Two hundred miles above her camp, an American KH11 satellite was peering through the dust columns with cold, expressionless eyes. Its infrared sensors were augmented by photomultipliers that could magnify starlight ten thousand light-years away or detect the glow of a cigarette on the desert below. Under cover of night the little orbiting robot saw vehicles from Baghdad and Basra etching unmistakable patterns on the ground, gravitating toward the Kuwaiti and Saudi borders; a deadly storm, incomparable even by the standards of the Mashkan-shapir fires and Leonard Woolley's flood-layer culture, was about to break.

8

THE
WORLD,
THE FLESH,
AND
THE DEVIL

What the Bible calls Paradise—Eden—was simply the Sumerian [or ancient Babylonian] word "Edem," the wild grassland of southern Iraq, the natural landscape before the arrival of the city. And picking over the debris of Paradise, it's hard not to see the psychological truth of the Bible's story: that the very beginning of our ascent to civilization was also the fall. When we tasted the fateful fruit of the tree of knowledge, [we at once acquired] the means by which we would become masters of the earth, and yet eventually gain the power to destroy it and ourselves. Truly a devil's bargain.

—MICHAEL WOOD, *Legacy*

SPRING A.D. 1991

As I write, black rain is falling across Iran and India—courtesy of Saddam Hussein and the first known use of ecologic warfare.* The fires burning in Kuwait are nudging an already alarming accumulation of greenhouse gases upward ever so slightly, creating a ripple in the biosphere that will be felt globally and for decades to come. The soot is raining acid upon India's rice fields even as the United Nations announces results of a disturbing study: 10.5 percent of the planet's most productive soils—an area the size of China and India combined—has been salted by unstable irrigation practices, overgrazed, deforested, acidified, and so seriously damaged since World War II that a global food crisis looms as the human population swells toward eight billion by A.D. 2020.

I fear that some of the worst scenes played nightly over CNN are a portent of things to come for much of our world during the next forty years. An elderly Iraqi engineer, nurtured by an electronic civilization, has found himself suddenly in a place where there is no food, no clean water, and always the sound of children crying. "You will hear the shouts of fathers and mothers," he says. "They are cursing, abusing. Cursing even God. And this is reality. This is the first time I hear Muslims and Christians who curse God."

If we are to have any future worth living, requiring every schoolchild to study the cyclic collapses of civilization in the riverworlds and then to

---

*There are in fact earlier, though relatively small, precedents, including America's use of poisonous herbicides to defoliate the forests of Vietnam, Genghis Khan's destruction of Mesopotamia's irrigation systems about A.D. 1200, and, in 146 B.C., Rome's destruction of the Carthaginians' fields by spreading salt on them. Nevertheless, Hussein's tactic stands as the first use of true ecowarfare, for in this case the consequences (deliberately so) reached far beyond the boundaries of conflict. Nothing in the history of war can even furnish criteria for comparison with the Kuwait fires.

compare what is learned with the modern-day republic of Haiti might be a good start. In A.D. 1975 more than 80 percent of the Haitian landscape was covered with rain forests. The people were fruitful and multiplied and subdued the land. Today less than 5 percent of the forest remains. The naked soil has washed down into the sea, choking the life-giving coral reefs to death. At 6:00 A.M., in scenes belonging to horror movies that were pure science fiction only two decades ago (*Soylent Green* comes quickly to mind), men and women gather with plastic jugs and fight bitterly around trucks bringing drinking water. All the farms have failed. There are no fish in the sea, no tourism, no exports. There is nothing except constant warfare. The fighting weakens the land still further, makes the people even more inclined toward fighting. The process becomes cyclic and self-sustaining, like a snake connected head to tail.

We know now that the people of the riverworlds were the first to learn that canal building and other new technologies permitting the beneficial exploitation of local environments could ultimately lead to overexploitation and collapse. For them, the cyclic decline of civilization, though it always seemed like the end of the world, was a relatively small event limited to the distance that one could travel by boat or horse in a few days.

In A.D. 1985, while working with prototypes for robots we one day hoped to launch into space (to explore new oceans under icy moons of Jupiter and Saturn), I sailed with explorer Robert Ballard and the deep-sea robot *Argo*. For me, all the present fuss over extinct civilizations began as just another interesting thing happening on the way to Jupiter, for *Argo* happened upon the *Titanic*, and the *Titanic* became my baptism in archaeology. I remember a shiny new beer can lying on the *Titanic*'s bow. I remember tar balls and plastic bags filled with garbage floating thousands of miles from the nearest city, and Ballard explaining to me that the mid-ocean waters were darker than they used to be.* But most of all, I remember that when I returned home, more and more of my work began to encompass archaeology, and as I began to see how past civilizations fell, so, too, did I find signposts pointing to the new millennium. I realize, now, that probing beneath the snows of Europa, hurling Valkyrie rockets toward A-4, and all my other dreams of planetary and extrasolar exploration will never come to pass if we do not take pause and get down to some basics right here on Earth. And the most disturbing realization of all is that in the

---

*"People are always griping about the plight of the forests—which is indeed cause for concern," Ballard said. "But the Amazon and Sequoia National Park are minuscule compared to the oceans. We don't breathe oxygen from the redwoods. We breathe it from plankton, from microscopic plants living in the first three inches of the ocean's surface!"

past the consequences of expansion beyond the limits of environmental stress have manifested themselves not as a science-fiction-like wave of starvation and plagues but as a more subtle pattern beginning with what is essentially a declining standard of living, rising food prices, and ultimate economic collapse followed by civil disorder and population crashes (cities are more fragile than we think, perhaps only two or three meals from rebellion). Seen in this light, the history of civilization's cyclic expansions and collapses is simply this: the desire of every empire and every city, every hamlet and every individual, to live beyond its means.

Indeed, the paleontologist in me begins to see the evolution of all creatures in this same light, casting the history of life upon this planet as the long march of every organism toward chaos (occasionally) forestalled by feedback. Yale University biologist Lynn Margulis might not entirely agree or at the very least might like to add a few details to the picture. She views living cells as collections of mitochondria, flagella, and other once free-living organisms whose ancestors colonized one another, and she sees in New Zealand's sea slugs and the lining of the human throat symbiotic linkages between once-individual, free-roaming cells. Seeing this, she proposes that humans, termites, and trees are part of a still-larger organization. Margulis is one of the framers of the Gaia hypothesis, in which the largest living organism on Earth is Earth itself, built from the interactions of whole ecologies.

In some quarters Gaia is taking on the qualities of a new religion, shifting the concept of "Mother Earth" from symbol and myth to a dawning reality. But the Earth is not a doting parent. The origins of hemorrhagic fever, the new leukemia virus (HTLV-1), and AIDS all can be traced back to a half dozen rain forests slashed down to accommodate expanding human populations. At the edges of the receding forests, farms have become bountiful food sources for monkeys and other carriers of diseases to which humans have never before been exposed. Fields of grain expand their numbers and bring them into close contact with man, whose airplanes instantly transport any new infection around the world. At brainstorming sessions with scientists from Brookhaven National Laboratories and the Rensselaer Polytechnic Institute (RPI), I once observed that if man had left the rain forests alone, some of these new diseases might have left man alone, to which RPI physicist Ed Bishop added a haunting and provocative question: "If the Earth's biosphere is viewed as a living, unified thing, could not the ever-expanding billions of us be the Earth's equivalent of a viral bloom?" If we answer yes to this question, then new diseases emerging from felled rain forests are simply the planet's immune system responding against us.

If Margulis and Bishop are right, then AIDS, ozone holes, and impending agricultural collapse are the feedback we earn by growing out of control and becoming increasingly harmful to our surroundings. Viewed in this way, Manhattan is a major infestation of *Homo sapiens,* an irritant. As we progress from an electronic civilization to a spacefaring one, Earth becomes potentially contagious to the rest of the galaxy, and if we are jolted by the arrival of AIDS and ozone holes, we should be equally alarmed that the conditions may be ripe for similar surprises. "I don't think these are going to be unique events," adds Joshua Lederberg, who holds a Nobel Prize in Medicine. "There will be more surprises because our fertile imaginations do not begin to match all the tricks that nature can play."

As ozone holes, acid rain, and other consequences of our expanding billions begin to take hold, people like Bishop and Lederberg, whose backgrounds are as varied as physics and genetics, have begun to see the one really relevant question for future man as this: How to save the ship from its passengers? As a paleontologist who wanders through the deeps of time, I ask a different question: Does the ship really need saving? The biggest difference between the scientific measure of time and the biblical measure of time is that the scientific measure teaches us that the ship has been here for billions of years and, no matter what, will be here for billions more. It's the passengers who are in trouble.

〜〜〜〜

Where once there were marshes and fertile grasslands on the Tigris-Euphrates Delta, Saddam Hussein's war has spread out sullen black lakes of oil covering a thousand square miles. Overcome by fumes, birds trying to cross Saddam's lakes have been seen to fall from the sky, forming new fossil beds. The latest footage of towns near Mashkan-shapir shows twisted bodies still lying in the streets after many weeks, and there are reports of attack helicopters sent by Hussein to dump sulfuric acid on rebellious citizens. The village Stone and Zimansky lived in was a Shiite area, right in the center of a postwar uprising against Hussein.

Watching from New York, Stone surveys the destruction and recalls her life at Mashkan-shapir, and the world tilts irrationally, like a dish, and she feels as if she were standing on its edge. Every new frame of film shows streets of broken glass and fallen walls, already being colonized by the first weeds. Everywhere the camera turns she sees scars: office buildings still smoldering, their former executives, with the look of madness in their eyes, offering for sale powdered milk looted from hospitals. Since early childhood Stone has been digging in ruins, often finding civilizations stacked on top of one another like the decks of a giant layer cake, the oldest, most romantic and mysterious of them always lying at the bottom of the mound, naturally (Hammurabi's time usually lay beneath more than twenty feet of earth and rubble). But now, for the first time, she sees the youngest, uppermost layer of ruins still gleaming in the daylight. Coming to terms with the destruction of ancient settlements has always been an excitement for her, even rip-roaring fun, but this is too immediate. This time it is her own home and the homes of her friends that have become the substance of archaeology. It is a painful and belittling thing to behold. There is no romance in it. No dignity either.

"I don't want to wax too pessimistic," her husband says, "but I was just thinking: Looking at modern Iraq, what do you suppose the population of Iraq a year from now is going to be in comparison to what it was before the invasion of Kuwait? I'll bet you'll find that there are a couple of million people missing."

Stone shakes her head, slowly. "Cholera epidemics . . ." Already, Red Cross and UNICEF investigators have reported that along the Tigris and Euphrates rivers, more children are dying from malnutrition and disease than all the Iraqis who were killed in the war. The discoverers of Mashkan-shapir begin to suspect that it has always been this way: masses of people just disappearing when authority collapses. "There is no electricity in Bagh-dad. No clean water, and these are populations who are as used to drinking purified water as we Americans. And the Tigris water is going to be polluted, and they don't even have fuel with which to boil that water. The only fuel Baghdad has right now is whatever wood its citizens can get by cutting down their fruit trees. They're going to run out of those real soon."

"Suddenly there is going to be a real decline in population," says Zi-mansky. "It's happened many times in Mesopotamian history."

The conversation becomes increasingly statistical and less immediate, slipping easily into the past. "It's difficult to know exactly what's happening when you lose a population," says Stone.

"As when Ur broke down," I interject.

"Yes. That was followed by what seems to us to be a universal disappear-ance. And the question is whether they died or whether a lot of them wound up practicing lifestyles which are now archaeologically invisible, which would be Bedouin nomadism . . . somewhat like your biblical wanderings of Abraham and Lot."

"But that necessarily is going to support a much smaller population," adds Zimansky. "The land is never uninhabited. People just go to a smaller, more pastoral system of living until civilization comes back again . . ." And it occurs to me that seeing broken television sets and trucks half buried in the desert, seeing the achievements of our present, electronic civilization swept into the dustbin of archaeology, reminds us of a conceivable, if not likely, future. It is plain that Stone and Zimansky would rather not dwell on that thought just yet. Me either. For us, the only quick escape from the future is into the past. And with strange ease does our conversation shift to more distant oddities, including population dynamics in the archaeological record and my recollection of a young RPI/CIA man poring over satellite images of southern Iraq during the Gulf War. The trainee had, at some point in his past, studied archaeology, and in one of the images he saw, targeted for bombing, a hidden bunker that was in fact an archaeological site. When he reported this to his superiors, the military wanted to know: "Are there any more of these?" He laughed. "Sure," he said. "It's the cradle of civilization. Most of the early chapters of the Bible take place there. The country is riddled with ancient buried structures."

Two concerns immediately arose among military strategists: (1) risking pilots' lives to bomb useless targets, and (2) anti-U.S. propaganda that might arise from going after archaeological sites.

"We sort of had this idea," says Zimansky, "that the bombing campaign was being organized the way we organized the campaign against Japan and Europe during World War Two, wherein ancient cathedrals and other cultural monuments were put on lists of places to be preserved and we tried to avoid bombing them. But not one of us archaeologists (despite a lot of agitation from us) was contacted until the bombing campaign was virtually over. Our troops are in control of Ur now. There was an incident involving Iraqi planes that were supposedly parked on or near the ruins. Dick Cheney was making a big deal out of it near the end of the war, saying at the time that the Iraqis were essentially using their archaeological sites as hostages."

"It doesn't make sense," says Stone. "Because you don't have a paved road to Ur. You have to go up a steep incline on a bumpy surface. You can't drive ordinary cars there, much less Soviet MIGs. I suspect somebody exaggerated, that somebody decided this was a good thing to say."

Zimansky nods in agreement. "All right. So three weeks after Cheney made this big fuss about it, we occupied Ur. And not one report came out as to what the status of that site was or where those planes were. Why didn't anybody follow up on that? I called a couple of news organizations and said, 'We're there now, why don't you send a reporter over in a helicopter and see what's going on?' Suddenly there was no interest in it at all."

"Before the war," says Stone, "we knew all the American Embassy staff, and they knew us. So they had every opportunity to know that places like Ur were important, every opportunity to know that if they were going to protect cultural monuments, they should get in touch with the experts, and they decided not to. There had to be a conscious decision somewhere that this was not sufficiently important for us to worry our little heads about."

~~~~~~

On January 2, 1991, on the very brink of the Gulf War, one of my more religious friends, a rabbi, found a crowning irony in how Westerners in general, and Americans in particular, knew almost nothing at all about the history and culture of Iraq. "They do not yet realize," he said, "that as they move into the southern desert and prepare to unleash the terrors of super-science, they are returning to the very source of their own civilization."

The Bible insists that a man named Abraham once lived, and that God sent him wandering out of Iraq, leaving behind what would one day be priceless reserves of oil (in response to this apparent mistake by God against his chosen people, Father MacQuitty argues, "The efficiencies of man are not the efficiencies of God"). On the eve of war I told Father MacQuitty that I had often wondered why, if anything like the Exodus ever took place, and if we were to suppose that this had anything to do with the will of God, the Hebrews were sent for forty years into the deserts of Sinai. And when at last they moved north into Israel, why were they sent into a land already occupied by people who fought them every inch of the way? It seemed to me that a more forward-looking God might have commanded them to build a fleet of the then commonly used reed boats and guided them outside the Strait of Gibraltar, where, in only four or five weeks, they could have sailed across the Atlantic to Florida or some other sparsely populated part of America.

This is the sort of question that one would normally expect to provoke anger among theologians. But at least among members of the Vatican's Jesuit order, questions—even the most challenging ones (especially the most challenging ones)—seem welcomed.

According to Father MacQuitty, it was important for the Hebrews to remain within the riverworlds, within the Crescent of Fire. "In that small corner of the world," he explained, "you have the epitome of every single dimension of human quest. The crescent . . . was the heartbeat of civilization, the heartbeat of the world. It is the place of revelation because directly underfoot we can trace back our history. Its roots are archaeological. We can identify with the people.

"And in the center of the crescent you had Jerusalem, which became the symbol of every city, the symbol of persecution, conflict, and restoration. It is the city of constant restoration and constant destruction. Right now, in Jerusalem and in the deserts beyond, the chaos of many religions is a perfect example of the babble that continues through man. The Babylonian captivities will continue.

"We have now, as I speak, three hundred thousand men and women—American Christians and Jews for the most part—who have been put into the land of Allah. Thousands of years have passed. An independent, electronic civilization has grown up all the way across the Atlantic, and still, we cannot get away from the crescent. We have come back to it. And now the amazing thing is that you have three hundred thousand representatives of variant religions. They are *in the desert*. The desert is the beginning of life. Even in the midst of conflict the desert will give new life. It is a strange sort of womb: sterile to man, but fertile to God.

"That's why Moses, John the Baptist, Saint Paul, even Christ and Muhammad had to retreat into that land. It forces meditation. And now we have the biggest desert retreat in the history of civilization. Under Islamic law our soldiers in the dunes cannot drink liquor. They have nothing. They can't get bombed out of their minds. They can't snort cocaine. They can't go into town cavorting and lusting because it is against the Saudi ethic. In other words, they are sober, looking at fields of rocks and sand as dead and dry as the surface of the moon, and at the absolute austerity of the blue sky and the stark sun. I've been there, and I find that the desert bothers people. The irony is that it is the beginning of spirit life. Removed from the fertilities of the world, what you're looking at is the austerity of God seen in a type of death-life.

"Those soldiers are being confronted. They are conscious of death, hope, fear, despair—all the dimensions. From August into January they have waited . . . helpless. I feel they're out there, all those people—and especially at night, with the cold stars burning in the desert sky—and they are being deified despite their frustrations. I feel that they will be conscious of utter loneliness. God is within that loneliness, and that is why so many of the founding prophets went into the desert seeking God. I hope they have sentinels out there who are just experiencing . . . that sky. They can't do anything except face themselves. And when you face yourself, you face God."

〰〰〰〰

As it turned out, the worst fears of the archaeologists were realized, while the best hopes of the theologian hardly materialized. Ur survived, barely, and only because most of it was still shielded under the ground. Other sites, including Mashkan-shapir, were not so lucky. The young soldiers Father

MacQuitty knew were indeed more religious upon their return from the desert. Some told him that after the rocket and tank battles had ceased, they were sent north and south to guard Kurds and Shiites from Hussein's vengeance, or east on giant hovercraft tank transports, bringing hundreds of tons of food and medicine to cyclone-ravaged Bangladesh. Father Mac-Quitty was impressed: "These were bodies trained for combat, and yet the war ended in India and Kurdistan with soldiers bottle-feeding babies." (MacQuitty had the curious distinction of being able to look into any evil event and find a way that human beings might somehow emerge enlightened and improved from it.)

The priest allowed himself to believe, for a fleeting moment, that the men had undergone a spiritual rebirth. But there is a true saying (this is how sayings become clichés; they're always true): "There are no atheists in a foxhole." MacQuitty expected the thin veneer of spirituality to erode with time, but he was astonished by the rapidity with which the men returned to their prewar attitudes. He would never cease to be amazed by the human mind's ability to adapt, to render even the strangest adventures commonplace. And thus did Iraq become many different things to many people in A.D. 1991, but mostly it was the shadow of Nergal, who held commerce with levels of darkness most people could hardly guess at.

III
COVENANT

9

YEAR
OF THE
DAMNED

When you stand back and look at a slice through the rings of a tree, you don't see people's lives; all you see is time, year upon year. . . . What seems like a long time to us is nothing but a tiny span in the life of a tree. Now, that same tree is nothing but a tiny span in the mind of God.
—ALEX HALEY, *The Search*

C. R. PELLEGRINO

There exists an archaeological Twilight Zone, where legend ends and history begins; and the two are sometimes hopelessly intertwined. Nowhere is this more true than in the fossilized city beneath Thera, with its connections to both the legend of Atlantis and Exodus.

In the year 1628 B.C., a Minoan city was crushed beneath nearly two hundred feet of volcanic ash on the isle of Thera. Discovered by Spyridon Marinatos in A.D. 1967, its multistoried apartment dwellings (of which only the lower floors have survived) would not have looked out of place on a modern street, and were equipped with such "modern" conveniences as bathtubs, flush toilets, and, apparently, central heating. A broken quartz lens hints that these people were inventing telescopes; and it becomes possible to believe that if not for volcanic upheaval, there might have been moon landings and television by the time of Christ.

C. R. PELLEGRINO

In the year Thera erupted, the pyramids were cased in a glass-smooth shell of polished white limestone and capped with sheets of gold. Thirty-three million years ago, the blocks from which the basic framework of the pyramids would be built were sediments forming on the bottom of a shallow sea, and today a close inspection reveals the blocks to be filled with clam shells and fish bones.

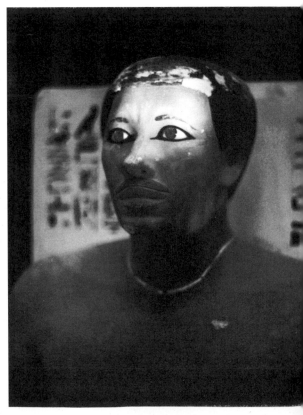

Egypt was a multicultural nation-state, a Bronze Age "melting pot" settled (not always peacefully) by Semitic tribespeople, merchants from the Minoan world, and, as demonstrated by this limestone funerary statue depicting an Egyptian prince, immigrants from the interior of the African continent.

Even by today's engineering standards, the Great Hall of Columns at Karnak is overwhelming. The temple complex was built as a copy of Egypt's legendary First Temple, where life was said to have crawled out of the sea. When the Greek historian Herodotus walked these avenues about 450 B.C., the columns moved him to write, "The Egyptians are religious to excess, beyond any other nation in the world."

After plotting her assassination and declaring the mere mention of her name a curse punishable by death, Tuthmosis III sought to destroy all traces of the pharaoh Hatshepsut's existence. He buried the obelisk depicting her deification by the gods, and in so doing unwittingly assured its preservation. Unearthed and reset about 20 B.C., it towered over Karnak as the tallest monument in all the world, until it inspired and was surpassed by the Washington Monument.

C. R. PELLEGRINO

A human figure stands in the ruins of Mashkan-shapir, whose mud-brick buildings are normally invisible to the unaided eye but stand out as clearly as blueprints when viewed from a kite at dawn. So long as the bricks still hold the morning dew, they stand out darker than the surrounding landscape, meaning that the city is visible for only a half hour each day.

Elizabeth Stone and Paul Zimansky excavate Mashkan-shapir, in southern Iraq, just prior to the Kuwait invasion. More than a third of the city, whose buildings lacked flammable materials, burned mysteriously to a cinder about 1895 B.C. Stranger still, no one ever settled there again, as if the land were somehow considered poisoned or forbidden.

Archaeologists excavate a cuneiform-inscribed tablet from the ruins of Mashkan-shapir. From such writings, the city is known to have been populated by judges and lawyers who worshipped Nergal, a prototype of "the Grim Reaper" and "Lucifer."

MALCUS WOODBURN

C. R. PELLEGRINO

Jericho is one of the oldest known cities in the world, dating back to at least 8000 B.C. Shown here is part of the city's defense system, a stone rampart that once towered over the land but was buried, inch by sequential inch, under the accumulated detritus of subsequent civilizations. Now, instead of soldiers climbing up the tower, archaeologists climb down to it on narrow stairs (right) carved into the detritus.

In the Bible, Genesis attempts to recount the early transformation of hunters and gatherers into settled farmers and city dwellers. The oldest mud-brick walls in Palestine are found at Jericho and were built by a farming community to keep the nomads out. Such fortifications required a complex system of social, economic, and political organization and teach us that civilization began at least ten thousand years ago.

The great Tell (man-made mound) of Bet-She'an marks the remnants of twenty cities, each built upon the ruins of its predecessor and the oldest of them dating beyond 4000 B.C. In the foreground, Byzantine bathing halls were built into an earlier Roman structure. On the horizon, near the top of the mound, Ami Mazar has penetrated below Byzantine and Roman ruins to uncover Bet-She'an's Canaanite and Israelite periods.

A panoramic view from atop the Tell of Bet-She'an, where we are able to stand upon the city's Twentieth Dynasty level and look down across its Canaanite period. According to biblical accounts, the Philistine rulers of Bet-She'an hung the bodies of Saul and his sons on the city walls after defeating them in the battle of Mount Gilboa. King David conquered the city, and his son, Solomon, made it an administrative center of his kingdom.

As if reading backward through the pages of a book, excavators at Ami Mazar's Bet-She'an site, moving backward in time as they move deeper into the Earth, patiently descend by inches into strata contemporary with Tuthmosis III, who was pharaoh during the time period covered in Exodus.

C. R. PELLEGRINO

The Temple Mount, in Jerusalem, is the burial site of sacred temples dating far beyond King Solomon of the tenth century B.C. The Bible tells us that Solomon's temple enclosed a splendid golden chamber that could only be entered by the High Priest on one day of the year. In the center of the room stood two *keruvim* statues, and beneath their giant golden wings stood the Ark of the Covenant.

C. R. PELLEGRINO

The Dead Sea Scroll caves, at Qumran, have yielded eight hundred documents hidden during the first century A.D., apparently as Roman legions approached the Jordan Valley. The scrolls, consisting mostly of contracts, lists (numbers of sheep traded, etc.), and deeds spanning more than three centuries, became a source of controversy the moment biblical texts were discovered among them.

C. R. PELLEGRINO

Nomadic tribespeople still inhabit the Judean Desert, living much as they did ten thousand years ago, except for the addition of electric generators and TV antennae to their tents. An important link in Jerusalem's antiquities market, they exchange archaeological artifacts for gas to run their generators, so their TVs can stay plugged in (the programs of choice are two old America soap operas: *Dallas* and *Falcon Crest*).

SPRING A.D. 1965

Just as Plato had described it, Spyridon Marinatos thought. Once there had been a circular island rising a mile out of the Aegean, but by comparison to what then was, there remained in small islets only the broken bones of the wasted body, strewn about a flooded hole in the Earth more than eight miles wide and a half mile deep. Thousands of years ago, the isle of Thera had simply disappeared. Fifty cubic miles of rock were hurled into the heavens and the depths, and Marinatos suspected that the force of the Theran upheaval might explain the strange displacements of foundation stones he had seen in the lost Minoan towns of Crete, only seventy miles away. If he could prove that the explosion had occurred during Minoan and Egyptian times, it might also explain the catastrophe Plato had written about in *Timaeus* and *Critias* or perhaps even the plagues of Egypt and the drowning of Pharaoh's army.

It was the biggest explosion human eyes had ever witnessed. Not even the USSR's fifty-megaton bomb, which had burst up through the Earth's atmosphere and into space when tested, could furnish criteria for the Theran upheaval. Harold E. ("Doc") Edgerton, the man whose photographic inventions had frozen in time the first few milliseconds of nuclear detonations, was now developing a new side-scanning sonar to peer below the mud of the Aegean seabed. Piece by piece, a mosaic of Edgerton scans was revealing that the Thera explosion had done more than dig out a volcanic blowhole. It had torn up the whole eastern Mediterranean. There were cracks in the Earth, radiating out from the island. One of them was more than a hundred miles long. The sound of the thing must have permanently deafened people three hundred miles away, and any surviving with their hearing still intact would have heard it again and again, as the noise shot around the world sixteen times.

If the explosion occurred during Minoan times, Marinatos theorized,

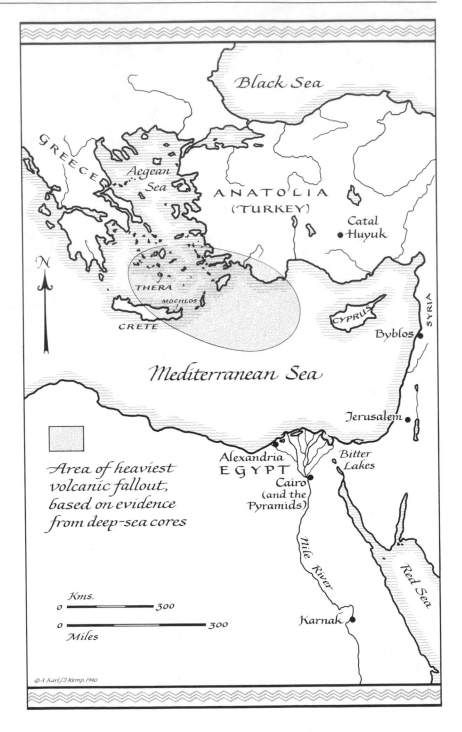

Black Sea

GREECE

Aegean
Sea

ANATOLIA
(TURKEY)

Catal
Huyuk

N

THERA

MOCHLOS

CRETE

CYPRUS

Byblos

SYRIA

Mediterranean Sea

Jerusalem

Area of heaviest
volcanic fallout,
based on evidence
from deep-sea cores

Alexandria
EGYPT

Bitter
Lakes

Cairo
(and the
Pyramids)

Nile River

Red Sea

Kms.

0 ———— 300

0 ———— 300

Miles

Karnak

© A. Karl/J. Kemp, 1990

then almost certainly it must have impacted on their civilization, perhaps even brought about its mysterious decline. But he had no proof. No one really knew when, in relation to before or after the Thera upheaval, or how many years from it, Minoan civilization had ceded to the Greeks. Nothing short of a town dating from the last days of the Minoans and buried conveniently at the bottom of the volcano's ejecta blanket would convince Marinatos (and his critics) that he was on the right track.

So Spyridon Marinatos went to Thera, hoping that before the final explosion, exhalations of volcanic ash had buried at least one town to its rooftops, protecting its buildings as if sealed in a cocoon. In some places the pumice and rubble were more than two hundred feet deep. In others nearly four thousand years of stream erosion had hewn the ash layers down almost to the original ground surface. These scratches in the Earth, Marinatos decided, would be his portals through time. From that moment his work was a matter of exploring the Theran wasteland, from end to end, and deciding which portal he would enter first. On the southern shore of a crescent-shaped rib of land, near a town called Akrotiri, some of the island's original limestone hills actually protruded into daylight, and atop one of these a farmer had discovered the foundations of an ancient lighthouse.

This is the spot I would pick, Marinatos thought, if I were going to build a town. In his mind's eye, he tracked the island back in time, placing a towering mountain where now existed only a quiet bay. Homes on the southern shore, near modern-day Akrotiri, would have been shielded by the mountain from prevailing winds and would have faced the Sun all winter long, whereas buildings on the north shore lay always on the windward and always in the shadow of the mountain. The south, he theorized, had been the choicest real estate around.

Spyridon Marinatos belonged to the Arthur Evans, Leonard Woolley, and T. E. Lawrence school of archaeologists, for whom intuition served instead of satellites and occasionally served even better than satellites. Marinatos had chosen wisely: There were in fact Minoan ruins hidden beneath Akrotiri, on the original ground surface of Thera; but they were part of a mile-wide city, not a town, and they were in a most remarkable situation. The same ash that cocooned the city had also collapsed roofs under its weight. Yet even in the most severely flattened rooms, vast meshworks of wooden beams and household furnishings had offered some small resistance, so that there were innumerable underground chambers and passageways. If, through some sort of magic, Marinatos could have entered one of these chambers through a door, he would have found signs of seemingly recent habitation all around him: an oil lamp with finger smudges upon its surface, rows of pottery jars half filled with walnuts, snail shells, and seeds (many of them still capable of germinating). Some of the very air was last breathed by the men and women who had made the finger smudges and filled the jars, thirty-six hundred years ago. But no magic doorways existed, and it was only through sheerest good fortune that one of the chambers revealed itself to him.

A notable feature of the city's cocoon was its permeability to rainwater. The land above was mostly desert, partly on account of poor agricultural practices but mostly because the water seeped through it with surprising speed. A severe autumn downpour would, in only an hour or two, transform the Akrotiri ravine into a foaming white hell, stripping away ash and rocks and carrying them into the sea. Yet within three or four hours of the storm's passing, the freshly scoured streambed, dampened here and there by diminishing trickles and stagnant pools, became the only evidence remaining of the surge, as the ground quickly reverted to its former arid aspect. From the year of the city's burial, each autumn squall sent little rivulets of rainwater down to the chambers of Thera. As the water seeped into the volcanic tephra, it gathered up sulfides and acids, so that when at last it oozed through cracks in ceilings and walls, spattering cedar panels, leather thongs, and rattan beds, it was received much as a newly ingested piece of meat might receive secretions of digestive juices. Unlike the deserts above, the atmosphere of the chambers was permeated to saturation with moisture. Even the bronze cups and frying pans began to turn green with oxidation. Of all the organic furnishings within the tomb, only the seeds of wheat and other grasses resisted the assault. During the first few decades after burial, a furry white film, interrupted here and there with splotches of black or pink, sprouted on all exposed surfaces: on carefully plastered and painted walls, blocks of fallen cement, pine beams, hanging plants, and screens of wooden doors that once slid in bronze tracks. In only a few centuries virtually anything that was not inorganic was reduced to cold black streaks of carbon. There must have been periods lasting many decades, in which leather, glue, the contents of the world's first flush toilets, and broken jars filled with prepared meats, wines, oils, fruits, paint, and paint thinner literally steamed from the forces of decomposition.

Chemicals in the atmosphere combined, fractionated, recombined, and sometimes evolved into strange new substances no one had ever seen before. The air of those chambers was like a very dense, very moist industrial haze. If, by some miracle, one chamber had been filled with leather, or imported tea, or some other source of tannic acid, and if the conditions of the surrounding ash had been just right, then papyrus scrolls (if such existed) and even bodies might have been preserved. But failing such improbable circumstances, supporting beams and carved wooden tables oxidized out of existence, leaving at best phantom fossil impressions, and leaving these only if they happened to be covered by a layer of tightly packed ash. As the beams vanished, some of the chambers caved in. Others endured even in the absence of timbers, for the ash, compressed by the ages and soaked in mineral-laden water, had taken on a claylike consistency. In such places the roofs, though crosshatched with nothing more substantial than fossil impressions of wood, could endure almost forever, if they remained deep underground and moist.

Three dozen centuries passed. A new civilization introduced irrigation and street lighting to the island. Grape vineyards were growing in the deserts, high above the rooftops, in what had once been the Theran sky. Some of the chambers, originally sheltered beneath a two-hundred-foot mantle of volcanic cinders, now lay within only ten feet of the ground surface. Annual rainstorms and flash floods had chewed away more and more of the overlaying tephra, until one day, as Spyridon Marinatos was deciding where in the southern ravine he would dig first, a farmer with whom he had struck up a friendship noticed that sinkholes were appearing in his fields, as if the soil were subsiding into hidden caverns. The man's name was Arvanitis, and what made his sinkholes worth mentioning to the archaeologist was their peculiar shape: not round, as any natural sinkhole ought to be, but perfectly rectangular, like rooms in a house. Arvanitis had a theory, and decided he would tell Marinatos that the sinkholes might be just what he was looking for: signs of an extinct civilization. But before they met again, a new and almost fatal sinkhole yawned open. The ground suddenly collapsed under the weight of Arvanitis' donkey, and he disappeared with the animal into the earth. He knew at once what had occurred, and as he regained his composure and the poor donkey thrashed and as sheets of volcanic dust continued to rain down, the farmer looked around and began to laugh. He had plunged through thirty-six hundred years of Earth history. Whole empires had come to this island and gone. Thera had ceded to the Greeks, the Egyptians, the Romans, the Turks, pirates, Christian Crusaders, and back to the Greeks again. None of them had discovered what lay only a short distance beneath their feet, and it filled Arvanitis with a joy that no one had ever described before, to know that he was the first in all those centuries to enter the lost city of the Minoans.

〰〰〰

At first Spyridon Marinatos tried tunneling into the city, but he discovered that once he exposed the volcanic ash to air and allowed the moisture to escape, it crumbled to powder, making every tunnel roof a trap looking to turn him into just one more permanent fixture beneath Thera. Moreover, it was always easier to excavate and document a room full of pottery, cooking utensils, and other household items if you entered from above, rather than burrowing in haphazardly through the doors and windows. So Marinatos decided to excavate the city like an open-pit mine, roofing it over with a tin rain cover as he progressed.

Here was a discovery that time would prove far more important than the tomb of Tutankhamen, which Marinatos could easily have hidden in a small corner of Thera's subterranean world. He knew that at least seven years would be required just to expose one city block to daylight. Trial trenches indicated that there were more than a hundred city blocks to be explored, meaning that failing the invention of machines enabling one to "CAT scan" the ground before excavating (combined with barely imaginable advances in robot technology), exploration of the entire city might require a thousand years. And as if that were not enough to keep him busy, while digging foundations for the tin rain shelter, Marinatos found evidence of streets and walls fifteen feet below what he had assumed to be the original ground surface—a buried city beneath his buried city, two thousand years older, and destroyed by the same volcano. It began to look like the archaeological realization of the Buddhist belief that life on Earth was multiple planes of existence.

The higher, terminal Minoan plane was wondrous enough for Spyridon Marinatos. He would leave the deeper wonders for future generations to explore, if technology ever provided a way of doing so without, at the same time, destroying the upper strata. And so he exposed multistory houses whose walls were honeycombed with systems of plumbing that made possible bathtubs, showers, and, apparently, central heating. Wall murals gave faces and costumes, if not names, to the people who had built the world's first navy. The houses themselves were of such sophistication that they would not have looked out of place among the seafront hotels then being built on some parts of Thera. When he first viewed the ruins, Marinatos thought Arvanitis had led him to a Minoan palace, but further excavation revealed great works of art, comfortable furnishings, hanging gardens, and advanced hydrotechnology in every house, and the houses had belonged not to kings but to shipping merchants, jewelers, bakers, and fishermen. It occurred to him that more than three thousand years ago ordinary Minoan citizens were living under conditions that a great many modern city dwellers would envy.

Some of us who were to follow in Marinatos' footsteps could not avoid wondering: If the Minoans had not been living in the most dangerous place on Earth, might there have been television by the time of Christ? And might interstellar voyages be something we were actually doing today, instead of merely watching on television? Marinatos himself could not avoid such thoughts, as he turned a rarely talked-about but much prized sliver of cut and polished quartz between thumb and forefinger. When reconstructed, the crystal, convex on both sides, hinted that someone on ancient Thera possessed a knowledge of lens cutting. Nothing short of an actual wood and quartz tube buried in one of the buildings would compel him to risk his reputation by publicly voicing his speculations about Minoan telescopes, but lacking such proof, he allowed himself the luxury of private speculation among friends. And the more one thought about it, the less strange the speculation became. Lens-cutting technology was not beyond Minoan reach, and the idea of ancient telescopes only seemed strange when modern uses were assumed for them. Even Galileo did not have stars and planets in mind when he assembled his first working models about A.D. 1609. There were much more immediate, much more lucrative applications. Galileo was living in the Age of Discovery. It was the time of the entrepreneur, and financially speaking, there was a very good reason for his obsession with long-range optics.

"Climb with me up the Tower of Saint Mark's and look through this," he suggested to a small gathering of friends, who also happened to be shipping merchants. Because his tube of lenses could magnify distant objects nine times, one could spot ships coming over the horizon hours ahead of the best unaided eye. This meant that Galileo and his merchant friends could review the commodity list and fix the market price accordingly because they knew the ships were coming while their competitors did not. For as long as their little secret lasted, the men profited from the world's first-known insider trading scam (it was only as an afterthought that Galileo pointed his telescope where his partners in crime had never dreamed of pointing it: toward the moon and Jupiter).* Further diminishing the strangeness of Marinatos' speculation, his quartz sliver, if actually part of an ancient lens, was hardly unique. There were other precedents for pre-Galilean telescopes, including accounts of King Kāsyapa, born about a hundred years after Christ, in Sri Lanka. It is said that when the women of his harem bathed in fountains at the foot of Demon Rock, he stood at the

*The official story, as presented today in advertisements for the optical company Bausch & Lomb, explains that "in 1609 Galileo was acclaimed by distinguished citizens of Venice when he demonstrated his first telescope from the Tower of St. Mark's." Now you know the rest of the story.

window of his citadel, high above, and watched them through a telescopic glass.

Thinking about the advances the Minoans had worked (and knowing that his magnificent city was a mere suburb of the world's first naval empire), Marinatos shifted easily into abstract fantasies about how far beyond telescopes these people might have advanced if left on their own. In the Minoan world every island had the Aegean for a moat. Competition was inevitable, universal control even more difficult to maintain than in the riverworlds. There was no reason for technological stagnation ever to set in. After a thousand years of advance, a golden age was about to dawn. And then, in the amount of time it took a man to draw in a breath, hold it reflectively, and exhale, the course of civilization forever changed. Like Galileo justifying his telescope, Marinatos could justify Thera beyond his fascination with antiquity because it probably marked the most powerful "what if" in the tenure of humankind upon this planet.

"My poor Minoans," he lamented. "If only they had not been living in the wrong corner of the wrong sea when the mountain exploded, the fires roared, the Earth darkened, the waters were poisoned."

And when might that have been? Marinatos wondered. His best guess was 1450 B.C. Other archaeologists were placing their bets on 1200 B.C., but the physicists who had examined charred tree stumps found at the bottom of Thera's ash layer came back with carbon isotope dates closer to 1640 B.C., give or take twenty or thirty years in either direction.* Marinatos did not know a great deal about atomic accelerators and the radiometric decay of carbon, but it seemed to him that the physicists were off by at least two centuries. The archaeological assignment of dates, by which changing pottery styles clocked the years, could not possibly be so wrong, could it? No, Marinatos decided. He would not believe it. For now, the pottery chronology provided enough fascinations. The style known as Late Minoan 1A, characterized by paintings of spirals and leafy vines on a pot or vase, was found everywhere in the buried city, and also in the ruins of Minoan Crete, and inside Egyptian tombs contemporary with the pharaohs Hatshepsut and Tuthmosis III. Life on Thera had therefore been cut short while the Late Minoan 1A pottery style flourished, which was also the time of Queen Hatshepsut. *And when was that?* The question nagged, but too many pieces of the puzzle remained undiscovered—*when?*—and he regretted, momentarily, that he would probably never know.

---

*Wood, much like fired pottery, will last almost forever once it is burned. Fire, the great destroyer, can also be a paradoxical preserver.

CAIRO, EGYPT, TIME PRESENT

About A.D. 1988, some fourteen years after Marinatos' death, the pieces started fitting together. On the Nile Delta a quarter-inch-thick layer of Theran ash was discovered (each volcanic eruption produces ash with its own distinct concentrations of metals and other elements, which can be matched to the eruption with the same certainty that fingerprints can be matched to individual people). Carbon dating of decayed matter lying immediately above and below the Nile ash layer bracketed both the Theran tree stump age and Marinatos' best guess (sometime between 1650 and 1350 B.C., according to the mud). From the Greenland ice sheet, in layers of ice that accumulate in yearly cycles, depositing distinct strata resembling the annual growth rings that give wood its grain, University of Copenhagen glaciologist Claus Hammer found one of the deeper, older horizons to have formed from acidic snow. Though similar acid signals in the youngest horizons of snow and ice can be identified as fallout from the carbon exhaust cloud that now shrouds the Earth, in ancient times volcanoes were the only source of acid snow. Counting backward through the ice horizons, and taking into account a few horizons where the transition from one year to another was not clearly defined, Hammer narrowed the Nile Delta and Theran tree stump brackets between 1664 and 1624 B.C.

An exhaustive study of overlapping generations of oak trees preserved in Ireland's peat bogs narrowed the brackets still further. Every century's pattern of dry and wet and somewhere-in-between summers was like a spectral signature written in fluctuating growth ring sizes. Careful comparison allowed the construction of an unbroken record of seasonal variation reaching all the way back to the last Ice Age. All trees dating from the seventeenth century B.C. displayed abnormally narrow growth rings sometime in the 1620s, suggesting that there were summer seasons, during that decade, too cold to support normal growth. Cornell University paleobotanist Peter Kuniholm (spearheading the emerging field of dendrochronology) was making similar discoveries in Greek and Turkish peat bogs (made possible because trees falling into peat bogs are soaked in tannic acid and do not rot).*

*To me, the most exciting thing about Kuniholm's Greek and Turkish tree ring chronology is that it preserves actual Mediterranean weather patterns, which will surely be reflected in wood grain impressions fossilized in Thera's buried houses, from which we should be able to date the entire construction sequence of the city right up to the day it died. (Unfortunately most of the wood grain fossils in the area thus far excavated have crumbled to fine powder during the nearly twenty years of neglect, and even pilferage, that followed the death of Spyridon Marinatos.)

The 1629–1620 B.C. brackets closed on a specific year once core samples were taken from the bristlecone pines of California's White Mountains, known as "the trees that live forever" (some are more than five thousand years old, meaning that they were alive when Leonard Woolley's flood layers and the first bricks of Mashkan-shapir's Cobra Gate were laid down). When paleobotanists counted the bristlecone pine's annual rings back to the year 1627 B.C., they found the wood there heavily peppered with unusual darkened cells: unmistakable signs of ice scarring during the summer season. Identical frost scars were found in rings dating to A.D. 1816, following the explosion of Tambora, whose shroud of stratospheric dust plunged global temperatures and ushered in what historians recorded as "the year without a summer." Evidently, something equally cold and dramatic happened to the world during the summer of 1627 B.C.

Across parts of Turkey, and two hundred miles east of Thera, almost as far away as Cyprus, drifts of volcanic ash several feet deep were deposited. The dust cloud spread fewer than one hundred miles west, suggesting that a powerful squall, blowing from the west, stopped it and hurled it east. Such Mediterranean squalls occur during the autumn months, meaning that the climatic upheaval seen in California's bristlecones began in the autumn preceding the 1627 B.C. freeze. Fifty cubic miles of Thera went into the heavens. The heavier debris rained down over the eastern Mediterranean; the lighter ash formed a canopy of atmospheric dust that circled the globe, intercepted some of the Sun's rays high above the ground, and eventually fell back to Earth as acid snow.

Chinese scribes provided eyewitness testimony to how truly global the Thera upheaval became. The oral and written histories of the Chinese emperors were always footnoted with records of solar and lunar eclipses. In 1991 NASA/JPL astronomer Kevin Pang, retracing the motions of Sun, Earth, and moon through the eclipses of 1302 and 1876 B.C., determined that the emperor Chieh ruled during the time of Thera. Of his reign, Chinese records say he "lacked virtue," and because of this, "the Sun was distressed . . . during the last years of Chieh, ice formed in [summer] mornings and frosts in the sixth month [July] . . . hot and cold weather arrived in disorder. The five cereal crops withered and died." The disorder lasted for seven years, and in the northern provinces whole populations starved, warred, and returned to tribal nomadism.

If the volcano produced such effects as far away as China, then if any one thing is clear, it is that no one in his right mind wanted to be on or around the Mediterranean the day Thera exploded. The tidal waves alone produced horrors. In some places, as in the Nile marshlands, the shape of the coast

The Theran tsunami of 1628 B.C. varied greatly in height, depending upon the contours of the coast and the incline of the shore. On the northern coast of Sinai, the tsunami stood "only" forty feet tall. At Argos and in parts of Turkey, peninsulas jutted into the precursor wave like the prongs of giant tuning forks, piling the waters eight hundred feet high and sending them thirty miles inland. The skyline of lower Manhattan is a familiar and instructive gauge for understanding what numbers alone cannot possibly convey.

and the slope of the beach (according to the latest computer models) piled the waters four stories high. But this was a mere ripple by comparison with what happened along parts of the Turkish coastline, where two large peninsulas jutted into the direction of the blast, trapping the shock wave as if in the mouth of a giant funnel. At the funnel's spout, the water shot thirty miles inland, dislodging house-size boulders as it went, scouring away every trace of civilization, and carving out channeled scablands that are still visible. To penetrate so far inland, the wave had to be eight hundred feet high when it reached the shore (it would have washed halfway up the World Trade Center's twin towers or two hundred feet over the Washington Monument). If the Babylonian floods of approximately 2800 B.C. could become a global legend of biblical proportions, then the upheaval of autumn 1628 B.C. was even more likely to echo down through oral and written traditions. Plato, writing about 350 B.C., described in *Timaeus* and *Critias* a civilization said to have existed long before his time and to have passed on to his Greek forebears much of the architectural and technological knowledge his brethren took for granted. They were the ancestors of King Minos, and according to legend, if he had traveled to the islands and dug beneath the streets of Classical Greek towns, he would have discovered remnants of this earlier civilization, which the Egyptians still remembered. Plato described their religious ceremonies in detail, practices that Arthur Evans and Spyridon Marinatos, excavating beneath Greek ruins, found depicted exactly as Plato had described them, on wall paintings and golden cups (the ceremonial cups, too, were described by Plato and were, like most other features of his lost civilization, distinctly Minoan).

The archaeological evidence suggests, as Plato had said, that some Minoan architects survived the Theran upheaval. On Crete we see them erecting new palaces and towns, but dedicating them to rulers from the Greek mainland, who apparently swept in and wrested control from the volcano-ravaged Minoans. Some archaeologists believe the takeover could have been severe, accompanied by murder or pillaging and the burning of what few buildings stood intact. Others believe that the survivors might have been so weakened as to welcome anyone who offered to govern them toward the resurrection of civilization, regardless of what new language, laws, and customs the governor planned to introduce. Plato, seeming to describe this transition, said that Greece made battle against the sea people, but sometime before, or during, or after the battle, the island civilization and many warlike men were overcome by thunderings, earthquakes, and floods, and "an island disappeared," presumed to have "sunk into the earth . . . in a single day and night of misfortune."

In 1992 Kevin Pang brought to my attention the work of Johns Hopkins University geologist Mott Greene, who had analyzed Hesiod's *Theogony,* a Greek poem of the eighth century B.C. It predates Plato by about four hundred years, heralding back to the wall inscriptions of Deir Alla in the Jordan Valley, the oldest references we have to the prophet Balaam and the dust cloud of Israel ("and they said [to the Sun]: Sew shut the skies with your cloud! Let there be darkness and no shining . . . for you will provoke terror by a cloud of darkness"). The first Old Testament texts and *Theogony* were, as it turns out, exact contemporaries.

Though traces of the Theran dust cloud have not yet been found in Israel, it must surely have passed over the Jordan, for Mediterranean sediments only two hundred miles west of Tel Aviv are thick with Theran ash, indicating that the cloud was still full of roilings and thunderings as it bore down on Deir Alla. Egypt was out of the direct line of fire, but we know that the explosion was seen and felt there, for it deposited a clearly defined layer of ash on the Nile Delta.

According to Plato, Greece was at war with the parent culture when a great national catastrophe struck. The *Theogony,* the earliest Greek poem that has survived to the present time, provides some corroboration of Plato's version: There was a war. One side was allied with the god Zeus; the other, with Kronos.* The battle was so fierce that the Earth shook, the sea boiled, fire and rocks shot into the heavens, farmlands were destroyed by drifts of ash. The crops failed. The people perished.

Greene, a geologist, does not draw his connections haphazardly. He knows, for example, that during the past ten thousand years there have been 5,564 eruptions of 1,343 volcanoes, 627 of them with clearly defined dates, but his detailed surveys of Mediterranean volcanoes have allowed him unambiguously to match the 1628 B.C. Theran explosion to an eruption described in *Theogony.* The poem details fifteen successive stages of an eruption that precisely matches the fifteen-step scenario written in Theran ash and Minoan ruins. "There is a complete one-to-one correspondence with no missing elements," says Greene. "And they are all in the correct order."

My own studies of the Thera debris fields had prompted me to write, in 1988, "that volcanic death clouds had surged out to all points of the compass, writhing and snakelike, clutching the ground . . . and in the paths of those snakes (each like a sea of fire cutting through billowing black smoke) . . . the air was hotter than live steam, hotter than molten lead, hotter than iron emerging white from a furnace."† Later, when Kevin Pang called my attention to the *Theogony* (excerpts 820–1022), I had the curious and impossible-to-describe feeling of hearing my own voice echoing from the past: "Monstrous Earth gave birth to . . . a hundred snake heads . . . with black tongues flickering and fire flashing . . . large tracts of the earth were set on fire by the prodigious heat and melted like tin heated in molded crucibles by skillful workmen, or like iron, the strongest metal, softened by the heat of fire in some mountain cleft."

Perhaps my experience of the echo has made me less objective than I ought to be about Greene's theory, but I know the existence of fiery,

---

*Most epic poems began as song legends, in which music became a mnemonic device for passing oral histories down through the ages; hence the rhythmic, almost poetic nature of *Theogony,* the *Odyssey,* and the Bible.

†The "snakes" struck forth over a wide area of the eastern Aegean, but miraculously missed the harbor district of Thera's buried city, which, although located right near the heart of the explosion, survived as if enclosed by a protective bubble.

snakelike clouds could only be concluded from careful geologic investigation, or be described by survivors who had been terrifyingly close to ground zero, and would not likely have been guessed at by a poet. I am thoroughly convinced that Greene has found (especially in *Theogony* 630–719 and 820–1022) eyewitness accounts of a stupendous volcanic explosion and that the explosion was none other than Thera:

> The limitless expanse of the sea echoed terribly; the Earth rumbled loudly, and the broad area of the sky shook and groaned . . . and a heavy quaking penetrated into the gloomy depths of Tartarus [a sunless abyss below Hades]—the sharp vibration . . . forming a solid roll of sacred fire. Fertile tracts of land all around cracked as they burned, and immense forests roared in the fire. The whole Earth and the ocean and the barren sea began to boil. An immense flame shot up into the atmosphere, so that the hot air enveloped the Titans. . . . The sight there was to see and the noise there was to hear made it seem as if Earth and vast sky above were colliding. As if Earth were being smashed and [as] if sky were smashing down upon her. . . . The winds added to the confusion, whirling dust around together with great Zeus' volleys of thunder and lightning. . . . [The gods] attacked relentlessly, throwing showers of three hundred stones one after another with all the force of their enormous hands, till they darkened the Titans with a cloud of missiles.

"While assessments of the violence of the explosion vary," says Greene, "it is generally believed to have been more violent than Krakatoa, and perhaps as much as fifty times as violent. If so, it would make [Thera] one of the half dozen most explosive events of the last ten thousand years. We must remember, in assessing the impact on observers, that this was no uninhabited Pacific island, but a thriving Minoan colony on an island little more than a hundred miles from Athens and seventy miles from Minoan Crete."

"When I read Greene's report," says Kevin Pang, "I was pretty convinced that this poem described a situation that had actually occurred. By itself I would not have put all my bets on the *Theogony;* but I have studied the Chinese literature, and I found that similar, related events were written down. China's *Bamboo Annals* were written during the third and fourth centuries B.C., describing phenomena that were observed more than a thousand years earlier, so I think the Greek poem has some weight. The events so greatly resemble what would happen after a catastrophic eruption, fifty times more powerful than Krakatoa (which is still legendary throughout the world, more than a hundred years after it occurred).*

---

*Most people today cannot remember the name of a single world leader from A.D. 1883, but they remember the name Krakatoa. The blast wave from the explosion caused permanent

"Thera was the nuclear winter of the prehistoric Greek world, and Greene's discovery is important because it shows us that the explosion was so unforgettable that detailed eyewitness accounts of great accuracy were preserved for nine centuries before being woven into Hesiod's mythological poem."

In the aftermath of Thera the Cretan mainland ceded to Greek rule. On eastern Crete (where the devastation was heaviest and the ash fell deepest), citadels were erected atop Minoan ruins, suggesting that post-Minoan society divided into armed camps and rival Greek kingships. A hundred twenty miles east of Thera, more than two feet of ash fell out of the cloud as it passed over the island Kós, thick enough to suffocate, if not for the fact that it was also hot enough to instantly scorch men's lungs and turn their tongues into charcoal. After the cloud had passed Kós, thousands of bodies lay bleeding under the starlight. The towns and farmhouses, those that were above the tsunamis, seemed to be sleeping peacefully under drifts of fresh-fallen snow; but it was not snow, and not even a spider stirred upon it.

On those islands where men and women did survive, there was no International Red Cross ready to move in with food and supplies, and there was no way to flee the devastation, for most of the ships and shipbuilders were probably caught in the harbors by tsunamis. Much like the Cretan citadels, the houses whose foundations appear in the next, post-Thera layer of construction (or, rather, reconstruction) resemble fortifications. One of the most haunting developments of recent years has been the discovery that on some of the islands, though cemeteries had been in use for hundreds of years before Thera and though archaeologically visible populations existed afterward, people suddenly stopped interring the dead. It conjures up images of survivors dining upon human limbs and throwing the bones over the walls, images that might explain passages in Homer's *Iliad* and *Odyssey* describing obscure Greek islands whose fields, strewn with bones, were inhabited by semihuman man-eaters called Sirens and Cyclopes.

When the first Greek and Egyptian ships sailed into the still-smoldering Thera Lagoon, their crews would not have guessed that a mountain had gone up into the atmosphere. To them, a mountain and most of an island were simply missing and presumed sunk at sea, like a lost Atlantis. There are passages in the Bible that may, like Hesiod's *Theogony,* preserve eyewitness accounts of the explosion, seen from perhaps a little further away. The

---

damage to people's hearing more than 500 miles away and was powerful enough to shatter windows and crack walls at a distance of 100 miles. More than thirty thousand were killed by tidal waves, as far away as 2,000 miles. Thera was as much as fifty or even a hundred times more powerful, and Egypt was only 430 miles southeast, Israel only 620 miles east.

Bible can be interpreted many ways by many people, but it is impossible for me to read the Forty-sixth Psalm and not see in it the peculiar horror of Thera's disappearance: "Will not we fear, though the earth be removed, and though the mountains be carried into the midst of the sea; though the waters thereof roar and be troubled, though the mountains shake with the swelling thereof. Selah . . . the earth melted . . . Selah . . ."

Oceanographer Daniel Stanley and I have been in contact almost from the moment his team found the Theran ash layer under the Nile Delta. He suspects that the Book of Exodus preserves an even more detailed, more dramatic account of the explosion, a belief with which I ally myself fully, given the evidence of the ash layer.

"What else can you say?" asks Stanley. "Flames raining down from the sky? A darkness so thick that people could not see one another? A pillar of fire lighting up the night sky? The Bible makes reference to all of these things, and all of these things could be descriptions of the ash cloud. This Nile ash layer is the first hard proof of anything like that in Egypt. We now have a record that statements given in Exodus which sound very strange most likely did happen."

"It is inconceivable what happened to the land," begins Egypt's Ipuwer papyrus, which may be the best-preserved nonbiblical memory we have of "the dust cloud of Israel" and the ten plagues of Egypt.

Ipuwer belongs to the Egyptian lament literature, which became very popular about the time of the pharaohs Hatshepsut and Tuthmosis III (the approximate period from which most copies of Ipuwer seem to date). The poem consists of the wise man Ipuwer's description of a mysteriously devastated Egypt and his rage against the Sun-god, who has abandoned the Nile and to whom he appears to be speaking:

> The land—to its whole extent confusion and terrible noise. . . . For nine days there was no exit from the palace and no one could see the face of his fellow. . . . Upper Egypt suffered devastation . . . blood everywhere . . . pestilence throughout the country. . . . The Sun is covered and does not shine to the sight of men. Life is no longer possible when the Sun is concealed behind the clouds. Ra has turned his face from mankind. If only it would shine even for one hour! No one knows when it is midday. One's shadow is not discernible. The Sun in the heavens resembles the moon. . . . Lo, the desert pervades the land, townships are laid waste, and a foreign bow people are come to Egypt! People flee . . . and it is tents that they make like bedu [nomads].

Perhaps coincidentally, and perhaps not so coincidentally, Ipuwer echoes throughout the Book of Exodus, sometimes with such spine-chilling fidelity that I am tempted to believe both accounts were copied from the same texts:

> The land—to its whole extent confusion and terrible noise (Ipuwer)/*And all the people perceived the thunderings, and the lightnings . . . and the mountain smoking* (Exodus 20:14–15). . . . For nine days there was no exit from the palace and no one could see the face of his fellow (Ipuwer)/*And there was a thick darkness in all the land of Mizrayim [Egypt] three days: they saw not one another, neither rose any from his palace for three days** (Exodus 10:22–23). . . . Upper Egypt suffered devastation . . . blood everywhere (Ipuwer)/*and there was blood throughout all the land of [Egypt]* (Exodus 7:21–22). . . . Pestilence throughout the country (Ipuwer)/*I smite thee and thy people with pestilence* (Exodus 9:15). . . . The Sun is covered and does not shine to the sight of men. Life is no longer possible when the Sun is concealed behind the clouds. Ra has turned his face from mankind. If only it would shine even for one hour! No one knows when it is midday. One's shadow is not discernible. The Sun in the heavens resembles the moon (Ipuwer)/*darkness over the land of [Egypt], darkness which may be felt* (Exodus 10:21). . . . Lo, the desert pervades the land (Ipuwer)/*And there remained no green thing in the trees, or in the plants of the field, through all the land* (Exodus 10:15). . . . Townships are laid waste (Ipuwer)/*Thunder and hail, and the fire ran down upon the ground . . . very grievous . . . smote throughout all the land . . . both man and beast . . . and broke every tree* (Exodus 9:23–26). . . . And a foreign bow people are come to Egypt! (Ipuwer)/*The children of Yisrael [people of the bow†] who came into [Mizrayim] Egypt* (Exodus 1:1). . . . People flee (Ipuwer)/*And it was told the king of [Egypt] that the people had fled* (Exodus 14:5). . . . and it is tents that they make like [nomadic] bedu (Ipuwer)/*And he [Moses] turned back to the camp: but his servant [Joshua] . . . did not depart out of the Tent* [Exodus 33:11].

*Though the events described here, and the language used, are virtually identical, the number of days differs (nine according to the Egyptian account, three according to the Israelites). Perhaps this is because no one was really sure how long people had stayed in the palace (only that it was within the range of one week) and because the number nine was sacred to the Egyptians, and three was sacred to the Israelites.

†"A foreign bow people are come to Egypt!" is the only independent Egyptian record we have associating Asiatic tribespeople (among them the Israelites) with the volcanic phenomena that appear to be preserved in Ipuwer's lament and the Book of Exodus. Because of their notoriety for resorting, at the slightest provocation, to quick and efficient use of the bow and arrow, they were sometimes referred to as "archers" or "people of the bow." From their hostility toward expeditionary forces, Egyptian propagandists called them "the wild men of Asia," a label that in meaning and tone approaches our current term *terrorist*. Because they lived beyond the Sinai Peninsula, they were also referred to as "those-who-are-across-the-sand."

Like *Theogony*, Ipuwer and Exodus describe what has all the appearances of a volcanic ash cloud passing over the land. When fifty cubic miles of rock are hurled at the stratosphere, the debris, ranging in size from boulders to microscopic flecks, immediately separates in two directions. The smallest particles remain in the upper atmosphere, intercepting sunlight and chilling the ground. The larger, heavier fragments, ranging down in size to grains of sand, fall to Earth, each trailing a little slipstream of air behind it, and the collective whole creating a mighty downblast, which, blocked by the Earth itself, flows out laterally, hugging the ground, surging impartially over islands, sea, and ships. The rock-heavy air acts more like a liquid than a gas: all horizontal velocities, confusion, and terrible noise. Laced with micro-shards of volcanic glass, the onrushing cloud becomes a tidal wave of cutting edges, speed-slung teeth, hundreds of billions of them.

Eyewitness accounts of the A.D. 1883 Krakatoa eruption describe such a cloud and echo Ipuwer and Exodus tales of a darkness that could not be cut by lamplight, of lightning striking down everywhere, jarring the ground, and a sky that deluged men with red-hot sleet instead of rain. In western Java, thirty miles east of Krakatoa, burning torrents of ash, blown with the violence of a hurricane, set homes afire even as tsunamis uprooted them. The stars went out. Night brought gusts of blistering heat alternating with hail and icy black rain. There was no dawn, no daylight. Ash blotted out the sun for two days, then scattered, rained down, or drifted west until it no longer filled men's lungs, until it no longer settled on the skin like micro-shards from a broken bottle. But still, where it fell, whether a quarter inch thick or ankle-deep, tens of thousands died from starvation and disease. There was pestilence throughout Indonesia. Fish died; ash-choked lakes and rivers turned from glass clear to an acidic murk, sometimes the color of drying blood. Fields of ash-covered grain yellowed and, for a time, took upon a desert aspect. Birds and livestock had suffocated in the storm, and everywhere there was the unsettling stench of decaying animals. Those creatures that were not killed outright by the ash fall were "made bold by hunger" and behaved in strange ways, reenacting scenes straight out of the biblical plagues of Egypt. Ants were driven from the banks of the rivers. With them came snakes and centipedes. They swarmed into kitchens and parlors, into bedding and nightclothes. Rats were seen to attack dogs and other creatures many times larger than themselves,* yet these phenomena

---

*The strange behavior of the animals was not unique to the Krakatoa eruption. Identical events were recorded during the Tambora (A.D. 1815) and Mount Pelée (A.D. 1902) eruptions. Only a quarter inch of ash was sufficient to create such havoc.

were of but trifling magnitude by comparison with the events of autumn 1628 B.C.

Thirty miles east of Thera, the ash cloud sluiced toward Kós at more than two hundred miles per hour, and where it touched the sea, it converted men and ships into gas.

The wave fronts—vast, globular, and black—spread out over the Mediterranean, weakening as they spread, capsizing ships and dislodging the lighter buildings, flinging their exterior walls indoors. By the time it struck Egypt, four hundred miles away, the winds were down to ninety miles per hour; but they blew for more than two hours, and though the dust particles had already shed much of their heat to the air through which they passed (on the Nile the air itself must have become intolerably warm but not necessarily lethal), random pockets of insulated, incandescent ash, still hot enough to raise blisters and ignite hair, were moving busily to and fro within the cloud. From excavations on the Nile Delta, we now know that a quarter inch of volcanic dust fell out of the storm as it rolled over Egypt, so surely there was darkness and pestilence throughout the land, and famine, and the death of cattle and men.

The Bible does not name the pharaoh of the Exodus, but we can determine with reasonable certainty who was ruling Egypt the day Thera exploded. In the tomb of Semut, architect and foreign minister to Queen Hatshepsut, a wall fresco depicts a procession of Minoan men wearing kilt styles identical to those painted on the walls of Theran homes immediately below the ash layer. The men carry vessels painted in the distinctive Late Minoan 1A style known to have been in wide circulation at the moment Thera's buried city was snuffed out. The implication is that Semut served Hatshepsut at a time before Thera exploded and that at least some portion of Hatshepsut's reign preceded the ash layer. Another procession fresco, discovered in the Karnak tomb of Rekhmire (a vizier to Hatshepsut's stepson, Tuthmosis III), identifies fourteen men bearing gifts as visitors from Keftiu (Crete). We know from a newer, second layer of paint (which has partly flaked away from its undercoat) that a gift bearer from Crete, originally shown wearing a traditional Minoan kilt (as depicted on Theran walls and in Semut's tomb) was overpainted during Rekhmire's lifetime to show a longer, more ornate kilt characteristic of mainland Greeks, who began to dominate Minoan Crete above and after the ash layer. By ordering kilts to be repainted, Rekhmire was giving diplomatic recognition to a change of leadership in the Minoan world. The painting suggests two things: (1) The change of kilt style and government, hence the Thera

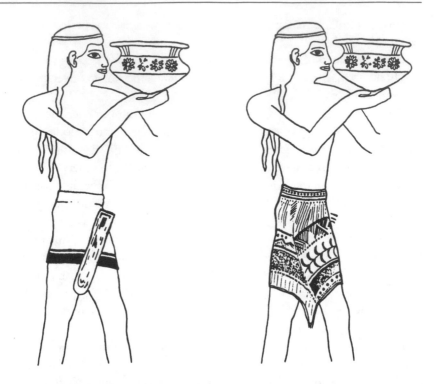

A man bearing gifts from Minoan Crete was repainted during the lifetime of Tuthmosis III's visier, Rekhmire. The original underpainting (left) shows the traditional Minoan kilt and codpiece known from Eighteenth Dynasty and Hyksos period frescoes on the walls of Thera in the Aegean and from Avaris on the Nile. The overpainting (right) displays a kilt design characteristic of mainland Greeks, who seized control of the Minoan world in the aftermath of the Thera eruption. By ordering kilts to be repainted, Rekhmire, acting as Egypt's foreign minister, was documenting for future archaeologists the fact that he was required to acknowledge a change of regime on Crete, and that the critical, Thera-induced change occurred during the reign of Tuthmosis III.

explosion of 1628 B.C., must have occurred during the reign of Tuthmosis III (possibly after a coregency with his stepmother, Hatshepsut), and (2) whatever the consequences of the explosion—including rival kingships and hints of sporadic cannibalism in the Aegean world—it did not completely extinguish the bright flame of Minoan civilization, for though shipping might have been temporarily interrupted, within a decade or two (still within Tuthmosis III's thirty-year reign), people who looked Minoan (who bore Minoan vessels and wore their hair in Minoan locks), yet dressed in kilts that were unmistakably Greek, began arriving in Egypt.* This is in agreement with what the Egyptian priests of Neith, according to Plato (in *Timaeus twenty-three c*), told his ancestor Solon: that much of Greek culture and technology could be traced back to a few learned survivors of a universal deluge.

∿∿∿∿

The Egyptians were not the only people leaving behind written records at the time of the Thera explosion. There were literate Canaanites beyond the Sinai Peninsula and the Hebrew tribespeople who lived, traded, and fought among them, and were then adopting or would soon adopt their writing. As with the tale of Cain and Abel, the biblical account of events leading up to the Exodus is a rare glimpse of the nomad's side of the story, as telescoped and edited by scribes who took great pride in their nomadic roots. The Exodus story was only a microsecond in the tribal and intercity struggles that have resounded throughout the riverworlds for more than eight thousand years. The city of Jericho was already more than four thousand years old when Thera spread its darkness over Egypt. Benjamin Mazar and the other Israeli archaeologists know that the Semitic people they call Canaanites had been living in the Jordan Valley at least since 3500 B.C., about seven hundred years ahead of Leonard Woolley's flood layers and sixteen hundred years before Hammurabi and the burning of Mashkan-shapir. In the Sorek Valley, midway between Tel Aviv and Jerusalem, archaeologists have discovered a fifty-five-hundred-year-old Canaanite temple edged by an arrangement of stone slabs that looks for all the world like a forerunner of Stonehenge. In layers of earth eighteen hundred years younger, fragmentary inscriptions tell us that the Canaanite alphabet was in use at the time of Thera, Hatshepsut, and Tuthmosis III. So were a furnace for smelting bronze, cisterns for trapping rainwater, and dozens of man-size storage jars imported from Egypt and Minoan Crete. Sorek is the legendary birthplace of Samson, biblical nemesis of the Philistines who, according to Scripture (particularly Genesis 10:14, Amos 5:7, and Jeremiah 47:1–4), were a remnant from the coast of Crete, whose cities were destroyed by "waters rising from the north," says Jeremiah. Modern Palestinians claim

---

*Ipuwer's lament describes an interruption of shipping like that which must surely have followed the Thera upheaval: "No one today really sails north to Byblos [a port of Lebanon, famous for both its cedars and a fine papyrus it exported; in time the port's name became synonymous with *papyrus scroll*, which lives on in the word *Bible*]. What shall we do for cedar for our mummies? . . . [M]en of Keftiu [Crete] come no longer. Gold is lacking. How important it now seems when the oasis people come carrying their festival produce."

their origins and early attachment to coastal Israel and the Jordan Valley from their Canaanite and Philistine ancestry.

Between 1900 and 1800 B.C., during the century of Mashkan-shapir's fall, the end of Hammurabi's Amorite dynasty, and the onset of another long night of Mesopotamian barbarism, a fresh wave of refugees-turned-tribal-nomads paraded out of what are now the oilfields of southern Iraq. Some began drifting into the Jordan Valley and pasturing their sheep outside Canaanite cities. In the strata above Sorek's Stonehenge, strewn with broken and discarded bits of Mesopotamian housewares, archaeologists can actually watch the influx of new peoples, among them the Hebrews, who were part of a larger cluster of skilled migrant shepherds and intercity traders known from cuneiform tablets as "Apirus" and "Habirus." They had been passing through Canaan for centuries before the Amorite collapse, and during this time, undoubtedly, some of them must have taken Canaanite husbands and wives, and their descendants were living behind city walls, peering down at their nomadic brethren, when the influx of approximately 1900 B.C. and its attendant pressures began.

Egyptian texts make no distinction between Hebrews and Canaanites. They are simply grouped together as "Hyksos and Asiatics," or "the people of the bow," or "the people beyond the sands [of Sinai]." Similarly, to archaeologists working in Israel, the distinction between Canaanites and the tribespeople who settled among them (and adopted their culture, pottery styles, and clothing while, at the same time, infusing the Canaanites with tribal legends and art styles) shifts and blurs until archaeologically there is often no distinction at all.

Genesis 17:8 tells us that Yahweh, the God of the Hebrew tribes, promised Abraham: "I will give to thee, and to thy seed after thee, the land in which thou dost sojourn, all the land of Kenaan [Canaan], for an everlasting possession; and I will be their god."* When he first arrived in the Jordan Valley, according to Genesis 12:6, "the [Canaanites] were then in the land." Most of the Canaanites spoke not only a Semitic language but what in time evolved into modern Hebrew. The Israelites who eventually seized control of ancient Canaan ended up speaking and writing in the language of the people they came to dominate. They wrote, in the first Old

---

*Actually, when God elaborates on his covenant with Abraham in Genesis 15:16–21, he promises more than Canaan to his descendants. If one decides to read the Bible uncritically as a historically accurate account of a God-given inheritance of land, and then to act upon it as law, there can be no end to four thousand years of bloodshed, for "the Lord made a covenant with Avram [Abraham], saying, To thy seed I have given this land, from the river of Mizrayim [the Nile of Egypt] to the great river, the river Perat [the Euphrates of Iraq]."

Testament texts, that Abraham begat Ishmael with Sarah's Egyptian slave girl and begat Isaac with Sarah. Ishmael's descendants moved into the deserts and founded new Arab nations. Meanwhile, Isaac begat Jacob, who won a decisive battle in Canaan and changed his name to Israel.

Then, in what by a Theran calendar would have been between 1900 and 1800 B.C., Israel's son Joseph reenacted a variation on the Cain and Abel story with his own brothers. Driven to jealous rage, the brothers sold him to "Ishmaelite" slave traders, coming full circle to an ironic recrossing of paths with their Arab stepbrothers. The Ishmaelites carried Joseph to Egypt, where he was promoted, step by sequential step, from a pharaoh's household servant to his chief adviser, foreign minister, and dream interpreter. The pharaoh and the Israelite are said to have gotten along so well that the rest of the Israelites decided to haul up stakes and abandon Canaan in favor of better opportunities in Egypt. A later pharaoh perceived them as an internal threat and "put slave drivers over the Israelites to wear them down with heavy loads." After two centuries of slavery a man named Moses stood before Pharaoh and demanded, "Let my people go!" Pharaoh refused, and in punishment for his refusal the God of the Israelites visited darkness and pestilence, famine and fiery hail upon Egypt and smote whole armies under towering walls of water.

Out of Egypt, the Israelites headed east and south, into Sinai, where Yahweh appeared to Moses on a mountaintop and renewed his promise of a new homeland that would range from the Euphrates River to the Sea of the Philistines (Crete and the Mediterranean). Construction plans for the Ark of the Covenant, through which the spirit of Yahweh vowed to spread panic ahead of Moses, to make his enemies turn and run, were detailed in Sinai, as if the Lord had in mind a long-range plan of conquest for his chosen people, who were, in the Book of Exodus, suddenly redefined. No longer did God's chosen, and the land promised to them, apply to all the descendants of Abraham; they applied only to Isaac's lineage. Somewhere between the time of Abraham and the Exodus, Ishmael's lineage (the Arab descendants of Abraham) had been disinherited.

Most people are surprised to learn that there is no historical mention of Isaac, Joseph, Jacob, or Moses outside Scripture (as I have been equally surprised to discover that, while much of the biblical story of tribal origins takes place in Egypt, there is absolutely no mention of the Pyramids in the Bible). Nor are there any independent accounts of a covenant between Yahweh and the Israelites. This is not to say that such texts were never written. The odds of any particular text surviving nearly four thousand years are probably less than one chance in five hundred. Given those same odds,

Isaac Asimov (who wrote nearly five hundred books during his lifetime) is probably the only modern author who has a hope of being read four thousand years from now. Future man may never know that Walter Lord and Arthur C. Clarke, Stephen King and Stephen Jay Gould, Charles Pellegrino and Harlan Ellison ever existed—which is probably just as well for future man.

Though written records tend to be as fragmentary as pottery shards, providing only random snapshots of the past, traces of what appear to be an Exodus tradition apart from the Hebrew version do exist. The Greek geographer and historian Strabo (who lived between 60 B.C. and A.D. 20) described an army—already an ancient legend by his standards—drowned in the sea; however, the drownings occurred on the coast of Canaan and "near Egypt," not in the Red Sea (as the story of Moses would have it). The memory of seas rising up and defeating a coastal military force is consistent with the probable effects of Theran tsunamis, especially if the volcano exploded during the reign of Tuthmosis III, who, Egyptian writings attest, invaded Canaan and Lebanon and placed sentries along the coast from Egypt to Canaan.

Phoenician legend (as recorded by Greek scribes, including Herodotus, the "Father of History," who lived between 484 and 425 B.C.) seems to preserve a nonbiblical memory of an Exodus out of Egypt. People of the eastern Mediterranean coast (particularly Lebanon and Canaan) descended to the Nile. Among them was Io, for whom a fiery moon of Jupiter would one day be named, and who, according to legend, married the reigning pharaoh and, like Joseph, ascended to prominence and power in Egypt. Four generations later, her great-great-grandson Danaos slew the sons of his brother Aegyptos and fled to the Greek coastal city of Argos, whose inhabitants offered him protection and were duly punished by a terrible thunder that boomed from within the Earth, heralding a wave that soared into the sky and came shoreward through the city. University of Toronto Egyptologist Donald Redford wonders if the slaughter of the sons of Aegyptos [Egypt] has anything to do with the slaying of Egypt's firstborn as described in Exodus 11:5–7. The connection does not seem at all unreasonable, since before his flight to Argos, Danaos is said to have led a sojourn of foreigners expelled from the Nile.*

---

*Argos, like the Turkish scablands (into which the Theran tsunami swept thirty miles), overlooked a harbor at the junction of two peninsulas. Like the scablands, the city stood in the mouth of a giant tuning fork whose prongs pointed toward Thera. When the wave came, Argos might just as well have been a flea located in the mouth of a cannon, for the waters were piled more than four hundred feet high when they passed through the center of town.

The story of the Exodus from Egypt, linked with empire-spanning destructions, was already more than eight hundred years old when Amos (in
3:1 and 9:7), Hosea (in 11:1 and 12:9–13), and other Hebrew prophets
of the eighth century B.C. began committing it to writing. There is only one
chain of historical events, recorded in Egypt, that can accommodate this
tradition, and that is the Hyksos descent to, occupation of, and ultimate
expulsion from the Nile.

During the century of Mashkan-shapir's fall and the Mesopotamian
displacements (and, as a result of those displacements, new population
pressures in the Jordan Valley), there began what Egyptian scribes called
"the time of the great humiliation." Between 1900 and 1800 B.C. the
Hyksos (Asiatic people who could fire arrows with frightening precision
from horse-drawn war chariots) swarmed west across the sands of Sinai and
into the Nile Delta. They stayed in Egypt, "like mice in the wheat,"
pillaging towns, murdering, and enslaving. They established Hyksos garrisons and inflicted upon Karnak and other centers of power a series of
puppet Egyptian "kings" manipulated by conquering armies. The invaders,
Canaanites mostly, were described as "an infestation of Egypt by plague-
ridden, unclean and leprous peoples" (which teaches us, not necessarily that
the Hyksos brought leprosy and bad table manners, but that the Egyptians
were no less adept at propagandizing their version of events than any other
culture).

The revolt began somewhere about 1730 B.C., led by Queen Hatshepsut's great-great-grandfather Sekenere III, a puppet king–turned-
revolutionary who charged headlong against his Hyksos overlords. His
mummified body, now on display in room 52 of the Cairo Museum, tells
the outcome. His lips are drawn back in a silent scream that never quite
finished, brought on by the first of five separate battleax blows that pierced
his head.

His mother, Tetisheri, survived to take over the education of his son,
who became "the avenger of the five wounds . . . the glorious liberator
. . . the unifier of Egypt . . . the great Conqueror." His name was Amose
I, first pharaoh of the New Kingdom and the Eighteenth Dynasty, grandfather to Hatshepsut. Somehow, without the Hyksos overlord in the delta
(near present-day Cairo) ever learning that a storm was about to break,
Amose I organized a mutiny of Asiatics, Africans, and Egyptians in the
south. Karnak was captured. Neighboring towns were liberated; garrisons,
destroyed. As word of Amose's victories spread, all Egypt was aroused to
a war for independence, and the Hyksos were routed back across the Sinai,
and southward into the Sudan. Driven by vengeance, Amose I marched

his armies after them, into Canaan, which he claimed as an occupied territory, a province of Egypt.

About 1710 B.C., "the glorious liberator's son," Amenhotep I, ascended the throne. The Hyksos, hearing of Amose I's death, began plotting uprisings within Egypt and Nubia and massing armies outside Egypt's borders. The mice were still "nibbling in the wheat." The pharaoh visited his enemies across the borders with preemptive strikes, and those still within his borders with public executions.

Amenhotep I died about 1690 B.C. His descendants were Tuthmosis I, II, and III. The first fathered Queen Hatshepsut. The second (the queen's

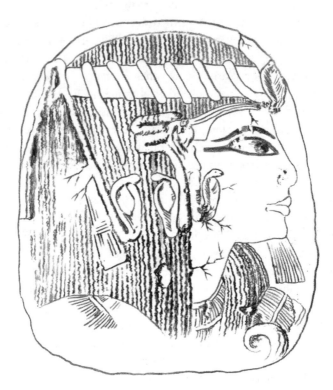

During the decades leading up to the Thera upheaval, Hatshepsut ruled as both king and queen of Egypt. In order to sit on the throne, her divinity as a man had to be declared by the gods, and she even wore a king's false beard. Ultimately, her rule was declared blasphemous by her stepson Tuthmosis III, who orchestrated her assassination. In an attempt to erase her memory for all time, he subsequently toppled and buried every monument and wall carving dedicated to her, including this limestone relief— which was, in effect, hermetically sealed for future archaeologists.

half brother) fathered her children. And the third (her nephew and stepson) buried her and despoiled her grave. Hatshepsut shared their ancestry, their hopes, and their thrones and, according to most scribal accounts, was obliged also to share their beds. She is said to have been the most beautiful woman who ever lived, from the time of Egypt's beginnings until Cleopatra and Livia.* The few sculptures of her that survive into our time bear out the stories of her remarkable beauty, which could only have complicated her life. Her father, Tuthmosis I, had two sons in line for the throne, and both died mysteriously in their youth, probably (according to Egyptian forerunners of gossip columnist Liz Smith) through the plotting of an ambitious concubine named Mutnefert, who was determined to clear the path of anyone who stood between her semiroyal son and the throne. At least two other half brothers who found favor with the pharaoh met with accidental death. Hatshepsut, evidently, was overlooked as a target, on account of her being a woman, who could no more inherit the throne than Mutnefert. Being harmless in Mutnefert's eyes, she was the only royal or favored child still living as Tuthmosis I neared the end of his thirty-five-year reign. Putting a protective wing over her, the pharaoh married his daughter and dressed her in manly robes and a false gold beard that she, as a woman, had no right to wear. In his will he proclaimed her both queen and king of Egypt, and on the walls of Karnak made this proclamation sacred with depictions in which the gods Seth and Horus baptized her and gave to her the symbols of masculine royal power. Thereafter she dressed as a man, and royal inscriptions prefixed her variously as "he" and "she."†

---

*The classic Elizabeth Taylor image of Cleopatra is entirely wrong: She was not Egyptian—she was a Ptolemy, a Greek—and she was a redhead, not a brunette. Augustus' wife, Livia, is said to have been even more beautiful than Cleopatra, so beautiful that she intimidated the emperor into impotence, so evil that when a snake bit her, it died.

†We know from Minoan frescoes that this appears to have been one of the rare periods in history when women came to positions of authority. (Writing more than a thousand years after the Thera upheaval, the Greek philosopher Aristotle, referring to historical texts extant in his time but now lost, noted that even the slaves of ancient Crete enjoyed the same rights as other citizens, except the right to bear arms.) Throughout most of the world, the women of Hatshepsut's time were treated as property—slightly more valued than slaves and cattle—and were traditionally beaten to death for even minor offenses. By contrast, a fresco from pre-Thera Knossos depicts women and men together at a public festival, extending hands to each other and engaged in animated conversation. Both men and women wear clothing suggestive of high office. In another fresco, women participate alongside men in sporting events. Both hold the reins of chariots. In contrast with the rest of the world, Minoan women seem anything but oppressed. At about that time, Hatshepsut had become one of the most powerful and respected rulers in the world. We do know that there was active trade

Hatshepsut was twenty years old when she inherited the throne about 1665 B.C. (give or take ten years). Mutnefert must have, by then, amassed political allies within Karnak, for (being politically aware that while friends may come and go, enemies accumulate*) rather than smite her father's concubine, the young pharaoh married Mutnefert's son, gave him the name Tuthmosis II, and took him as coregent.† He was a weak and effeminate man, precisely what the queen sought in a husband. Trying to rise above the graffiti on the town walls (which poked fun at his plump body and dainty fingers), he launched punitive expeditions against remnants of the Hyksos, who continued to "nibble in the wheat." Hatshepsut encouraged his attempts to shake off the weakling image, but unlike his predecessors, he did not lead his armies into Nubia or Canaan. Instead he *sent* them, favoring the safety of Karnak's walls over the field marshal's tent. Unofficially, Mutnefert and Hatshepsut ruled Tuthmosis II. Officially, Tuthmosis II ruled the army. And yet officially, Hatshepsut wrote of herself:

> Thou shalt triumph . . . flex thine arm in smiting the bow people . . . and my power overwhelms them that are across the sand [across Sinai]. . . . The Asiatics . . . bring me the choicest products, namely cedar [from Lebanon] and all the fine woods of God's land . . . the banks [of the Nile] are united under my authority, and the black land [Nubia] and the red land [across Sinai] are in terror of me. My power causes foreign countries to kneel. The northerners [Syrians, Hyksos, and Canaanites], their gods are fashioned as my amulets.

And thus reads Egypt's side of the story. The Canaanite version of the Hyksos oppression, told in a way calculated to put forth the best possible face, depicted the bow people's one-time rise to power in Egypt not as a conquest but as a peaceful migration west by pastoral nomads who rose quietly through the ranks to political control. Their oppression and depar-

---

between Crete and Egypt, and along with trade there might also have occurred some measure of cultural cross-pollination. Until a Minoan library is found with its records intact, I will always wonder if Hatshepsut's rise to power ushered in similar opportunities for Minoan women or if perhaps the reverse is true: if Minoan equality made Hatshepsut's ascent more palatable in Egypt (or if, in fact, there is simply no connection at all).

*As so often cited by Brookhaven National Laboratory physicist James Powell, this is one of the most sensible pieces of advice we can carry with us through everyday living, as applicable to office politics today as it was to a young pharaoh of 1665 B.C.

†In biblical times, social norms regarding marriage between close relatives were vastly different from today's. It was not quite so shocking, in those days, to read that Hatshepsut married her father, her brother, and her son, or that Lot slept with his daughters.

ture came not as the result of a humiliating defeat but as a voluntary and spiritual stand taken against Pharaoh. By about 900 B.C. the legend of Israel's coming out of Egypt was already a tradition of long standing, constantly alluded to by early biblical prophets. It had been incorporated into the origin stories of all the Semitic-speaking enclaves in and around the Jordan Valley, including the Canaanites and the Hebrews who dwelled among them. Blurred, telescoped, and consciously edited over time, the biblical account that reaches us today does not tell us that there were Pyramids in Egypt or that there were probably several pharaohs of the Oppression and that one of them was a woman.

By about 1645 B.C. (possibly ten or fifteen years earlier) Tuthmosis II was dead, probably poisoned by his concubine Isis, who was determined that her son, Tuthmosis III, must first rule beside "King" Hatshepsut as co-regent and ultimately rule over her as Pharaoh. Hatshepsut protected her position by creating a new religious doctrine and placing herself at its center. Somehow, she accomplished this with the full support of Karnak's priests and politicians. Inscriptions on the temple walls told of her divine birth by immaculate conception, depicted her baptism and purification by the gods, and proclaimed her the salvation of Upper and Lower Egypt, who would return after death, on Judgment Day, to lead the souls of her chosen people to Paradise. Politically it was a brilliant maneuver, for in theory, one does not want to assassinate his savior. And in practice, her claim to divinity did in fact prove effective, at least in the short term. Why else would her stepson and coregent have copied it? When a giant comet appeared in the sky (orbital motions, tracked backward in time, suggest that this could have been an early, close flyby of Halley's comet), Tuthmosis III sensed an opportunity and grasped it. He quickly assembled his own enclave of priests and political allies, and through his vizier and foreign minister, Rekhmire, he sent forth word that the pillar of fire was a pronouncement of his divinity, a sign put in the skies by the gods themselves. Writers of his day began praising him as "a circling comet which shoots out flames and gives forth its substance in fire." Suddenly Egypt had the peculiar distinction of being torn apart by two living gods competing for the same throne. By what was to follow, it becomes easy to understand how early civilizations came to regard comets as heralding both the birth of kings and terrible misfortune.

∼∼∼∼

While Hatshepsut lived, at least one military incursion was made into Canaan, to subdue the remnants of the Hyksos, with Hatshepsut herself leading chariots into the field. In her own account of the expedition, recalling the time of Egypt's "great humiliation," she described "barbarian nomad groups [among the Hyksos, who had] destroyed what [Egyptian monuments] had been made," and she vowed to "restore that which was in ruins . . . since the Asiatics."

It is clear that, even as Asiatic rebellions occasionally had to be put down, Asiatic people lived among and were tolerated by and even accepted as true Egyptians. Nowhere is this more clear than in records attesting that Amose I's mutiny against the Asiatic Hyksos included mutineers who were themselves Asiatics. Hatshepsut herself boasted of wearing the amulets of northern gods, and it was during her reign that Egypt began to accept the gods of resident aliens, apparently under the rationale that the gods of the vanquished had finally recognized whose kingdom was stronger and "more right," and now extended to the goddess-king their loyalty and protection. Thus did the Canaanite Lord Baal (a forerunner of Moses' all-powerful Yahweh) gain a foothold among the Egyptian and Asiatic merchants of Memphis, becoming as much a god of the conquerors as of the conquered. And under the goddess-king, Minoan ships carried Egyptian trade goods north to Canaan and Syria, then west to Cyprus and Crete, then turned south again toward Egypt, with holds full of cedar from Lebanon, fine Canaanite oils, and Syrian gold. There are indications that in Hatshepsut's time, overland trade routes from as far away as India linked up with the Minoan fleet. A hunting leopard was imported for Pharaoh, and chickens (which, like the hunting leopard, had evolved in India) made their first appearance in Egypt shortly before Thera exploded.

About the twentieth year of her kingship, curses went out from the temple of Karnak, and it was prophesied by Hatshepsut's priests that any who betrayed her, or despoiled her tomb, or broke any engraven image of her would be damned for all time, would live in infamy for as long as human beings dwelled upon the Earth. There must have been a reason for the curses. She must have seen the storm coming. She would probably have been in her early forties when, sometime between 1635 and 1630 B.C. (though possibly as early as 1645 B.C.), Tuthmosis III orchestrated the murder of her closest advisers and defaced the interiors of their tombs. He then either buried (and unwittingly preserved) or mutilated every likeness of her, built walls around the monuments she had erected, and placed himself on record as sole ruler of Egypt. There is no record of how the end came, but by these actions we know that Hatshepsut was dead.

Within three months of crowning himself pharaoh, Tuthmosis III renewed punitive attacks against the Asiatics, both within and without the country. Officers and soldiers were promised shares in the loot of victory, and he gathered them together on the Mount of Megiddo, fifty miles north of Jerusalem, into a place called in the Hebrew tongue Har-Megedon, which has come down through the Bible as Armageddon. The pharaoh confronted armies massed under the three kings of Palestine, Syria, and Kadesh.* The king of Kadesh alone had brought to the field of battle three hundred Asiatic princes, each with his own army. Tuthmosis III defeated them all. He took the sons of Asiatic kings to Egypt and held them as hostages in his Karnak palace, where they were educated and catered to as equals to his own sons.†

Then, in the autumn of 1628 B.C., Thera exploded, and Tuthmosis III's world, which up to that point had confronted him only with a straightforward struggle for power, was never so simple again. Whether or not an exodus of Hebrew slaves occurred at this time, the Theran ash layer, the California bristlecones, astronomical dating of China's *Bamboo Annals,* and the inscriptions of his own foreign minister mark him as the Pharaoh of the Plagues.

North of Egypt, nearer the epicenter, an even mightier, more refined civilization went into eclipse. Amos and Jeremiah tell us that remnants of a lost empire, reduced to refugee status, sailed to Canaan and that the origin of the Philistines was, at least in part, a remnant from the coasts of Crete. When we look at Philistine ruins in Israel, we indeed see the influence of Minoan pottery styles, and on occasion we find actual Minoan temples and wall paintings.

At a shrine near Karnak (built to honor "the Great Speos Artemidos"),

---

*Kadesh, a region in the northeastern corner of the Sinai Peninsula, means "holy," and was probably so named for some as yet unknown event that became part of the pre-Thera, pre–Tuthmosis III oral history of the area. It is, relative to the rest of Sinai, a fertile region, into which Moses and his people (according to the Exodus account) are said to have settled for thirty-eight years.

†If one is permitted to speculate on this arrangement, the historic reality (as recorded in the annals of Tuthmosis III) of Semite children reared and educated under a pharaoh's roof, coupled with the generally accepted view that the name Moses is a variant of Tuthmosis and/or Amose, is hauntingly reminiscent of the biblical account of Moses' youth in the house of Pharaoh. Additionally, the original Hebrew name of Moses was Mosheh. When the Greeks later translated the Old Testament into the Septuagint, their language had no letter for the *sh* sound, and it was thus substituted with a simple *s*. Another *s* was added because Greek names traditionally ended in *s*. The original pre-Greek pronunciation of Tuthmosis also varies, from Thutmose to Thutmoshe. In Egyptian, Moses (Moshe) means "son." Thus Thutmose means "son of Thoth," the Egyptian god of wisdom, learning, and magic, who to the Greeks became the scribe and messenger to all the gods and took the name Hermes (the Romans later called him Mercury).

Johns Hopkins University Egyptologist Hans Goedicke has translated an inscription dating from the time of Tuthmosis III. It reads, in part: "And when I allowed the abominations of the gods [immigrant peoples] to depart, the earth swallowed their footsteps. This was the directive of the Primeval Father [water, according to Egyptian beliefs] who came one day unexpectedly."

A fragment of diary (age uncertain) recorded on the back of the Rhind Mathematical Papyrus, and "lamentation inscriptions" on a stela of Amose I (possibly one of several known overcarvings by later, nameless scribes) describe a terrible storm in which "the sky came on with a torrent of rain, and darkness covered the western heavens [in the direction of Thera and Crete] while the storm raged . . . [louder] than the noise at the Cavern that is in Abydos. Then every house and barn where they might have sought refuge was swept away, and [they] were drenched with water . . . and for a period of [illegible] days no light shone in the two lands [Upper and Lower Egypt]."

I used to agree with Benjamin Mazar that accounts of cities going up like the smoke of a furnace, of darkness throughout the land of Egypt and towering cliffs of water crashing down upon armies were all sheer mythology, with no basis in fact. One of the most fascinating surprises of my life has been the dawning realization that some of the more dramatic episodes of the Bible, things that appear very strange to most of us living today, perhaps even miraculous, seem actually to have occurred. The stories might have been distorted by successive retellings, with moral lessons and judgments stitched in, but I think we can begin to read in them, at the very least, snippets of geologic and historic reality.

And the closer I look, it seems the less astonished I should have been in the first place. I used to live in New Zealand, where Maori talking chiefs could recite royal ancestries going back two thousand years and peek beyond the first kings to the voyages of discovery from distant islands. They gave descriptions of giant birds, long extinct, that once roamed the land. Archaeology and paleontology have corroborated much of what the talking chiefs were telling us. Yet the oral transmission of New Zealand's history goes back more than twice as far as the span separating Thera from the writing of *Theogony* and Exodus. So why, then, should I have been so surprised?

If, in the time of Thera and Tuthmosis III, there did occur a migration of people out of Egypt, it was probably not so much a matter of a patriarch standing before Pharaoh and demanding, "Let my people go," as the pharaoh, pondering what to do in time of famine, deciding that dwindling resources might best be preserved by jettisoning his most expendable subjects (immigrants) into the eastern desert, and then demanding of them: "Get your people out of here!"

Is it possible that two groups of refugees (Minoan-Cretans-turned-Philistines-turned-Palestinians, and Asiatics-turned-Israelites) were created by the same volcanic catastrophe? Is that how it all began, all those thousands of years of bloodshed?*

History so changed the day Thera exploded that it is possible to believe we would long ago have landed on Halley's comet, explored oceans under the ice of Europa, begun to colonize worlds around distant suns, and Egypt would now be a mere footnote in Minoan history books, if only Minoan civilization had survived. But whenever we consider such alternate histories, we must be very careful about using the word *we*. As way leads unto way, and roads never traveled become the mainstream, our very existence may be erased. Minoan bull worship becomes the new Bible. The development of rapid global transportation two or three thousand years early so scrambles and resets the pattern of marriages and births that there is no way my grandmother, Bible in hand, could have been an Irish-born immigrant bound for Ellis Island and a country called America in A.D. 1921, there to meet my grandfather. In a world where Thera never exploded, none of us should exist today. The Earth would be inhabited by total strangers, but for a geologic time bomb hidden in her mantle long in advance of the dryopithecine and sivapithecine ascents.

---

*The Bible appears to be peppered with echoes of Thera's destruction and Crete's dispersal into Philistine settlements. As is common in biblical texts, events in the past are often pulled into the writer's forward field of vision and given prophetic significance. Thus we read in Ezekiel 27–28: "Zor [a onetime neighbor of Sodom that came to a similar end], O thou that dwellest at the entry of the sea, that art a merchant of the people of many islands . . . they [like Minoans] have taken cedars from Lebanon to make masts for thee. . . . Dan [a Philistine city] gave yarn for thy wares . . . thou wast upon thy holy mountain of GOD; thou hast walked up and down in the midst of the stones of fire . . . but . . . thou hast sinned: therefore I have cast thee as profane out of the mountain of God: and I will destroy thee . . . and all thy men of war that are in the midst of thee, shall fall into the heart of the seas in thy day of ruin . . . therefore I have brought forth a fire from the midst of thee, it has devoured thee, and I have turned thee to ashes upon the earth in sight of all them that behold thee. . . . Now thou art broken by the seas, in the depths of the waters; thy merchandise and all thy company are fallen in thy midst. . . . Thou hast come to a dreadful end and thou shalt exist no more for ever."

10

ON THE TRAIL OF THE LOST ARK

As meteorites are a poor man's space program,
archaeology and paleontology are a pauper's
time machine.
—PELLEGRINO'S FIRST LAW

Why meet we on the bridge of Time to exchange one
greeting and to part?
—THE KASIDAH OF HÂJÎ ABDÛ EL-YEZDI

A NY VANISHED PEOPLE will keep far more information to themselves than we are ever likely to discover in the fallen walls, housewares, and random scraps of writing they have left behind. But that should not discourage us from trying. If we believe that the second and third books of the Bible actually echo some geologic and historical truths, then following the trail of the Exodus tribes promises to be a complicated, if not futile, endeavor because as Elizabeth Stone and Paul Zimansky's work in Iraq has taught us, people returning to tribal nomadism tend to become "archaeologically invisible." If, as a starting point, we accept the twin theories that (A) the story of the ten plagues is linked to the Thera upheaval and (B) the story of the Oppression and ultimate flight from Egypt grew out of the Hyksos oppressions and expulsions (covering the century from Hatshepsut's great-grandfather, Amose I, through her stepson Tuthmosis III), then we are looking at multiple oppressions and expulsions—at least three exoduses—telescoped into one account, culminating with darkness, plague, famine, and a parting of the waters.*

The most dramatic of the exoduses would of course be the most indelibly burned into tribal memory, and there was nothing in human experience quite so dramatic as Thera. References to a pillar of cloud and a fire in the sky, thundering mountains, and other volcanic motifs appear to have

---

*From a strictly dramatic and literary standpoint, it seems sensible that an epic tale of deliverance, meant to be read by every succeeding generation, should have but one climax. Even if the compilers of Exodus had all the facts at hand, the notion that they would follow the miracle of the waters and the escape from Egypt with another oppression in Egypt by another pharaoh, requiring yet another escape, defies both logic and good storytelling. They would be no more likely to tell of multiple exoduses than to make Noah build the Ark twice, no more inclined than Hollywood producers would have been to put Clark Gable under Charles Laughton's command a second time, requiring a second *Mutiny on the Bounty,* just as he was about to ride off into the sunset with his Tahitian bride.

become so symbolic that we find them spread throughout the books of Exodus and Numbers. Even the mountain where Moses made his covenant with God (in Exodus 19:16–20, on the Sinai Peninsula, where no volcanoes exist) is described thus:

> And it came to pass . . . that there were thunderings and lightnings, and a thick cloud upon the mountain . . . exceeding loud; so that all the people in the camp trembled. . . . And Mount [Sinai] smoked in every part, because the Lord descended upon it in fire; and the smoke of it ascended like the smoke of a furnace, and the whole mountain quaked greatly . . . and the Lord called [Moses] up to the top of the mount; and [Moses] went up.

The cloud from Thera was certainly seen and felt in Egypt. Actual, close-up eyewitness accounts of the eruptions leading up to the final explosion, as evidenced by *Theogony,* had become legendary, and these, too, must have been received in Egypt. One of the strangest coincidences of all, when we read about Moses and the miracle of the sea, is that like the *Theogony* account of a volcanic upheaval, the Book of Exodus provides what looks for all the world like a scientifically accurate description of the tsunamis that radiated out from Thera. The Song of the Sea, in Exodus 15, is one of the most ancient passages in the entire Bible (much like the Book of Genesis, Exodus appears to be woven together from four different and sometimes contradictory textural strands). In this song story, we read, in celebration of the Lord's deliverance from Egypt: "Lord is His Name! Pharaoh's chariots and his army . . . the deeps covered them . . . And with the blast of thy nostrils the waters were piled up; The floods stood straight like a wall."

Burton Rudman (of the Archaeological Institute of America) has pointed out that the October 22, 1883, issue of *The New York Times* used almost identical language to describe (from eyewitness accounts) the tsunami that struck Java in the wake of the Krakatoa explosion: "Far out to seaward a piled-up wall of water, standing like a high column and coming in upon the shore with inconceivable swiftness."

To judge from models using both the contours of Egypt's northern coast and the slope of its shore, walls of water did indeed pile up and crash down upon the rim of the Nile Delta, where a major trade route followed the north shore of Sinai to Canaan and where Tuthmosis III had, at intervals of several miles, placed his sentries. The wave from Thera probably towered some four stories over the shore road.

The Theran tsunami, however, was not tall enough to burst southward through the Sinai Peninsula and make its power known in the Red Sea (where most modern sources allege the miracle of the waters to have taken place), not unless one expands the Red Sea to include possible damage to the northern reaches of the first Suez Canal, an ancient Egyptian construction which, like the modern canal, extended the Red Sea to the Mediterranean. Ever since Cecil B. DeMille and Charlton Heston made the Red Sea parting into a cultural icon, their version became widely regarded as a reasonably accurate portrayal of what the Bible was really trying to tell us, but the words *Yam Sup,* the original Hebrew location of the miracle of the sea, has been translated variously as the "Red Sea," the "Reed Sea," and the "Sea at the End of the World." In both Egyptian and Hebrew, the word *yam* means "sea" (no argument there), while the word *sup* derives from an Egyptian word for "papyrus reeds," and a direct biblical linkage between *sup* and reeds can be found in Exodus 2:3–5 (when the baby Moses is hidden among the *sup* [papyrus reeds] growing along the banks of the Nile) and Isaiah 19:6 ("and the canals of Mazor shall be diminished and dried up: the *sup* [reeds] and rushes shall wither"). The Hebrew word for "red" is *adam,* meaning that if the biblical scribes had meant to tell us that the miracle of the waters took place on the "Red Sea," they should have written it *Yam Adam,* not *Yam Sup.* Northern Egypt's papyrus marshland, known since ancient times as the Reed Sea (Yam Sup), is on the Nile Delta, right where Tuthmosis III had placed his sentries, and right where the Theran tsunami struck.*

*Biblical scholar Bernard Batto points out that the "Reed Sea" solution is not quite so simple as it seems, and he calls attention to Numbers 33:8, which describes the Israelites fleeing Egypt, then passing miraculously through the sea (*yam*) into the wilderness. Which "sea," we are not told. Afterward the Israelites travel east for five days, setting up camp at Marah, Elim, and finally by the *Yam Sup*—which here refers to the Gulf of Suez, the northwestern finger of the Red Sea. But *this Yam Sup,* though referring to the Red Sea, not the Reed Sea, is five days' journey east from the sea at which the parting of the waters is said to have taken place (Egypt's "Reed Sea" is approximately five days' travel west from the Gulf of Suez). Batto notes that the confusion may arise from the fact that *sup* may be connected to the Semitic root *sop* (meaning "the end of the world"), as well as to the Egyptian word for "papyrus reeds." "Sea of Sup" could thus mean "Sea of Reeds" and "Sea at the End of the World" at the same time. "Sea at the End of the World" is an ancient symbol for "the great unknown," an abyss leading to the underworld. Geographical knowledge in those days was such that any sufficiently large body of water was potentially a pathway over the edge of the world and into the infinite. Thus does the Dead Sea Scroll text known as the Genesis Apocryphon (21:17–18) refer to the Tigris-Euphrates as emptying into the *Yam Sup* (not the Persian Gulf), and the third century B.C. Book of Jubilees associates the *Yam Sup* with the Indian Ocean. In other words, the biblical location of the "Red Sea" is not as clear-cut as most people think. Like the city of Sodom, it moves around a lot.

Contrary to popular belief, tsunamis are not great cresting waves that crash and foam across the ocean surface, overturning ships as they go. In fact, one of the safest places to be during a tsunami is on the open sea, where it manifests as a gently rising and falling long wave, scarcely three feet high. The shore is the worst place to be, for here the wave is reflected, compressed, and enhanced by the ever-diminishing distance between the bed of the sea and its surface. Its first breath is felt as a sudden retreat of the water, plunging over a matter of minutes deeper than any tide can ebb. Tuthmosis III's north shore sentries could not have known what was happening to them. Such things are rarely seen in a human lifetime. Because they did not know what they were seeing, there was no cause for fear, only astonishment and curiosity, even joy at whatever miracle was turning the seabed into dry land. Mud dunes that had lain under thirty feet of water were suddenly climbing into open air. In the spaces between the dunes, puddles glistened. Fish thrashed in them. On a tidal plain such as the Sea of Reeds, the Mediterranean would have drained a half mile out, making hissing and sucking noises as it went. Driven by wonder and curiosity, at least some of Tuthmosis III's sentries must have followed the retreating coastline, perhaps even pausing to drag prized fish from the puddles. Then the sucking noises abruptly ceased, and a new sound was in the air, and even those who had remained on the higher shore knew suddenly, and too late, that where they thought themselves to be utterly safe, their high ground was but an illusion. It rippled away in that moment of awful realization when they saw the black shapes stirring out there, rising up on their haunches. When the sea returned—faster than any charioteer could hope to flee—it was forty feet above the high-tide mark. As told in Exodus 14, "And the Lord caused the sea to go back by a strong east wind* . . . and [He] made the sea dry land. . . . And the Lord said to [Moses], Stretch out thy hand over the sea, that the waters may come back upon Mizrayim [Egypt] . . . and the waters returned, and covered the chariots, and the horsemen . . . and the waters were a wall to them. . . . Thus the Lord saved [Israel] that day out of the hand of Egypt."

The dust cloud from Thera was itself a driving wind. Probably this arrived even before the tsunami, and continued to roar through the floods, and lingered afterward. Occasional thinnings in the cloud would have allowed daylight to break through, and if eyewitness accounts downwind of the A.D.

---

*Analysis of ash deposition on the Mediterranean floor, in the Nile Delta, and across Turkey tells us that in the aftermath of Thera there was indeed a strong wind blowing from the volcanic ground surge and that the surge drove in an easterly direction.

1883 Krakatoa and 1980 Mount Saint Helens eruptions provide reasonable clues to life under a volcanic shroud, then the Sun in the sky gave little more illumination than the full moon. Interestingly, Exodus 9:8 tells us that at God's instruction Moses held out his arm toward the sky and sprinkled soot from the furnace heavenward "that it shall become a fine dust all over Egypt." Today, on the Nile Delta, in a time horizon coincident with Tuthmosis III, we scientists can actually see, and feel, and probe with our instruments a layer of fine volcanic dust. We do not know that a man named Moses ever existed, but black, ground-hugging clouds, easterly winds, walls of water, and dust there were, in great abundance. The rocks tell us so.

What the rocks cannot tell us are the specifics of the Egyptian sojourn of a people who came to call themselves Israelites (like the postcollapse nomads at Stone and Zimansky's end of the riverworlds, the early Israelite warrior tribes, if we assume the Bible is correct about their existence, left few archaeological traces of themselves). According to Egypt's Leningrad papyrus (1116A), some immigrant peoples were subjected to compulsory labor during the century after the Hyksos expulsion leading up to and through the reign of Tuthmosis III. In a condition of slavery they were recruited and forced to work in state quarries, in gold mines, and in the building of sacred temples. The Egyptian record is consistent with the Exodus account of Hebrew servitude, in which the Israelites were drafted for slave labor in connection with public building projects. However, contemporary Egyptian records do not even mention the Hebrews, who, rather than being specifically singled out for forced labor, were, if present, just one more little minority group caught up in a wider oppression of Asiatic Semites (the minority would thus have been made large by the fact that it did, in time, come to write a tribal history, now immortalized as Scripture). I have a nagging suspicion that if we had access to the ultimate archaeological tool—a time machine—and if we could arrive some years after the Thera upheaval and ask Tuthmosis III, face-to-face, what happened when Thera exploded, he would tell us about the ash and the darkness and the need for rationing grain. He would remember the cold summer season that followed the days of darkness and the danger of hunger and poverty-induced rebellion. He would probably remember the jettisoning of Asiatic immigrants; but to him the fate of the Asiatics themselves would be of far less significance than his successful implementation of history's first-known rationing. If we could press him about the Hebrews and Moses, we might gasp at his total, almost contemptuous lack of interest ("A Hebrew exodus? Israelites? I don't remember them!") in events sacred to modern Jews, Christians, and Muslims.

That we do not read of Hebrews and Israelites in Eighteenth Dynasty Egyptian documents no more implies that they were not present than the fact that Egypt's Pyramids are not mentioned in the Bible may be taken as proof that the kingdom's most obvious architectural features never existed. Absence of evidence is not always evidence of absence: Pyramids and Hebrews were simply not important, depending on whose side was telling the story.

The actual events were, without doubt, much more complex than the Exodus narrative suggests, with multiple migrations and traditions knitted into a single, comprehensible Egyptian sojourn. Even the biblical account, taken as a whole, speaks of more than one migration into and out of Egypt (Abraham in Genesis 12, Isaac in Genesis 26, Jacob in Genesis 42, Joseph in Genesis 37 and Exodus 1, Levi in Exodus 1, and Moses in Exodus 15), indicating that the flight from Pharaoh leading to the miracle of the sea is but a single "snapshot" recorded during what must be viewed as a continuous process of migration, settlement, displacement, and resettlement taking place at different times and along different routes by small and large armed bands passing back and forth between Egypt and Israel. While the descendants of Hebrew tribes did eventually come to dominate the Jordan Valley, the biblical account, which portrays the events culminating in Canaan's conquest as a single mass exodus from Egypt, followed by one refugee tribe's forty years in Sinai, followed by the migration of a Hebrew fighting force from Sinai into the Jordan Valley, almost certainly comes down to us in an oversimplified, all but freeze-dried version. Still, events that make archaeological sense shine through. The forty years in Sinai ring true to the fact that we would expect some decades (in all probability considerably more than forty years) to pass before the post-Theran mosaic of Semitic tribespeople could amass sufficient numbers (including, perhaps, covert alliances with nomadic Semites who never left Canaan) and hope to challenge the established might of the Canaanites and the Philistines. Among their allies would have been the people of Kadesh, in northeastern Sinai, whose king, according to Egyptian records, had gone north to Canaan and fought Tuthmosis III at the Mount of Megiddo.

The biblical path taken by the people who left Egypt—south into Sinai—makes perfect sense if indeed, one of the migrations out of Egypt took place during the time of Thera and Tuthmosis III. Whatever the physical and economic consequences of Thera, Egyptian civilization soon recovered, and Tuthmosis III continued, for the rest of his life, to pursue campaigns into Canaan and Lebanon. Anyone leaving Egypt, and not in the pharaoh's good graces, would have wanted to avoid both Canaan and the shore road leading to it. The only safe route was south, into the Sinai Peninsula (whose coasts, incidentally, provided some of the best fishing in the world and could easily have sustained a population of migrants). Because Tuthmosis III had placed military outposts along the north shore of Sinai and along the coast of Canaan, some of his soldiers would inevitably have come to harm from Theran tsunamis. Word of Egyptian soldiers crushed under waters that first retreated, then "stood straight like a wall" was bound to reach the pharaoh's adversaries in Sinai and Canaan, and it is easy to understand how an event so extraordinary as to be "miraculous" should be interpreted as a judgment by the Lord against an oppressive empire. In succeeding generations the story might have grown to include, on the very day of judgment, a direct confrontation between the pharaoh, his fleeing victims, and their God. It then became but a small step from interpreting the tsunami and the ashfall as divine punishment on Egypt to construing it as pressure exerted to affect a release from bondage.

~~~~~~

"Man's concept of God evolved as man evolved," Father John MacQuitty has said. As he sees it, if one or more groups of people did in fact leave Eighteenth Dynasty Egypt and hunker down, for a time, in the deserts and along the southern shores of Sinai, they would have entered the peninsula thoroughly Egyptianized, in most cases after many generations of residency on the Nile. "The Old Testament tells us that it took the Hebrews one day to get out of Egypt. But it took forty years just to begin to get Egypt out of the Hebrews: including their tendency to worship gilded carvings of sacred bulls and a sense of vengeance that proclaimed, 'If you kill one of my kin, I must kill seven of yours.'

"With the very first civilizations came a series of sacrificial gods to whom people made blood oaths. By the time of Exodus there had evolved a single, law-giving creator. The evolution of God had a lot to do with attempts to set standards of morality, to come to terms with right and wrong, light and darkness, sin and judgment, but mostly it had to do with man's becoming conscious of himself."

And thus, according to MacQuitty, arose the Mosaic law, strange to us today, strangely merciful to those living in a time when it was socially and morally acceptable for Hatshepsut and Lot to share beds with and marry their children, or to execute a man (and sometimes his entire family) for withholding a single egg from the tax collector, or to whip a slave to death for spilling a handful of barley. According to Scripture, the new laws were (in Exodus 19–23) given to Moses by direct contact with God in Sinai.

They have by now become very familiar to us, and they read, in part: "Thou shalt not covet thy neighbor's house . . . nor his manservant, nor his maidservant, nor his ox. . . . If the ox was wont to gore with his horn in time past, and his owner had been warned, yet he had not kept him in, but it killed a man or a woman, the ox shall be stoned, and its owner also shall be put to death . . . if any harm ensue, then thou shalt give life for life, eye for eye, tooth for tooth."

On first reading, the judgments set before Moses' people are one of civilization's earliest attempts to make the punishment fit the crime, to guarantee equal protection under the law. But as we read on, we begin to see hints that even in earliest biblical times, some men were protected more equally than others. According to Exodus 21, if your ox killed a slave, the rule of giving life for life did not apply. You paid your debt to his master in thirty shekels of silver and your ox was stoned to death. If you killed a man's ox, the penalty was the same as if you or your ox had killed his servant: You paid him in silver. If you plucked out your servant's eye, your own eye was not forfeit, as in the blinding of a freeman. Your punishment was to set the slave free. If we find the Mosaic laws brutal or unfair, that is because we make the mistake of contaminating them with the norms of our own civilization, without truly understanding that the Covenant was framed by an Egyptianized tribal people, so different from us as to be almost an alien civilization. And thus do we inherit from them Ten Commandments, four books full of laws, and not a single word against slavery (indeed, long into the nineteenth century A.D. Scripture was actually quoted as a justification for slavery and all its associated cruelty):

And if his master has given him a wife, and she has born him sons or daughters; the wife and her children shall be her master's, and [if he is sent out free] he shall go out by himself. And if the servant shall plainly say, "I love my master, my wife, and my children, I will not go out free": then his master shall bring him to the judges; he shall also bring him to the door, or to the door post; and his master shall bore his ear through with an awl; and he shall serve him forever [Exodus 21:4–7].

"We must understand that the patriarchs and the priestly writers of the law, while trying to move away from their Egyptian roots, could not revise the old traditions in too extreme a fashion," according to Boston University biochemist and biblical scholar Isaac Asimov. "And so you could still sell your daughters into slavery without any penalty at all, or beat a woman to death without necessarily suffering much greater punishment than if you'd destroyed your neighbor's ox, for your wives and your daughters were, according to tradition, your property—just like oxen. Even the holy days of obligation were carryovers from earlier traditions. In all probability the Passover was an Egyptian agricultural festival long antedating the time of Moses. Such festivals are common in all settled agricultural societies. (Americans have even invented one for themselves—Thanksgiving.) Usually such festivals, even among the early Israelites, were thoroughly pagan in inspiration. Some of the old laws, festivals, and customs were too popular and too deeply ingrained in tradition to be done away with completely. The best that could be done was to associate them firmly with some legendary event in biblical history and divorce them from idolatry. Passover, the most important of the agricultural festivals, came to be associated with the most important event in the early legends—the Exodus."*

The God of Moses was, in many ways, a warrior god who promised his people conquered lands if they kept his laws. The patriarchs and priestly scribes must have known that it was the only language their people were willing to understand. In Deuteronomy 7 the Lord tells Moses that the Canaanites and the Philistines are to be shown no mercy—"thou shalt devote them to utter destruction"—because they serve other gods. God suggests to Moses that all human beings are lowly (scum of the earth) and that his chosen tribe is neither larger nor better than any other and must indeed be punished, from time to time, for cursing him and the patriarchs and for breaking the law; but for reasons that seem to have as much to do with a game of chance as anything else, "the Lord thy God has chosen thee

*Asimov has also pointed out that shifting associations were common in the origin and evolution of religions. This explains how, in the early history of Christianity, the pagan celebration of the winter solstice (the shortest day of the year, on which the Sun in the sky has wandered as far south as it will go and begins to drift north again) was converted into Christmas, in celebration of the birth of Christianity's founding prophet. Nor is it any coincidence that Christmas and Hanukkah are celebrated at the same time of the year (although never on the same day) or that Passover and Easter coincide with the spring equinox (the halfway point of the Sun's journey north). After the Babylonian Exile (about 500 B.C.) the Passover became a festival during which all pious Jews attempted to make a pilgrimage to Jerusalem and worship at the Temple of Solomon, atop the Temple Mount. According to the Christian calendar, it was during the course of one of these Passovers that Christ was crucified. The word *Easter*, celebrating the Resurrection, derives from an ancient Teutonic goddess of spring (borrowed by the Romans during the Germanic wars). A pagan spring festival became the Christian Easter festival, retaining even the Germanic name to make the transition as easy as possible. Other additions and shifts accumulated throughout the centuries, and it is anybody's guess what the Jesus we read about in the Book of Matthew would have made of his modern-day associations with Santa Claus and the Easter Bunny.

to·be a special people to himself, above all peoples that are upon the face of the earth. . . . And thou shalt consume all the peoples which the Lord thy God shall deliver to thee . . . thou shalt smite them."

The conquest epics of the Old Testament provide graphic glimpses of what it must have been like when settled city dwellers were overrun by Bronze Age nomads. In the case of the children of Israel, God himself is said to have given specific instructions on how to loot and burn a city and to exterminate the conquered population down to the last child, except for special cases in which the Israelites were told to kill only the men and boys, and all the women old enough to have slept with a man, but to carry away young girls as slaves and concubines:

> And the Lord said to Moses [Moshe], Fear him not: for I have delivered [the king of Bashan] into thy hand. . . . So they smote him, and his sons, and all his people, until there was none left of him alive: and they possessed his land [Numbers 21:34–35]. . . . And the Lord spoke to Moses saying, execute the vengeance of the children of Israel [Yisrael] on the Midyanim. . . . And they slew all the men. And the children of Israel took all the women of Midyan captives, and their little ones. . . . And they burnt the cities in which they dwelt. . . . And they brought the captives . . . and the spoil [loot], to Moses. . . . And Moses said to them, Have you saved all the women alive? Behold, [surviving pagan women of an earlier conquest] caused the children of Israel . . . to revolt against the Lord . . . and there was a plague upon the congregation of the Lord. Now therefore kill every male among the little ones, and kill every woman that has known a man by lying with him. But all the women children, that have not known man by lying with him, keep alive for yourselves [Numbers 31:1–19]. . . . And then Sihon came out against us . . . and we smote him, and his sons, and all his people. And we took all his cities at that time, and devoted to destruction every city, the men, and the women, and the little ones; we left none remaining: only the cattle we took [Deuteronomy 2:32–35].

On eastern Crete, Nicholas Platon has found the city of Zakros suddenly, completely, and mysteriously destroyed. The archives and even the treasure rooms were found crushed flat and burned, but all their contents were still in place. In a shrine, scores of religious artifacts, lying in the same positions their owners left them in thirty-six hundred years ago, had just been used for a ceremony. In the kitchens, kettles and cooking pots were on the hearths, the eating utensils laid out for use, and meats in the process of being served when something interrupted the Minoan chefs. Platon believes the shock wave, ash cloud, and tsunamis from Thera caused the destruction;

and similar destructions on the coast of Israel, dating to approximately the time of Tuthmosis III, have been attributed variously to Thera, Tuthmosis III's incursions, and Israelite conquests.

At the Second International Congress on Thera and the Aegean world, archaeologist Sinclair Hood, perhaps carrying a valid counterpoint too far, suggested that we need not raise the specter of Thera to explain Canaanite destruction layers (a point on which he is probably correct) or even Minoan ones (a point on which he is almost certainly wrong). To him, a reading of Exodus, Numbers, Deuteronomy, and Joshua underscores the extreme inhumanity that humanity is willing to inflict upon humanity.

"I would like to comment," he said, "on the reluctance among archaeologists today to accept wars and invasions as explanations for archaeological phenomena. Must I remind you of what we and our fathers and grandfathers have achieved in the way of slaughter and destruction in two world wars? I venture to suspect that all the natural cataclysms, earthquakes, volcanic eruptions, and tsunamis that have occurred since the dawn of historic times have not achieved such destruction and slaughter as this; and I cannot help wondering whether a reluctance to accept the destruction for which we have been responsible in our own age, combined with the threat which hangs over us of a further and perhaps final slaughter with the use of atomic bombs, might not make us unwilling to consider war as an explanation of the archaeological evidence in those early times."

"I thank my dear friend Sinclair Hood for his excellent way of painting this picture," said the Swedish archaeologist Arne Furumark. "I agree with him entirely, except for one point. The destructions which took place on Crete [and elsewhere during the decades about 1628 B.C.] were child's play compared with modern wars. . . . The spears and arrows of that time were much less dangerous than modern weapons . . . they were simply not powerful enough [to account for the destruction at Zakros, and at some of the Philistine sites, directly in the path of Theran ash and tsunamis]. . . . Besides, a conqueror had only to cripple the central administration, and with that gone the small communities could not have resisted any longer."

"The view that primitive weapons were not capable of producing mass slaughter is not really true," countered Hood. "Far more people were killed in World War One than in World War Two. In ancient wars there was unbelievable ferocity against the losers, and there was the added mortality due to disease and the lack of medicine." He then pointed to a well-documented example from the Byzantine reconquest of Crete in the tenth century A.D. The Byzantines had captured the Arab capital, near the ruins of ancient Knossos. A contemporary poet, echoing Moses' concerns in

Numbers 31, praised the Byzantine general Nicephorus Phocas for the slaughter of the whole Muslim population, including the women, so that they should not tempt and contaminate his Christian soldiers.

The laws received by Moses at Sinai are astonishingly violent to those who have never before made a thorough reading of the Bible (which is so often referred to as "the Book of Love"). Most people find it difficult to believe that the Bible's Covenant between God and Moses contains instructions for setting cities aflame, looting their valuables, carrying off their young girls to servitude, and killing everyone else. These instructions are difficult to reconcile with commandments against coveting thy neighbors' property and killing. Yet if we consider John MacQuitty's suggestion that man's concept of God has been a process of continual evolution, we see hints that, in contrast with Egypt's pharaohs, who celebrated the cruelties of warfare on temple walls (at Karnak, for example, Tuthmosis III proudly displayed stone reliefs of foreign king's severed heads, and for this the gods praised him), the Hebrew patriarchs, early in their history, were beginning to view killing, even in victorious warfare, as unclean. Numbers 31:19 is the first ancient religious text we have that suggests, to however small a degree, that killing somehow stains a man, so that after destroying a city, one must "abide outside the camp seven days: whoever has killed any person, and whoever has touched any slain, and purify both yourselves and your captives [the virgin girls] on the third day, and on the seventh day."

This early tradition of washing or purifying the blood from one's hands resurfaces in the story of King David (who killed so many in his conquests of Philistine and Canaanite cities that he could never again be pure in God's eyes, no matter how much he washed, and was therefore obliged to cede the building of the Temple in Jerusalem to his son, King Solomon) and in the New Testament story of Pilate (who, upon sending Jesus away to be crucified, washed his hands).

In addition to receiving laws regarding killing and the handling of one's slaves, Moses, according to Exodus 25, was given directions for the manufacture of a tabernacle (or portable sanctuary modeled after Egyptian houses of worship), golden dishes, spoons, candlesticks, and other structures deemed necessary for the worship of God. The most sacred of these was a holy "Ark" (which, translated, means "trunk" or "coffer") in which the Hebrew army would carry before it the original stone tablets proclaiming the Ten Commandments (according to legend, Moses either cast the commandments in stone at God's command or saw them engraved directly by God's own hand), the books in which he wrote all the words of the Lord, one omer (approximately one quart) of the mysterious, life-giving manna that is said to have fed the Hebrew tribes, and Aaron's rod (which, though hewn from dead wood, is said, in Numbers 17:8, to have somehow bloomed flowers).

Over a squat, golden "throne" atop the Ark, whose seat was symbolized by the outstretched, feathered arms of two *keruvim*, the very presence of God himself was supposed to hover:

> And they shall make an ark of [acacia, a resilient, oily wood]: two cubits and a half [four feet] shall be its length, and a cubit and a half [about two feet] its breadth, and a cubit and a half its height. And thou shalt overlay it with pure gold, inside and outside. . . . And thou shalt make poles of [acacia] wood, and overlay them with gold. And thou shalt put the poles into the [golden] rings on the sides of the ark, that the ark may be carried therewith. . . . And thou shalt make two *keruvim* of gold . . . at the two ends of the [ark's lid]. . . . And the *keruvim* shall stretch out their wings on high, overspreading the [lid] with their wings, and their faces shall look one to another. . . . And there I will meet with thee, and I will speak with thee from above the two *keruvim* which are upon the ark.

The *keruvim*, which translates down through Greek as "cherubim," figured in later legends as being among the higher orders of angels, including the childlike cupids (or cherubs). The Bible fails to tell us exactly what the *keruvim* meant to people about 550 B.C. The men who compiled the Old Testament texts knew that contemporary readers were familiar with and would understand, without any explanation beyond the positions of their wings and faces, exactly what *keruvim* were. What they could not know was that the meaning would be lost during the next two and a half millennia, leaving us puzzled. The *keruvim*, misty and vague, reappear about six hundred years after the Theran upheaval and the apparent origin of the Mosaic laws, when (about 1000 B.C.) King Solomon built his Temple in Jerusalem. According to I Kings 6:23–29:

> Within the sanctuary he made two *keruvim* of olive wood, each ten cubits [sixteen feet] high. . . . And from the tip of one wing to the tip of the other was ten cubits . . . both the *keruvim* were of one measure and one form. . . . And he set the *keruvim* within the inner house: and the wings of the *keruvim* were spread out, so that . . . the wings which they stretched toward the midst of the house touched one another. And he overlaid the *keruvim* with gold.

If only we did possess that ultimate archaeological tool, and could travel back in time to the Temple Mount, and explore sacred chambers now gone so utterly that one cannot guess within a hundred feet where they once stood. From the description given in Solomon's Book of Kings, his *keruvim* were a newer version of God's throne, as originally rendered atop the Ark of the Covenant; but this time projected large. As the Hebrew tribes were no longer nomadic herdsmen and conquerors of cities but had themselves become settled city dwellers, the throne no longer needed to be portable, and the *keruvim* could now grow to occupy an entire room. Some scholars have suggested that the *keruvim* were shaped like hideous demons, in keeping with their role as guardians of places particularly holy and unapproachable, as suggested, perhaps, by God's placement (in Genesis 3:24) of *keruvim* about the Garden of Eden, and with them the bright blade of a revolving sword to prevent any return by man to the tree of immortality. The mention of wings suggested to Asimov a composite creature, familiar in the forms of Minoan and Egyptian sphinxes, which generally had the head of a woman, the wings of an eagle, and the body of a lion. But the stance given in I Kings 6:23–24 suggests to me an even more familiar form: A *keruvim* stands ten cubits high, and its outstretched wings (arms), from wingtip to wingtip, also measure ten cubits. I stand seventy inches high, and my arms, from fingertip to fingertip, also measure seventy inches. The *keruvim*'s dimensions are distinctly human (lions, for example—and also my two cats—are a third longer than the spread of their forelegs), suggesting, but by no means closing the argument, that the figures in Solomon's Temple, and atop Moses' Ark, were understood by biblical scribes to be represented as men or women with wings instead of arms (and the wings had the same dimensions as human arms).

The role of *keruvim* as protectors of Eden suggests that great and fearsome powers were associated with their presence; this probably explains why carvings of them were represented on the Ark's lid.* When God showed his favor and manifested his presence above the *keruvim*'s wings, any army carrying the Ark before it was invincible.† The Ark had ancestors

---

*In Ezekiel 27–28 God sends *keruvim* down to explode a mountain on an Eden-like island that in all essentials echoes the volcanic destruction of Thera and its neighbors.

†During World War II Adolf Hitler actually sent his men searching the world for religious artifacts, including the Ark of the Covenant and the Ark of the Cross. No one ever learned why. Among the treasures he managed to acquire was the holy spearpoint, said to be (but not known with any certainty to have actually been) the very spear with which Jesus was stabbed on the cross. It also became known to the Allies, late in the war, that Hitler was consulting astrologers and planning battle strategies accordingly. The British responded to this news by calling in the best astrologers in all England to determine what the best astrologers in Germany might be telling Hitler. They predicted correctly that the German

and parallels among other nomadic tribespeople, who also carried portable sanctuaries before their armies. The tradition of carrying sacred artifacts in a gilded chamber continued through the Christian Era of the first millennium A.D. The Ark of the Cross was smaller than the Ark of the Covenant: only about two feet long and less than a foot wide. In it were contained the books of the apostles and what was said to be a piece of the true cross. Carried into warfare, it was ultimately captured and ransomed, apparently several times. Oral tradition suggests that on at least one occasion it was either seriously damaged in battle or returned from ransom in many pieces. To reduce the probability of total loss, three separate fragments of the holy log were eventually placed in three identical Arks. When a battle was won, the Christian soldiers were said, after the fact, to have been virtuous, and the presence of the Lord had therefore hovered over the Ark for that reason. When a battle was lost, it was presumed that the Lord had, for one offense or another, forsaken the defeated army. After one such defeat a rival chieftain stripped an Ark of its gold, burned whatever remained, and returned the ashes to the Christians. No one knows what became of the other two Arks. By A.D. 1200 they had simply disappeared from history.

According to Exodus, Numbers, Judges, and Samuel, the Ark of the Covenant had a similar record of capture and recapture, with similar reasons given for victory and defeat. Moses' people did not always march with God's favor. When he came down from the sacred mountain, Moses found that his (still-Egyptianized) followers had returned to pagan bull-worship, and in holy wrath he smashed the tablets upon which the Ten Commandments were written. Then, by earnest pleas and direct bargaining with God, he obtained forgiveness for his people and was even given a new copy of the commandments. Judging from the Exodus 32–34 account alone, we can read, between the lines, hints of the fragmentation, duplication, and multiplication events that are bound to accompany any sacred object carried into war. Even by a strictly biblical accounting, we are told that there existed at least two copies of the Mosaic law and that at least one of these somehow came to be divided into several pieces (to those of us who tend to view the Bible as a compilation of orally transmitted histories, it is easy to imagine the tablets simply being dropped during a migration or a battle, with the story of Moses himself smashing them being added centuries later). The

chancellor would be inclined to dismiss any attack on Normandy as a decoy operation and would therefore leave that portion of the French coastline less heavily defended than any other. Almost to the moment that the Allies recaptured the holy spearpoint, Hitler committed suicide in his bunker.

obvious next question is: If we posit multiple copies of the Mosaic tablets, then why not also multiple Arks?

This, of course, is preceded by a more fundamental question concerning the reality or fiction of the tablets and the Ark. Accounts, outside the Bible, of tribes carrying sacred chests before them, and gilded thrones over which their gods were to appear, tell us that such customs did indeed exist among nomads and suggest that something very much like the Ark described in Exodus must actually have been built by the Hebrews.* From medieval times we have examples of Torah Shrines that, according to texts dating from about A.D. 1100, were built to house handwritten Torah scrolls and were regarded as symbolic descendants of both Solomon's Temple in Jerusalem and the more portable Ark of the Covenant. It has been known for some time that the modern tradition of housing sacred chests in synagogues and churches (in the Roman Catholic church, a chest similar to that found in synagogues houses sacred cups and wafers of holy communion, much as the Ark of the Covenant was said to have held wafers of holy manna) could be traced back to medieval times, and it was long suspected that since Christianity and Judaism shared a common cultural ancestry, the custom of building sacred chests did not simply arise coincidentally and fully developed among two separate cultures in the Middle Ages but must have gone back to high antiquity, probably predating the split, by Jewish reformers, of one religion into two. Support for this theory emerged A.D. 1981, when Duke University archaeologists Eric and Carol Meyers, working at Nabratein in the upper Galilee, unearthed a half-ton stone from a lost synagogue Ark dating to approximately A.D. 200.

"It is a kind of missing link," says Eric Meyers, a link poised some fifteen hundred years after the first Bronze Age descriptions of nomadic armies carrying Ark-like figureheads and more than a thousand years before the oldest-known medieval Torah Shrine. Working from a rough estimate that only one in about five hundred examples of a given artifact is likely to be preserved for two thousand years, the Meyers' Torah Shrine should not have been the only one existing at that time. There must have been hundreds of them. And if there were hundreds of them, then by implication, if we imagine ourselves following the chain of cause and effect backward

*Nearly sixteen hundred years after Thera, Roman armies would preserve a version of the ancient tradition of carrying powerful symbols before them into battle. Augustus Caesar's legions went into Germany with golden eagles mounted atop long staffs. Later armies carried flags before them. In a very real sense, the classic photograph of American troops raising the flag atop Iwo Jima is part of a tradition going back to the time of Exodus.

through history, from the first tangible remnant of the Ark tradition, we behold, in reverse order, an ever-expanding pattern of duplication that must inevitably funnel down to one or two forebears. But even these remotest forebears could not have sprung out of nowhere, fully developed and without ancestors. Probing back to the time of Tuthmosis III, we find a long, preexisting tradition of Ark-like structures—"the holy of holies"—in Egyptian temples (the empty granite chambers where the chests were housed still stand today, at Karnak, at Abydos, and in the Temple of Hathor at Dendera). As for the stone slabs on which God's laws were engraved, there are examples of this, too, long in advance of the Old Testament. The Minoans are said (by Plato in *Critias* 119c) to have inscribed their laws on a sacred stone pillar. In similar fashion, the most important artifact of Hammurabi's reign is a black cuneiform tablet standing just over seven feet tall. On it are inscribed more than two hundred laws, some of which are so remarkably similar to the Mosaic codes ("If a man shall put out the eye of another, then let his own eye be put out. If a man shall knock out the teeth of another . . . then let his own teeth be knocked out") that it is difficult to believe they were not borrowed and simplified by the authors of Exodus 21 ("And if any mischief follow, then thou shalt give . . . eye for eye, tooth for tooth"). The Exodus account tells us that Moses was called to the mountaintop, where he met the Lord face-to-face and received from him laws inscribed in stone. The uppermost panel of the Hammurabi tablet tells the same story some three hundred years before the time of Thera and Exodus. The cuneiform text is accompanied by an illustration showing Hammurabi with a long beard, sitting opposite the Sun-god, who is placing the tablets of the law in the king's outstretched hands. The Ark of the Covenant and its contents may be lost, but in the Paris Louvre, just a few paces from the *Mona Lisa*, we can see on display what is probably the direct ancestor of the Mosaic tablets.

The Hammurabi tablet, though carved by settled city dwellers, without portability in mind, was nevertheless captured by a rival king and transported to the city of Susa, where archaeologists found the multi-ton slab of diorite (a type of volcanic rock) broken into three pieces. According to Exodus 32, the original Mosaic tablets also ended up in pieces on the ground. Moses' stay on the mountaintop continued for so long that the Israelites in the camp below began to fear that he had met with misfortune and might never return. They reverted to their older, Egyptianized habits of animal-worship, and Aaron asked the people to bring him gold: "And he . . . fashioned it with a graving tool and made it a molten calf: and they said, These are thy gods, O [Israel]."

The idea of worshiping an animal image is not so bizarre as it may sound on first hearing to those of us looking on from the twentieth century A.D. Even in the Bible some animals are given human qualities (as in the cases of the serpent, and the donkey who speaks to Balaam in Numbers 22:27–34), while others, including lambs and doves, become sacred. In Egypt, animal representations of the gods were believed to embody some part of the spirit of the patron they represented on Earth, a belief that made falcons, cats, crocodiles, baboons, and bulls subjects for mummification. Cats were regarded as associates of Pasht, the goddess of war (from whom the name Puss is derived). Every falcon was believed to contain a tiny particle of Horus' spirit; every baboon, a particle of Thoth. At Saqqara in Lower Egypt, a subterranean tomb, only now being explored, has so far yielded the mummified bodies of five hundred baboons (each with a stone inscription noting its age, together with prayers for its spirit) and five million birds (among them, in separate clay coffins, were placed the embalmed bodies of shrews, presumably so the spirit falcons would have spirit prey in the afterlife). According to Egyptian tradition, those who donated mummified representatives of Horus, Thoth, or Pasht to the temple gained special grace in the gods' eyes; but deliberately killing these sacred animals was regarded as a crime so serious as to be punishable by death, yet natural deaths could not possibly have accounted for the demand expressed by millions of mummified birds. In archaeology, as in politics, when we encounter something totally strange, we can count on one thing: Somehow, somewhere, money had to be involved. The priests must have found a theological loophole permitting special dispensations against killing, and more than a few of us have begun to suspect that in what is now a dried lake bed near the tomb, the priests of Saqqara might have created a lucrative industry raising the birds, which they later killed, embalmed, and sold ready-wrapped to the devout. It would appear to be a practice that became widespread. A tomb at Beni Hasan, near Karnak, was found to contain nearly a half million mummified cats (in A.D. 1880, twenty-eight tons of them were dumped into the hold of a ship, sent to Liverpool, ground up, and sold as fertilizer).*

The holiest of all Egyptian beasts was the royal Apis bull, raised in special

*Recent lakeside excavations near the Saqqara bird tomb have yielded what appear to be the remains of mud-walled rookeries, possibly supporting the theory that birds were raised for eventual burial. There are also traces of a pottery factory, where little clay coffins were produced. The picture begins to resemble an ancient Egyptian, Ford Motors–style assembly line for the mass production of animal mummies, through which anyone with enough money could buy grace with the gods. If true, this was not a unique development. In A.D. 1490, when printing presses had begun to proliferate throughout Europe, the Vatican came up with the idea of mass-producing certificates, called indulgences, with which the wealthy

quarters at the Memphis Temple of Ptah, and anointed daily with perfumes. On the anniversary of Pharaoh's ascent to the throne, the bull accompanied the monarch in a procession through the countryside, in theory renewing the fertility of the fields. And when he died, his body was mummified with all the care accorded the king himself, so that his spirit would be resurrected in the next world. At Saqqara, in a tomb called the Serapeum, more than a hundred bull-gods were buried. When the underground chambers were discovered about A.D. 1860, all but one of the huge granite coffins had been broken into and plundered, through entrances so small that they could only have been planned for a child to slip through, and pass out anything of value. Only one burial had remained hidden from looters. In it, beside a jeweled and mummified bull, excavators found a statue standing four feet high at the shoulder. It was a calf carved from solid gold.

While orally transmitted histories and traditions tend to involve a great deal of telescoping, the one thing I have found them to preserve best is religious ritual. An ancient Maori chief or shogun warrior might have become a composite character writ larger than life, but archaeology teaches us that the same Maori ceremonies, using the same jade figurines being used today, were enacted two thousand years ago. The same may be said of the Japanese tea ceremonies that sometimes became a backdrop for shogun sagas, and it can probably be said as well about the story of the molten calf: At some point in the ancient past, bull-worship must have been common among the Hebrew tribes. That large a golden calf has been found in an Egyptian tomb provides a clue to where the biblical reference to bull-worship might have originated, although the practice might also have been assimilated centuries after the Thera upheaval, through Hebrew contacts with Cretan Minoans displaced to Canaan. In the Minoan world, as in Egypt, bulls were worshiped above all other symbols of power, and vestiges of Bronze Age bull rituals persist, not only in the pages of the Bible, but in Italian jewelry's lucky charm—a golden bull's horn—and in Spanish bullfights (at the core, bullfights are religious rituals probably dating back to Minoan Crete, where we find the first-known bull arenas).

The story of the golden calf was probably handed down by the Joseph

---

could buy forgiveness for their sins. They were an instant best-seller, through which the pope was able to build St. Peter's Basilica and pay Michelangelo's bill. In October A.D. 1517, an East German professor of theology used the new printing technology to mass-produce his own papers: protests against the pope for, among other things, suggesting that people could buy their way into heaven. His name was Martin Luther, and his reformist movement against Rome was to become the origin of *Protest*antism.

tribes of the upper Jordan Valley and the Sea of Galilee. When, between 1000 and 900 B.C., Jerusalem became the center of worship under David and Solomon (both strict believers in their one and only true God), it almost certainly became politically, militarily, and economically unwise for the northern kings to allow paganism to continue. Nevertheless, the old roots ran deep, and there are hints that to one degree or another, they endured. One of the northern kings was Yarov'am (pronounced Jeroboam), who ruled over the Efrayimites, one of the Joseph tribes. In the first Book of Kings (12:28–32), in what by a biblical accounting was several centuries after the Exodus, Yarov'am resurrected the ancient animal symbol of his tribe:

> Whereupon the king took counsel, and made two calves of gold, and said unto them, It is too much for you to go up to [Jerusalem]; behold thy gods, O [Israel], who brought thee up out of the land of [Egypt]. And he set the one in Bet-el, and the other put he in Dan. And this thing became a sin: for the people went as far as Dan, to worship before that one. . . . And Yarov'am ordained a feast in the eighth month, on the fifteenth day of the month . . . and he offered upon the altar. So did he in Bet-el, sacrificing to the calves that he had made.

This continuance of pagan rituals among the Hebrew tribes is a recurrent theme, ranging from the first books of the Old Testament through the Dead Sea Scrolls (which were written during a period spanning 300 B.C. to about A.D. 70). The Joseph tribe in particular seems to have some special and long-standing association with Minoan and/or Egyptian bull cults. Thus, in Deuteronomy 33:16–17, when Moses is near the end of his life and gives his special blessing to each of the tribes of Israel, his legacy to the Efrayimites is: "Let the blessing come upon the head of [Joseph]. . . . The firstling of his [bullock] herd, grandeur is his, and his horns . . . with them he shall push the peoples altogether to the ends of the earth: and they are the ten thousands of Efrayim. . . ."

The persistence of paganism can be seen even today. In the walled city of Jerusalem one will occasionally meet Orthodox Jews who practice crystal healing and embrace California New Age cultism.* Such practices have also

---

*I urge every visitor to Jerusalem to spend a few days within the Old City. It is a place of contradiction, cultural tides, and conflicting undercurrents like none other on Earth. At Christchurch compound, near the Jaffa Gate, I attended a Christian Mass given in Hebrew, at which fundamentalist Christians sat beside Orthodox Jews and two Palestinians sat between the Jews and the Christians while, outside, Israeli police clashed with Palestinians

been attested to archaeologically. In A.D. 1990 workers building a water park in Jerusalem's Peace Forest broke through the roof of a burial cave whose limestone caskets were inscribed with the name Caiaphas (a name known to us from both the New Testament and the first century A.D. Jewish historian Flavius Josephus). One bore the bones of a sixty-year-old male, believed to be the Caiaphas known from Matthew 26:57–68 as the Jewish high priest who interrogated Jesus before handing him over to Pontius Pilate for trial as a heretic. In one of the caskets, a coin minted by Herod Agrippa I was found inside the skull of an adult woman.* There is little doubt that the coin had been placed in the woman's mouth (as was the custom among Roman and Greek polytheists) to pay the ferryman Charon for safe passage across the river Styx, meaning that as late as the first century A.D., a pagan custom was being practiced at a Jewish burial in Jerusalem.

According to the Book of Exodus, it was attempts to adopt and preserve such rituals that got Moses' followers into trouble at the sacred mountain. God told Moses (in Exodus 32) to go down from the mountain, for the Israelites were celebrating the golden image of a young bull—"they have turned aside quickly out of the way which I commanded them"—and God called them a stiff-necked people who must be destroyed utterly. But Moses changed God's mind, reminding him of the great deed he had demonstrated to the world in bringing his people out of Egypt through darkness and parting waters, asking if he would forever after have Egypt mock him for bringing his chosen people out, only to slay them in the mountains.

Then, when Moses descended from the mountain, the tablets of the law were smashed, and more than three thousand were put to the sword by the

---

over a recent stabbing and two hundred random arrests. One of the Christians turned out (by sheer coincidence) to be a New Zealand radio personality who had been instrumental in my 1982 exile for writing about evolution and debating Creationists on the air (and who, between episodes of speaking in tongues, asked for and received my forgiveness). The Jews (calling themselves Messianics) feared that if the beliefs they were embracing became known, their shops would be boycotted. The two Palestinians had blood oaths against them for their conversion from Islam but feared not. One explained that the angels would protect him (he had already survived a hand grenade attack). At every turn of the corner in Jerusalem, something fascinating is always afoot, and I know I shall return to the city, again and again.

*Herod Agrippa I grew up in the same house with Caligula (who believed himself to be the Messiah the Hebrews had been awaiting). Before his death in A.D. 44, he sent Jesus to Caiaphas and then, at a later date, proclaimed himself Messiah. He was in the process of organizing a revolution against his old friend, the Emperor Claudius, when he died mysteriously. Legend has it that when he dressed himself in a silver robe, and proclaimed himself king of the Jews before a whole stadium of onlookers, an owl landed before him, hooted three times, and he fell to the ground rotting with sores.

tribe of Levi, in what reads to some scholars as the memory of a civil war in which the Levites, siding with Moses, slaughtered the leaders of a rebel bull cult, and reaffirmed Moses' authority.

Still, scattered temple remains (and biblical references to Efrayimite bull cults) tell us that long after the time of David and Solomon, bull-worship and other pagan rites continued in the northern kingdom at least through the Babylonian invasions and exilic period (about 587 B.C.). However, while exerting at least some lasting influence (as seen in the Caiaphas tomb), the idea of multiple gods never again obtained a strong foothold in Jerusalem and the southern provinces, and it was from the south that the history of later Judaism and the reformist movement that became Christianity descended.

~~~~~~

About 1510 B.C., some 115 years after Thera, the monotheist pharaoh Akhenaton triggered a civil war in Egypt that ended with his assassination and the ascendancy of the child-king Tutankhamen ("King Tut"). The chaotic time had begun during the reign of Amenhotep III (approximately 1575–1550 B.C.), who had taken the heretic pharaoh (also known as Amenhotep IV) as coregent. It was a time during which Egypt stumbled and (temporarily, at least) weakened from within; a time during which Asiatic tribes (among them some ancestral Hebrew clans), hunkering down along the southern coasts of Sinai, could begin to move more freely. If the books of Joshua, Judges, and Samuel are based upon historic reality, then this is also the approximate time frame to which a Thera-based calendar assigns them.

During the confusion of Karnak's internal upheaval, Canaanite city-states managed what appears to have been a successful but ultimately short-lived breakaway from Egypt. Meanwhile, increasing numbers of nomadic tribes (some, possibly, swarming up from Sinai) began assailing the gates of Canaanite cities. Somewhere between 1550 and 1450 B.C., as legend and the Thera calendar would have it, the Israelite warrior Joshua swept in from the desert east of the Dead Sea, and according to Joshua 16:21, his army massacred every Canaanite in Jericho: "They utterly destroyed all that was in the city, both men and women, young and old, as well as oxen, sheep, and [donkeys], with the edge of the sword."

The Book of Joshua begins where the first five books of the Bible ("the law") leave off. Moses is 120 years old, and near death, when he appoints Joshua (who up to that time served as the patriarch's military adviser) commander in chief. Joshua, like everyone else who traveled from Sinai to the Promised Land, was born after the Exodus. According to the Bible, as punishment for worshiping the golden calf, the passage to Canaan could not occur until the entire Exodus generation had died of old age.

The first of the twenty-one Books of the Prophets (Joshua) attempts to describe a simplified version of the conquest of Canaan, with a clear beginning, middle, and climax, thus bringing to final and triumphant realization the inheritance promised by God in the five Books of the Law. The archaeological picture suggests chaotic and almost random destructions of Canaanite cities, coming from many different sources over several centuries. And when we look more closely at the Bible itself, we find corroboration of the archaeological record: The destructions, if they did occur mostly or even occasionally at the hands of Hebrew tribes, were not quite as simple as the Book of Joshua suggests. The Book of Judges, for example, appears to have been compiled from a diverse collection of ancient documents, not necessarily very closely related to one another. Its version of the conquests is nowhere near as idealized as the Joshua account, and is sometimes contradictory and often unflattering of the Hebrew tribes. Here we do not find a well-organized fighting force sweeping in under Joshua's command to a quick and clear-cut victory. Rather, we begin to see hints of a loose coalition of guerrilla factions fighting mostly on their own, sometimes against one another, and, in at least one instance (King David's brief alliance with the Philistines), joining the other side. The scenario duplicates more recent recollections of tribal warfare against city dwellers, as in the cases of the Arab tribes (at one point unified under Thomas Edward Lawrence), New Zealand's Maori and America's Plains Indians against European settlers. As the span between a king's victory and the year in which the event is finally transferred from oral tradition to written documents steadily diminishes, we begin to get glimmers of increasingly reliable history. The Judges account would thus appear to be younger than the Joshua account, and while we may have no record of Joshua outside the Bible, we do know where Jericho was and what happened to it.

Fortified (like most other cities of its time) by walls and guard towers, Jericho stood near a freshwater spring in the Jordan Valley, five miles north of the Dead Sea and seven hundred feet below sea level. The spring is still in use today, supplying a local reservoir, as it did during the time of Joshua and Judges. The life-giving water hole has been pressed gradually east as one civilization after another, each building upon the wreckage of the old, thrust up a mound fifty feet high on the valley floor. The city that stood at the time of Joshua was the fourth major settlement to rise from the edge of the spring. Each predecessor was, in its own turn, destroyed, but when we dig beneath the first city, tracking the spring's sediment layers deeper and deeper into the cellars of time, we find that the water attracted settlers the way a magnet attracts iron filings. There are traces of mud-brick towns going back as far as 8000 B.C. By comparison to Jericho's deepest, oldest levels, Karnak and the Pyramids are barely old enough to be called antiques, the Roman Colosseum is a modern structure, and Gaius Suetonius' history of the first Caesars describes current events.

About midway up the mound, the fourth city, which probably still stood intact about 1550 B.C., was surrounded by not one but at least two concentric walls. It was not a city in the modern sense of the word, like Paris or New York, and by our standards it might not even have qualified as a hamlet. It was simply an aggregation of storehouses, administrative build-ings, shops, and mud-brick houses crammed behind a defensive barrier and taking up barely as much space as one of today's ocean liners (given such small dimensions, the description, in Joshua 6:15, of soldiers and priests bearing the Ark of the Covenant and circling the city seven times in a single day does not seem at all out of place).

University of Toronto archaeologist Bryant Wood, who began his career as a nuclear physicist, has assembled a body of evidence from fields reaching far beyond the strict boundaries of archaeology—to military history, seis-mology, and theology—with which he forces us to rethink all that is "known" about the fourth city.*

After the destruction, Jericho remained abandoned for several centuries before anyone built there again. Earlier researchers had dated this destruc-tion to the Middle Bronze Age, about a century before Hatshepsut and Tuthmosis III, and on this basis it was concluded that the city must have still been abandoned in 1500 B.C., meaning that there was little at Jericho to conquer when the Bible says Joshua conquered it. Although graves near the city are known to contain scarabs (small hieroglyphic stones cut in the shape of a sacred beetle) bearing names of the (Late) Bronze Age pharaohs Hatshepsut, Tuthmosis III, and Amenhotep III, the prevailing view, based on the premise that scarabs were used as symbols of good luck and were often manufactured long after the monarchs themselves had died, holds that the graves date from the *fifth* city and the Early Iron Age, centuries after the pharaohs reigned. This is Wood's first point of contention: While it may

---

*An astonishingly large number of archaeologists have backgrounds in physics and theol-ogy, much as those of us who have worked in the area of antimatter propulsion and SETI (the Search for Extraterrestrial Intelligence) invariably work in many different scientific disciplines at once and are either only children or firstborn sons. Similarly, a great many hyperactive and reading-disabled children eventually gravitated toward paleontology—not just any field of paleontology but toward the study of dinosaurs. No one has yet explained why.

be true that with regard to scarabs of Tuthmosis III and Amenhotep III, there is no telling how long after death the men's names remained sacred, and continued to be used on scarabs, Hatshepsut was a very different matter. Unlike her successors, she was reviled from the moment of her death. Her name was systematically obliterated from monuments, and to bear an amulet of hers was punishable by death. A grave with a Hatshepsut scarab must therefore have been dug while she was still alive and revered, meaning that the scarabs were buried not centuries after Egypt's Eighteenth Dynasty and the years of abandonment that followed Jericho's fourth city but *during* the Eighteenth Dynasty. If we accept this conclusion, then the fourth, pre–Iron Age city was still alive while Hatshepsut ruled; its streets still carried a noisy traffic; its plastered and whitewashed walls were still gleaming in the Sun. That Amenhotep III's name appears in the same graveyard probably tells us that the city continued to thrive at least through his reign, approximately a hundred years after Hatshepsut, Tuthmosis III, and Thera. According to a dating system based on the Theran ash layer, Amenhotep III died about 1515 B.C. (give or take twenty years). Political chaos and an eclipse of Egyptian power were not far behind him, meaning that conditions in Canaan were just right for some of the events described in the Book of Joshua.

In one of the fourth city's rooms the floors are piled three feet deep with collapsed roof beams, severely burned furnishings, and fallen bricks. A piece of wood has yielded a radiocarbon date placing the building's destruction somewhere between 1540 and 1450 B.C., lending further support to the view that Jericho's City Four stood during and after the time of Hatshepsut but did not last more than a few decades beyond the death of Amenhotep III. This was the period we call the Late Bronze Age, the two or three centuries just prior to the development of iron technology. A three-foot layer of erosion-borne silt on top of the destroyed room (requiring centuries of rains washing down from points higher on the mound), followed by Iron Age ruins on top of the silt (City Five), testifies that Jericho lay unused and unlived in for many years after the end of City Four.

"So there was an active population at Jericho during the time period covered by the Books of Joshua and Judges," says Bryant Wood. He draws further support from a broken water jug decorated with a distinctive pattern of stripes. It was found in the wreckage of City Four, and much as plastic Coca-Cola bottles can be used to date strata in a landfill as having been laid down during or after the mid-1970s A.D., the water jug was painted in a style, known from other sites, not to occur before the first decades of the Late Bronze Age.

"What this means," Wood says, "is that there *was* a city for Joshua to conquer when the Bible says Joshua conquered it. Moreover, it would have been a prime target, for . . . the site is strategically located. From Jericho one has access to the heartland of Canaan. Any military force attempting to penetrate the central hill country from the east would, by necessity, first have to capture Jericho. And that is exactly what the Bible (in Joshua 3:16) says the Israelites did."

The only written record to survive concerning the history of Jericho in the Late Bronze Age is that found in the Hebrew Bible. When we compare the remains of City Four with the biblical narrative describing the Israelite destruction of Jericho, we enter a fascinating convergence of archaeology and Scripture that includes warehouses whose stores of grain were left unplundered and, most interesting of all, tumbled walls.

Those who would like to place the end of City Four in pre-Theran times (the Middle Bronze Age) attribute the destruction and abandonment to punitive expeditions under Hatshepsut's great-grandfather, who overthrew the Hyksos and then pursued their remnants all the way into Canaan. Opposing this view, Wood points to the storage jars in the destruction layer. They all were filled to their brims with grain when the end came, and this, he believes, mitigates against a siege on City Four by Egyptians, no matter what part of the Bronze Age one posits for the siege, "for when the city met its end there was an ample food supply. This flies in the face of what we know about Egyptian military tactics."

Egyptian campaigns were customarily and diabolically mounted just prior to harvesttime (in the tropical Jordan Valley, harvesttime has always been in the spring), when food supplies within the cities were bound to be at their lowest level. Camped in the fields outside the city walls, the invaders could harvest the ripening crops to their own advantage. As with Tuthmosis III's seven-month siege at Megiddo, the primary strategy was to starve the enemy into desperation. This was clearly not the case at Jericho. The abundant food supply indicates that City Four succumbed quickly, not after a long siege, and that the destruction came after, not prior to, the spring harvest. Indeed, the most abundant item found in the ruins, apart from pottery, is grain. Once wood and other organic substances are burned, they, like oven-fired clay, are essentially fossilized and cease to decay, and charred grain is scattered throughout the rooms of all the houses.

"This is unique in the annals of Palestinian archaeology," says Wood. "Perhaps a jar or two might be found, but to find such an extensive amount of grain is exceptional."

Along with bronze tools and objects of gold, triumphant invaders normally spirited away a city's grain stores (which then, as now, formed a major part of the commodities market). The mystery of the unplundered grain invites speculation, which Wood provides by calling our attention to Joshua 6:17–18, in which the Lord tells the Israelites that (except for gold and silver, which are consecrated and can be taken from the ruins and indeed are absent from them), "the city and all that is in it shall be devoted to the Lord for destruction," and they were commanded to "keep away from the devoted things [things devoted to destruction], lest you make yourselves accursed, when you take of the devoted things." This warning might explain why so much grain was scattered and put to the torch with the rest of City Four. That the buildings were destroyed shortly after the spring harvest is also consistent with the biblical account: In Jericho a woman named Rahav was drying freshly harvested flax on the roof of her house (according to Joshua 2:6) and the Israelites had just celebrated Passover before crossing the Jordan to attack the city (Joshua 5:10).

The Joshua account describes the march on Canaan in vivid detail, highlighting the story with unusual phenomena. The Sun stands still in the sky (Joshua 10:12–14*), and there is (in Joshua 3:17) a reenactment of the miracle of the waters that occurred during the Exodus from Egypt, in which the Jordan River parts to let the Hebrews cross over its bottom. God then commands Joshua's army to circle the city for seven days, with seven priests bearing before them the Ark of the Covenant and blowing seven trumpets of ram's horns. And on the seventh day, according to Joshua 6:1–20, the priests circled the city seven times, blowing the horns, "and it came to pass,

---

*"And the Sun stood still in the middle of the sky" may be an apt description of a total eclipse, in which the Sun is covered and essentially ceases to give off light. Totality lasts only a few minutes, but the darkness that comes before and afterward covers a significant part of the day (the approximate span of time suggested in Joshua 10:13). The moon is always new on the day of totality, meaning that it is nowhere to be seen in the night sky, and as if to more tightly link the event to a solar eclipse, the moon is described, in Joshua 10:12, as staying (hidden) in a valley. Totality is a rare and memorable event, and it is possible to believe that an eclipse recalled from antiquity was incorporated into the Joshua story (a sixth century B.C. scribe, upon reading an ancient account of totality, or hearing an oral recitation, might simply have decided, "Oh, that must have happened at the time of Joshua!"). In those days Hebrew scribes might actually have thought an eclipse resulted from the Sun stopping in the sky. The mechanics of the stoppage and darkening were easy to understand if the Sun was viewed as a reasonably small object embedded in a firmament only a few miles above the ground. If the Sun had actually stopped in the sky, this would have happened because the Earth itself had somehow stopped rotating. Under this "slamming of the brakes," as it were, inertia would have taken over, piling billions of tons of the Earth's atmosphere into mighty shockfronts, and producing windspeeds twice the velocity of sound in the Jordan Valley. Joshua's army would have had time enough for a very short scream.

when the people heard the sound of the horn, that the people shouted with a great shout, and the wall fell down flat, so that the [army] went up into the city, every man straight before him, and they took the city."

Isaac Asimov has said that "if the biblical account is taken literally, then this is a miracle, but those who seek natural explanations often suggest that it was an earthquake that did the trick. If so, it was a most fortunately timed earthquake," leading Asimov to suppose that the circling of the city had a carefully designed, two-pronged tactical purpose. The first (with which there is no argument) would have been the demoralization of Jericho's inhabitants, as they were forced to watch the Hebrews spinning a religious web about the city, bearing before them the Ark of a powerful God who might be expected to do almost anything—especially if the Hebrews had been spreading stories of parting seas and God's deliverance from Egypt during the decades since the memorable events surrounding Thera. Asimov's second and more speculative prong involves the noise created by the trumpets. While the city's defenders watched nervously and listened to the loud blasts, they might never have heard the more mundane activity of sappers sent by Joshua to tunnel up to the wall and undermine it.

The latest archaeological evidence suggests that sappers, though commonly used by invading forces, might never have been needed at Jericho's City Four. More than just the defensive wall fell on that last day. The buildings behind the wall were destroyed by a massive conflagration, but the collapse of the rooms (in which large quantities of unburned wood were buried under tons of fallen mud brick, and charred sticks were deposited on top of the fallen bricks) seems to have taken place before they were affected by fire, suggesting that an earthquake preceded the fire.

"This [observation] may be compared with the biblical account," says Wood. "According to the Bible [in Joshua 6:24], after the Israelites gained access to the city, they burned the city with fire, and all that was in it. In short, after the collapse of the walls—perhaps by earthquake—the city was put to the torch."

The parapet wall, which surrounded the city, also suffered damage. Archaeologists have cut three trial trenches, much as one cuts slices through a layer cake, all the way down to the base of the city's perimeter. In one of these trenches, on Jericho's west side, the guard towers and other upper structures are seen to have tumbled in an unusual way: out from the city and down flat. A huge volume of mud-brick came to rest in a heap, forming an inclined plane reaching to the very lip of a ruined rampart. The wall was not merely breached but had actually collapsed in a manner that invited any army arriving on the scene up a ready-made ramp and into the city. Hauntingly, the Joshua account claims that the wall fell down flat and that the army marched single file, up a flattened wall, into the city.

The Bible also tells us that the Lord said to Joshua [in 3:7–16, just prior to the march on Jericho]:

> This day will I begin to magnify thee in the sight of [Israel], that they may know that, as I was with Moshe [Moses], so I will be with thee. And thou shalt command the priests that bear the Ark of the Covenant saying, when you come to the brink of the water of the [river Jordan], you shall stand still in the [river]. . . . And it came to pass, when the people removed from their tents to pass over the Jordan. And the feet of the priests that bore the Ark were dipped in the brink of the water (for the Jordan overflows all its banks throughout the time of the harvest), that the waters that came down from [upriver] stood and rose up in a heap very far from the city Adam . . . and those [waters] that came down toward the . . . salt sea [the Dead Sea] failed, and were cut off, and the [Israelites] passed over opposite [crossed the river from the east bank, opposite] Jericho. And the priests that bore the Ark of the Covenant of the Lord stood firm on dry ground in the midst of the [river], and all Israel passed over on dry ground.

The march on Jericho is recorded in very explicit language, and Bryant Wood has correctly pointed out that historians and biblical scholars have focused on (and often dismissed as total mythology) the "miraculous"

Keruvim, *represented as angelic human figures through the time of Solomon, spread their wings to form "the seat of God" atop the Ark of the Covenant.*

The origin of the angel-like keruvim *mentioned in stories about Eden, Solomon's Temple, and the Ark of the Covenant can probably be traced back to Egyptianized Hebrew scholars of the seventeenth century* B.C. *In the ruins of Abydos, near Karnak,* kerivum-*like figures predating the Exodus legend adorn temple capstones. These Egyptian symbols of power and holiness took on the shape of lions in Torah Shrines of the first century* A.D., *but survived in winged human form as a Roman tradition (including the* keruvim-*derived word* cherub, *with its modern-day Saint Valentine associations) and were adopted as angels by the Roman Catholic Church.*

nature of the event, with little regard for the seismology of the continental seam that has created the Jordan Valley.

The Joshua account describes what appears to be a blockage of the Jordan River's flow, extending from somewhere far upstream and to the north of the city Adam (known since the first century A.D. as Damiya, which is some eighteen miles upstream of Jericho and from which the word *dam* is derived) all the way down to the Dead Sea. The Jordan River flows down the center of a crack in the earth, a rift valley where frequent earthquakes mark the inch-by-inch parting of two continental plates. Throughout recorded history, tremors have been known to bring down the valley walls, throwing natural dams across the river, typically near the village of Damiya. Citing the work of Stamford University geophysicist Amos Nur, Wood notes that in A.D. 1927 Damiya was the site of an earthquake that brought a 150-foot-high cliff tumbling into the river, damming the waters for twenty-one hours. Other quake-triggered blockages of the Jordan were recorded in A.D. 1906, 1834, 1546, 1267, and 1160. During the 1267 earthquake, according to the Arab historian Nowairi, a hill on the west bank, near Damiya, suddenly quivered and leaped like a living thing. It stumbled into the river, cutting it off. For sixteen hours no water flowed south from Damiya to the Dead Sea.

The combination of destruction at Jericho and natural dams flung across the Jordan is so much a geologic reality of the rift valley that when we read of such events in the time of Joshua and Judges, there can remain but the faintest doubts as to their historical reality. If the geologic account is taken literally, then we need not resort to sappers whose tunneling activity was hidden by loud trumpets, or to a fortunately timed earthquake that happened to occur as the Ark was displayed and the trumpets sounded. Rather, a powerful quake and the stoppage of the Jordan must have signaled to the nomads that cities across the river had come to harm and were now more vulnerable than they were ever likely to be for a very long time (add to this the advantage of being able to move a large fighting force quickly across the river's bottom to preselected points on the other side without having to worry about ferrying small numbers of men in sluggish boats subject to unpredictable drift and easy attack by archers on the shore). When scouts returned with news of fallen walls and weakened defenses, taking advantage of a rare, almost miraculous opportunity was simply the next logical move. The earthquake would thus have precipitated the attack on Jericho, and postquake events, including the parading of the sacred Ark around the city, might have been pulled forward in time by later retellings, and given prophetic significance.*

As his army crossed the Jordan, Joshua (in 4:3–20) ordered the men to pick up twelve heavy stones from the dry riverbed, so that in generations

---

*The time of the attack on City Four, as attested to by archaeology and alleged by Scripture, suggests that the invaders took advantage of surprise opportunities arising from a natural disaster. Invading immediately after the harvest, while the warehouses behind the city walls were fully stocked for a long siege and the fields outside the walls could provide no food for the invaders, was at best illogical and at worst utterly senseless under normal circumstances. As for Joshua himself, he appears to have been enlarged by legend. While discussing the City Four period with the Israeli archaeologist Ami Mazar, I once compared some of the Joshua legends with the story of America's first president, George Washington, miraculously throwing a silver dollar across the Delaware River (a feat made all the more miraculous because America had not yet begun to mint silver dollars). "The miracle is true," I explained to Mazar, "only to the extent that the dollar used to go a lot farther." Mazar agreed. The Bible describes Joshua's march through ancient Canaan as a wave of monumental destruction moving from city to city. Mazar suspects that in reality only one or two cities (including Jericho) were destroyed: "Archaeology does not prove such warfare over the whole country during a very short time. Joshua was probably a local hero, of one or two tribes, and the Hebrew tradition just blew this figure up. As with the legend of George Washington, he became larger than life."

Between 1540 and 1450 B.C., a powerful earthquake rumbled forth from the great rift of the Jordan Valley, collapsing Jericho's walls into ramps of rubble upon which invaders could easily have ascended directly into the city, as described in the Book of Joshua. A trench cutting through the ruins reveals that during the 3,500 years following the earthquake, the city lay abandoned and eroding for several centuries, until Iron Age people built upon layers of silt and gravel. Behind Jericho's earth and stone ramparts lie the residua of nearly ten thousand years of civilization.

to come, when children asked their fathers about the meaning of the stones, they would answer that when the waters of the Jordan were cut off before the Ark of the Covenant, the stones were collected as a lasting memorial to the children of Israel.

Then, after the priests who bore the Ark came up from the bed of the river, we are told (in Joshua 4:18) that the waters of the Jordan quickly returned to their normal level. This is consistent with the recent history of the Jordan's natural dams: Within forty-eight hours (and typically within as few as sixteen hours), the waters piling up behind an earthquake-made barrier overflow the mound, tearing great holes in it as they spill forth. If City Four's attackers did indeed come across the Jordan in the aftermath of an earthquake, they had to complete their reconnaissance, plan their strategy, and move very fast.

The Bible follows Israel's twelve memorial stones from the temporarily dry bed of the Jordan to a place called Gilgal and then loses track of them forever. The word *gilgal* means "a circle of stones," and throughout Europe and the Mideast, circles of large stones have been found dating back a quarter million years or more, to *Homo erectus* times. The most famous stone circle is Stonehenge in England. A virtual miniature of it surrounds the ruins of a thirty-three-hundred-year-old temple in Israel's Sorek Valley, near a town known from antiquity as Ir-Shemesh, meaning "town of the Sun" and believed to be an ancient center of Sun-worship.

Archaeology tells us that circles of stones long antedate the time frames of Joshua, Exodus, and the Canaanites. It is quite likely that the circles came down through Canaanite tradition playing a role in Sun-worship (England's Stonehenge and at least two or three similar structures appear to have been sophisticated sundials, aligned and calibrated to mark eclipses and solstices). The Bible's reference to a miracle of the Sun occurring shortly after Joshua's collecting of the stones may preserve an assimilation, by the Hebrews, of Canaanite Sun-worship and the sanctity of stone circles. As with the persistence of bull-worship, a tendency to adopt bits and pieces from older religions is thematic throughout the Old Testament, and God is repeatedly shown venting anger at his unruly children for doing this. In contrast with the molten calf in the aftermath of Exodus, this time the priests got away with the assimilation, apparently by incorporating at least one stone circle into the priestly view of history.

After the miracle of the waters, the collecting of the stones, and the destruction of Jericho, Joshua (according to 6:26) charged that the city had been destroyed for all time and bound the Israelites to an oath, saying, "Cursed be the man before the Lord, that rises up to build this city [Jericho]."

With perhaps the exception of Mashkan-shapir (whose ground is also rumored to have been either poisoned or cursed), cities located at strategic choke points, or near major watercourses and the junctions of trade routes, are never abandoned for all time. Hiroshima and Nagasaki were rebuilt within a decade of their destruction in spite of the curse of nuclear death that rested upon them (manifested as unusual blood disorders, microcephalic children, and a whole new, ostracized class of atomic bomb survivors). Jericho, because of its strategic location, was no more likely than Hiroshima to spend eternity in a vacuum. But something unusual happened there nevertheless. As if the curses described in the Bible were actually spoken and enforced, cross-sectional slices through City Four's ill-fated walls reveal that the fallen bricks lay weathering in the sun and the wind for many decades until finally they were covered by a steady downwash of gravel from somewhere above the ramparts. Then a layer of fine silt accumulated several feet deep, uninterrupted by foundation stones for almost three hundred years until, in the reign of King Ahab, at the beginning of the Iron Age, a new Israelite Jericho (City Five) arose directly upon City Four. The debris of successive construction periods raised the mound nearly twenty feet higher during a period extending through New Testament times. In the seventh century A.D., Arab invaders once again burned the city to the ground. It was resurrected by Byzantine empire builders, then conquered and rebuilt by Christian Crusaders about A.D. 1050. This last construction period has continued virtually uninterrupted into the present-day town known by the Arabic name Eriha. Its two thousand inhabitants make their living from small farms that thrive on the oasis and from concession stands for the small but constant stream of tourists seeking the world's oldest man-made mound.

〰〰〰

After the destruction of Jericho's City Four (between 1540 and 1450 B.C.) the Israelites entered into nearly three centuries of hit-and-run guerrilla warfare. Mostly, they camped in the hills, for it was, under even the best of circumstances, nearly impossible to occupy and hold the walled Canaanite cities and the fertile fields that surrounded them—which probably gave rise to the sack-and-burn tactic evidenced at Jericho. The nomadic Israelites were still challenging (very poorly, it seems) the settled Canaanites for control of the Jordan Valley when another displaced tribe, known to the Israelites as Kaphtorim settlers (from the island of Crete) or Philistines, and to modern scholars as a probable remnant of Minoan Crete, came to dominate the coasts of Canaan, and from walled cities sent forth armies of their own. In the ruins of Canaanite towns, traces of Philistine frescoes, architecture, and pottery (still preserving many of the earlier, Minoan stylistic quirks) suggest that the Philistines must have allied themselves with the Canaanites. None of this was good news for the Israelites, for whom the odds against victory or even survival were being decreased by a discovery made around the time Jericho fell.

Somewhere in Syria, or Turkey, or along the shores of the Euphrates River, an unknown hero of the declining dynasty of Hittite kings developed a method for smelting iron ore. The ores had been deposited more than two billion years earlier in much the same fashion that flint nodules were formed near Karnak (by microorganisms concentrating minerals in the mud that enclosed decaying worms and crabs). For hundreds of millions of years, microscopic photosynthetic plants had been pumping out vast quantities of oxygen until, about 2 billion B.C., a sort of atmospheric flash point was reached. Rocks from this time tell us that life on Earth had created conditions under which free oxygen began to combine with and precipitate iron from seawater. More than 90 percent of the world's iron deposits were laid down during the years surrounding 2 billion B.C. Seen from the surface of the moon, the Earth was blue ocean, swirls of white clouds, and continents as red as the sands of Mars.* Today, iron oxides are found everywhere on the Earth's surface, whereas copper and tin (the two metals which, when alloyed, make bronze) are relatively rare. The Egyptians, throughout much of their history, were able to corner the copper market in Sinai, and the presence of Egyptian beads in graves near Stonehenge suggests that Egypt's trading partners, the Minoans, dominated southern England's tin outcrops thousands of years before the Romans got there.

As with the rise of the flint cutters, the stage for human history was once again set by microorganisms that had lived and died long before man arrived on the scene. Iron ores were so widespread that they could never be monopolized, and properly treated, iron blades were harder, and in the heat of battle held a sharp edge much longer than bronze. Iron technology did

---

*A similar precipitation of iron was occurring on the planet Mars about the same time, but for different reasons. Today the Martian soil is full of ozone, peroxide, and other superoxides, not because life evolved there (from all indications, Mars is totally lifeless) but because the planet's gravity (one third of the Earth's) is too weak to maintain a hold on its hydrogen. When solar radiation splits atmospheric water molecules into separate hydrogen and oxygen atoms, the lighter hydrogen floats away into space, while the heavier oxygen falls to the ground and binds to iron, carbon, or even other oxygen atoms. Mars' peroxide-laden soil tells us that if life ever did make a start there, it could not have lasted very long. Dust there is, in enormous abundance, but no allergens.

not come soon enough, or spread fast enough, to save the Hittites from ultimate decline; but it did slowly filter down from Syria to some of the larger Canaanite cities.

Iron requires hotter, larger, and more sophisticated furnaces than bronze, and such furnaces require a settled, city-dwelling civilization. The Israelites in the hills had to remain constantly on the move and, living as they were near the dividing line between the Late Bronze Age and the Early Iron Age, were probably required more and more frequently to fight iron with bronze. Whatever they could accomplish with sheer force of will, and a little desperation, in addition to the natural advantages of being a mobile guerrilla force whose position and motion were impossible to predict, they did. But anyone trying to fend off iron swords with bronze quickly learns the limits of conquest; one might just as well try smashing diamonds with glass. During the approximately three hundred years between the fall of Jericho and the approach of the Iron Age, the Bible tells us that the tribe of Benjamin failed to take Jerusalem (Judges 1:21), the tribe of Manasseh failed to take Bet-She'an (Judges 1:27), the tribe of Ephraim failed to take Gezer (Judges 1:29), the tribe of Zebulun failed to take Kitron (Judges 1:31), the tribe of Asher failed to take Zidon (Judges 1:32), the tribe of Naftali failed to take Bet-Shemesh (Judges 1:33), and after the tribe of Dan was prevented from even descending into the Jordan Valley, we find (in Judges 2:11–15) accounts of a terrible rout in which the Israelites were delivered into the hands of their enemies. This, we are told, occurred after Joshua died at the age of one hundred ten and a generation that had not known Joshua began worshiping pagan gods. Judges 2:13 uses provocation of God's wrath to explain the Israelite defeats, but Judges 1:19–20 preserves hints that a dawning technology played a key role: "and he [Judah] drove out the inhabitants of the mountain; but could not drive out the inhabitants of the valley, because they had chariots of iron."

The situation grew so bleak that when Samson's mother was (according to Judges 13:5) visited by an angel of God, instead of total rescue, the Lord promised only that he would "begin to deliver" Israel from the hands of the Philistines.

In historical records outside the Bible, the Philistines (a Greek derivative of the Hebrew word *phelishtim*) appear to be first mentioned by the name Pulesati at the beginning of Egypt's Twentieth Dynasty, approximately two hundred years after the fall of Jericho. About this time (1300 B.C., give or take thirty years), the pharaoh Ramses III ascended the throne and began minor incursions into Canaan. He was the last powerful native monarch of Egypt, for the horrible spasm of intercity warfare and economic collapse that had spread out of Greece and the Aegean islands about 1500 B.C. (130 years after Thera), disrupting maritime trade, collapsing the Hittites, and forcing chaotic migrations upon once-settled people, was now about to impact on the Nile. In one last shining moment, Ramses III won a decisive battle against the Sea People (among them the Philistines), and then the bright flame of Egypt was upon the wane.

Ramses III's invasion of Canaan would have worked much to the advantage of any trained guerrilla warriors who might have been hiding in the hills. If the pharaoh had any interest in them at all (which is doubtful), constantly shifting tribespeople were a far less easy target than Philistine and Canaanite cities. It seems likely, under these circumstances, that Egypt dealt severe blows to the Philistines, and as their cities lay bleeding and diminishing in vigor, the nomadic Israelites, seizing the opportunity to descend upon already wounded adversaries, became little more than a final *coup de grace.*

From this point forward the distance between events and their commitment to writing has so narrowed that the convergence of archaeology and Scripture becomes easy and produces a sense of actuality: You can actually touch the Temple walls. You can almost touch Saul, and David, and Solomon.*

The Book of Judges ends with Saul's accession to kingship over the Israelites. Probably, this occurred between 1080 and 1030 B.C. The tribes had by now begun to master the new iron technology and must have been evolving more and more toward a settled, city-building existence. From an

---

*Until the summer of 1993, many scholars regarded King David, the man who founded the city of Jerusalem, with its legendary Temple, and laid down the foundations of the Davidic Dynasty, as "a sort of fairy tale concocted during the Babylonian captivity to justify the Jerusalem Temple and the priesthood." Until eighty-four-year-old Avraham Biran, an associate of Benjamin Mazar, excavated an ancient wall near the Jordan River, no one had seen an independent reference, outside the Bible, suggesting that "the House of David" or the Davidic Dynasty ever existed (reports, and leads into the first technical literature on this subject, can be found in the February 1994 issue of the *Biblical Archaeology Review*). Over the remains of the Philistine-turned-Israelite city of Dan, about 850 B.C., the Syrian king Benhadad erected a stone monument proclaiming his victory over the towns of northern Israel. A century later, the Israelites had recaptured Dan and, apparently angered by King Benhadad's victory stela, smashed it into dozens of pieces and buried them within a new wall, intending them to be forgotten forever, but unknowingly preventing their deterioration. Thirteen lines of Aramaic on the broken tablet describe the defeat of Baasha, King of Israel, by the King of Syria and the Judean "King Asa of the House of David." The tablet not only corroborates the very same inter-city battle described in I Kings 15 and II Chronicles 16:1–2, but the House of David and thus the existence of King David himself.

Israelite perspective, the roles were slowly reversing: It was the Philistine remnants, not the Semitic tribespeople, who had become a lingering nuisance, ratlike and "nibbling in the wheat." The gathering strength of the tribes was the birth of a kingdom, but the Philistines, though growing increasingly weaker, were nevertheless the most technologically advanced, the best organized, and the most dangerous of Israel's early enemies.*

According to I Samuel 4:1, Israel went out against the Philistines at Afeq and lost four thousand soldiers. The elders, hoping to alter their fortunes, called forth the Ark of the Covenant and rejoined the battle in the belief that the physical presence of the Lord would guarantee victory. The Philistines are said to have believed it, too—to the point of total desperation, which in the end strengthened their resolve and made superior fighters out of them. In I Samuel 4:8–9 the Philistines lament: "Woe to us! . . . these are the gods that smote [Egypt] with all the plagues in the wilderness. Strengthen yourselves and act like men, Oh [Philistines], lest you fall slaves to the Hebrews, as they have been slaves to you: quit yourselves like men, and fight."

And the Philistines fought, we are told, and Israel was beaten. Thirty thousand soldiers are counted dead in I Samuel 4:10, including the old high priest Eli's two sons, and—worst of all—the Ark of the Covenant was taken. On hearing that the Ark was lost, Eli, who was, according to I Samuel 4:14, ninety-eight years old, died of shock.

For somewhere between seven months and several decades (biblical accounts conflict), the next two chapters of Samuel trace the Ark's movements through Philistine territory.

It was first brought to the temple of Dagon in the city of Ashdod, whereupon the golden idol was found one morning fallen upon its face before the Ark. The priests set the statue back in place and the next morning found it completely dismembered, whereupon they fled the temple; but this did not save them. From the abandoned temple the Lord smote Ashdod and several nearby towns with plague and swellings. Then mice descended upon the fields and destroyed the crops. The Ark of the Lord was beginning to lose its charm, so the survivors of Ashdod delivered it to Gat, which promptly suffered similar misfortunes (since plagues have always tended to spread as people move to and fro, and because in ancient times infection had

*By the end of the Book of Judges, and the beginning of the books of Samuel and Kings, the biblical scribes came to regard Canaan as essentially conquered territory, even if they continued to mention Philistine strongholds and the land was in actual fact not completely under Israelite control. Nevertheless, the scribes had, at this stage, changed the name of the country from Canaan to Israel.

a capacity for death exceeding most major battles, the story of the pall of death trailing behind the Ark may have genuine biological overtones, especially as the first Book of Samuel tells us that not even the Israelites were spared). Plague-ravaged and desperate, the city of Gat tried to pass the Ark on to 'Eqron, whose people immediately refused it.

Calling their priests and magicians to a meeting, the Philistines decided to send the Ark back to the Israelites. The priests advised, however, that it should not be sent back empty, that the "Gods of the Israelites" must be appeased with offerings of gold: five golden sculptures of the blisters caused by the plague and five golden carvings of the mice that were destroying the fields. The tribe of Levi found the Ark near the Sorek Valley's Stonehenge, being hauled aimlessly by two cows tied to a cart. But when they retrieved the golden mice and blisters, they looked inside the Ark, and for this reason (according to I Samuel 6:19), the Lord smote fifty thousand of them (sixteen thousand more than had been lost in the crushing defeats at Afeq and a number probably matching, if not exceeding, Philistine losses to the same plague).

King Samuel subsequently obtained an oath from his people to once again, and with all their hearts, abandon the foreign gods of fertility and the Sun, and especially to forsake the fire-god Baal, to whom children were sacrificed. After sacrificing a suckling lamb by fire to the Lord, Samuel received a promise that the Philistines would be subdued, and during a time of plague, we are told in I Samuel 7:13–14, "they came no more into the territory of Israel: and the hand of the Lord was against the Philistines all the days of Samuel."

As Samuel grew old, there arose a new king whose name was Saul, and who (in I Samuel 10:1) was baptized king of the Israelites by Samuel himself. This is generally believed (according to the modern Jewish calendar) to have taken place about 1030 B.C. Saul's crowning is associated with a triumphant battle against a people called the Ammonites, who dwelled in the Jordan Valley. During his reign the Philistines, though clearly on the wane, were nevertheless able to put up a strong resistance, as evidenced by the fact that the Bible continues to devote dozens of chapters to them. In I Samuel 27, when a new national hero, David, became a contender for Saul's throne and whole armies were sent to hunt him down, he found shelter among the Philistines and actually became a mercenary for them. The Philistines not only were able to hold Saul's armies at bay but in I Samuel 31 attacked his soldiers on Mount Gilboa and won a complete (albeit their last) victory. Saul's sons were killed in the battle, and Saul is said to have committed suicide.

The Battle of Mount Gilboa is thought to have taken place about 1010

B.C., and the extent of the Philistine victory can probably be read in the contemptuous way they treated the corpses of Saul and his sons in I Samuel 31:8–11: "And it came to pass on the morrow, when the Philistines came to strip the slain, that they found Saul and his three sons fallen on Mount Gilboa. And they cut off his head, and stripped off his armor. . . . And they fastened his body to the wall of Bet-She'an."

Bet-She'an was a city almost as old as Jericho, a man-made mound on the west bank of the Jordan River, about six miles north of Mount Gilboa. Formerly a Canaanite center, it had, during the past century, ceded to Philistine control. When word spread that the headless, naked bodies of Saul and his sons were being left to rot on the walls of Bet-She'an, one of the Israelite tribes (according to I Samuel 31:11–13) sent a task force to fight the Philistines and to remove the bodies from the wall for proper burial.

Free of both Saul's interference and Philistine hostility, David was able to assume kingship over the Israelites about 1004 B.C. He established his capital on the hill of Jerusalem and (in what may be interpreted as a maneuver calculated to win over the dead Saul's followers) betrayed his former Philistine protectors, declared them his enemies, and sent an army north to seize Bet-She'an (after David was gone, the mound of Bet-She'an would become one of the major administrative centers for his son, King Solomon). David then routed the Philistines out of the Jordan Valley and beat them all the way back to the Mediterranean coast, from which they would never again exert even the most marginal control over Israel's interior.

Early in the second Book of Samuel, David realized that the twelve Hebrew tribes, whose politics and even religious practices differed widely, somehow had to be united under a single flag. Some idea of the multitude of factions that must have challenged him can be gained from an inscription dating from this time (found on a goblet unearthed twenty miles south of Jerusalem), which mentions the one and only Hebrew God Yahweh paradoxically taking the Canaanite fertility goddess Astarte as his consort, and as accounts both within and without the later books of the Bible attest, the Hebrews were a quarrelsome people, always on the verge of civil war.

In II Samuel 6, King David turned to the Ark of the Covenant as a means of bridging all tribes and centralizing the new kingdom. The object was distinctly Israelite, sacred to the twelve tribes since before the time of Joshua and Judges. Since its return by the Philistines, it had been kept near the Sorek Valley, about ten miles west of Jerusalem. David ordered a new cart built for the Ark, and in a great celebration it was carried by oxen to Jerusalem, whereupon the king sacrificed the oxen and made burnt offerings before his Lord.

The unification did not last. By 786 B.C. the tribes were divided into two fiercely independent kingdoms, the north (Israel) and the south (Judah), each dissipating its strength against the other in civil war. The books of Chronicles (written about 400 B.C.) describe battles in which the north eventually vanquished the south, setting towns afire and taking the king of Judah captive. It was the literal, biblical realization of nation against nation, brother against brother; and while havoc reigned supreme along the river Jordan, the cities of the Tigris-Euphrates riverworld were rising once again to prominence. In Babylon, Sargon II knew that the quickest route to expanding his nation's economy and keeping his people loyal was to march into and claim new and potentially prosperous territories. The divided Israelites were the easiest prey in sight. In 722 B.C. Sargon II's armies carried the northern Israelites off into exile and reduced Judah (what was left of it) to a tribute-paying state. According to II Chronicles 33:1–20, the Hebrews began to backslide, accepting the gods of their conquerors. In Jerusalem, King Manasseh ordered the building of giant idols dedicated to Babylonian gods. One of these is said to have been so large that a thousand men were needed to transport it. The writers of Chronicles blamed a return to (what else?) paganism and the provocation of God's anger for Babylon's ultimate destruction of Judah, and in adopting this view, history became fractal to them. Adding a brilliant and haunting irony, the writers introduced the phrase *king of the Chaldees* (rather than the more widely used "king of Babylon") to identify the man (Nebuchadnezzar) whose army looted Jerusalem and burned Solomon's Temple in 586 B.C. In II Chronicles 36:16–17, as we read of survivors shackled and carried off to Babylon, the language echoes back to the Book of Genesis, in which God first promised Jerusalem and all that surrounded it to Abraham, who had begun his journey from Ur of the Chaldees and whose distant descendants are, in the last book of the Old Testament, being hauled in chains out of their promised land by the king of the Chaldees. Seen against a backdrop of the parting Sea of Suf, the parting river Jordan, and the cyclical return of the Hebrews to paganism and punishment, the chroniclers seem to have anticipated a fractal geometry in which no matter how far forward we advance, identical shapes and events are bound to reappear. The snake consumes its tail. The universe achieves symmetry: "But they [the people] mocked the messengers of God, and despised his words, and scoffed at his prophets, until the wrath of the Lord mounted against his people, till there was no remedy. So he brought upon them the king of the Chaldees."

About this time, the Ark of the Covenant made its famous disappearance, much as the Ark of the Cross, more than a thousand years later, would simply vanish from oral and written traditions. Both disappearances occurred without mention of any great calamity or fuss. On the matter of the Ark of the Covenant, the Bible simply became strangely silent after Solomon's Temple was built, and no one can be sure, working from the biblical account alone, if the Ark actually existed or where it eventually went.

The fact that duplicate, Ark-like chests were proliferating in Israel by about A.D. 200* suggests that making multiple copies of the sacred scrolls and their container was a custom of long standing. Even when the Mosaic laws were first written, the Book of Exodus tells us that the prophet received at least two copies. The obvious next question (would they carry both copies in one chest?) hints, perhaps, at an early Hebrew tradition of building duplicate Arks, of not putting all of the most irreplaceable relics in one container. In traditions outside the Bible, suggestions of multiple Arks become more clear-cut. The Ethiopians have claimed since antiquity that King Solomon possessed at least one duplicate of the Ark, which he gave to his son by the queen of Sheba.

According to I Kings 11:3, Solomon had three hundred concubines and seven hundred wives. His most adored bride was the queen from the land of Sheba. Arab tradition identifies her by the name Queen Balkis, and she is called by that name in the Koran. African tradition claims Balkis as an ancient queen of Ethiopia, which lies twenty miles across the Red Sea from Yemen, a country known to the ancient Arabs as the land of Saba, to the Greeks as Sabaea, and to the Israelites as Sheba. The identification of Yemen with the land of Sheba would seem, at first, to preclude Ethiopian claims of kinship with Queen Balkis and King Solomon until one takes into account some basic facts about Ethiopian history, not the least of which is that during the first millennium B.C. Ethiopia was a wealthy nation that, from time to time, extended its rule out of Africa, across Yemen, and over sections of southwestern Arabia.†

---

*The Holy Ark found by the Meyers team, near the Sea of Galilee in A.D. 1981, was made of stone instead of acacia and gold, and the two *keruvim,* facing each other, were represented in the then-popular form of lions (instead of winged human figures). The decision to use stone instead of expensive wood and gold might have been a protective measure against being carried off and stripped out of existence, and this may explain why, though ultimately destroyed by vandals, pieces of the Meyers' multiton Torah Shrine entered the archaeological record of the Galilee. It is possible that total plunder of the original Ark and its early duplicates led to a policy of not using precious metals in later copies.

†As metallurgists the Ethiopians were unsurpassed. Their furnaces were sophisticated enough to smelt platinum, meaning that they were very close to being able to transform bauxite into aluminum. Had history continued along this course, the discovery of electricity might not have been very far behind (indeed, there are hints that in Sheba and Babylon some gold trinkets were actually electroplated).

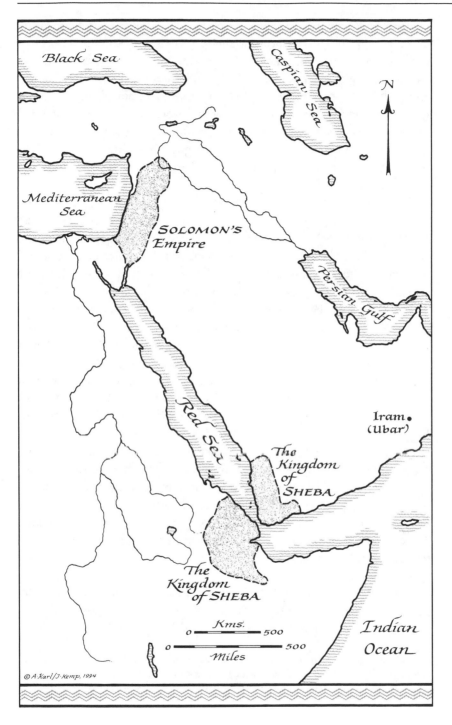

© A. Karl / J. Kemp, 1994

The Bible's Song of Solomon, one of literature's most ancient and erotic love poems, portrays the queen of Sheba as the most beautiful woman to walk the Earth since the time of Hatshepsut. The love between Solomon and Balkis begins (in I Kings 10:1–2) when, upon hearing of the king's impeccable wisdom, the queen travels to him with a train of camels bearing spices, gold, and precious stones, and with a list of hard questions intended to confound his legendary good judgment. In the Song of Solomon 1:5–6, an African origin becomes apparent in the queen's self-description: "I am very dark, but comely . . . as the curtain of Solomon. I am black, because the sun has looked upon me."

Ethiopian legend corroborates the Bible's Song of Solomon so perfectly (perhaps too perfectly?) that the two halves of the story seem to fit like matching pieces of a jigsaw puzzle. According to Ethiopian tradition, Queen Balkis had a son by King Solomon, and the son came to inherit a duplicate Ark containing copies of the Mosaic laws, a wedge from the original broken tablets, a portion of manna, and a sliver from the staff Aaron had thrown down before Pharaoh. The son of Balkis and Solomon was named Menelik, and from him descended a people who still abide by the Hebrew religion and who, in A.D. 1990, were recognized by the Israeli government as true descendants of David and Solomon (and therefore entitled to immigration privileges). As far back as Ethiopia's oral and written traditions extend, one of the honors given the king was the title "Lion of Judah," apparently in reference to an ancestry that could be traced back to King David.*

Menelik's Ark, like the original (although current Ethiopian oral history claims that his Ark *was* the original, owing to a cleverly arranged substitution), simply vanished from history. Some biblical scholars claim that it remained in Ethiopia, while at least one Arab tradition tells that it was sent by caravan across Yemen to present-day Oman, whereupon it was hidden in the uncharted miles of caves that lay beneath Thomas Lawrence's "Atlantis of the Sands," the lost city of Iram.

A widely told Israelite account of the Ark's final resting place ties the disappearance to either the 722 or 586 B.C. lootings of Solomon's Temple. Rabbinical tradition holds that sackings of Jerusalem had been anticipated by King Solomon, who built, beneath the Temple, a maze of hidden

---

*As evidenced by the representation of lions as *keruvim* atop the Meyers' Ark, by the second or third century A.D., the lion had become a dominant symbol in Hebrew tradition and continued as such even into modern times, as when Golda Meir, describing the loss of her soldiers during the 1973 Yom Kippur War, said, "They fought, and fell, like lions."

passages to rival the system of tunnels that honeycombed the Earth beneath Egypt's Pyramids. In advance of either Sargon II's or Nebuchadnezzar's troops, the priests took the Ark to "Solomon's Vault" and, to guarantee that its location would not be revealed by capture and torture, sealed themselves in the tunnel and committed ritual suicide.*

In A.D. 1981 three rabbis claimed to have tunneled under Warren's Gate (one of the original entrances to the Temple Mount) and discovered Solomon's hidden passages. They took a handful of pictures which, though very fuzzy, were clear enough to reveal that the corridors were nothing like the rectilinear chambers, passageways, false walls, and traps I had seen beneath the Pyramids in the Valley of the Kings. The Temple Mount passageways were tube-shaped, and their walls were sheathed in a mantle of large bricks, as if intended to conduct water. I came to suspect that the three men had simply entered an ancient water conduit, but Rabbis Shlomo Goren and Yehuda Getz insisted that during one of their explorations of the tunnel system they actually saw some of the lost Temple treasures, including the Ark of the Covenant, when they broke into a secret vault and pointed a flashlight through the opening. There were objects of wood and gold among blocks of fallen stone rubble, and what appeared to be the lid of the Ark—crushed, but with one of its *keruvim* still intact.

"Unfortunately," Rabbi Goren recalled, "when we came so close, the Arabs started rioting, they would not let us return to the tunnels. And the government became afraid. And they stopped us, to build a wall [blocking off the entrance to the tunnel system], and we had to [stop] our digging."

For years, some of Jerusalem's more radical Christians and Jews have been calling for destruction of the Muslim Dome of the Rock, which stands atop the Temple Mount, and for its replacement by a reconstruction of Solomon's Temple (of which all that is presently believed to remain are underground passages and a few stone blocks at the bottom of the Western Wall). I was once walking with a very devout and kindly Christian lady who preached God's lasting mercy toward all men, when suddenly she turned and pointed toward "that abomination standing where the Temple ought to be," and as if to teach that her human mercy was far shorter-lived than her God's mercy, she declared to me that "the abomination should be torn

---

*By contrast, the architects of the Great Pyramid of Cheops, who were supposed to be entombed with the pharaoh after they voluntarily released a cascade of giant granite slabs, had built for themselves a narrow and apparently secret passage to the outside world. It was through this escape route that the tomb was eventually looted, rendering the millions of tons of protective limestone blocks and the labor of decades immediately useless.

down, and the Temple restored, and all who try to defend the mosque must be destroyed with it.''

During my first visit to the Temple Mount, two rabbis were being escorted from the Dome by soldiers, who had interrupted their attempt to excavate a new foundation. This was nothing new. For more than two years, rabbis and ministers had repeatedly been arrested trying to lift old foundation stones from the Muslim shrines and to commemorate new stones of their own. It was easy to understand why, given this background, the Muslim Supreme Council, which oversees the Temple Mount, viewed the story of Solomon's lost treasures as a fiction concocted to permit ingress that, once sanctioned, might become permanent. Fearing both the theological and literal undermining of their shrines, and hoping to avoid a holy war between Muslims, Christians, and Jews, the council members (with support from the Israeli government) sealed the tunnel entrance in concrete and posted guards.

My own suspicion is that, although the Ark probably existed, it (and any of its contemporary duplicates) will not be found inside the Temple Mount, or in Ethiopia, or in Iram, or anywhere else. Unless buried someplace where, like the tomb of the child-king Tutankhamen, it was accidentally and, against all odds, totally forgotten, it must long since have been dismantled, probably in the time of Sargon II or Nebuchadnezzar, and probably for the same reason that two other treasures of the Exodus period—the golden death masks of Hatshepsut and Tuthmosis III—will never be found. The Ark of the Covenant was essentially a wooden chest covered inside and out with sheets of gold. Two golden *keruvim* knelt upon its lid, and any who captured the Ark would have been inclined (unless it could be held hostage or bargained for ransom) to strip off the gold and discard the wood.

And yet . . . at least one other Israelite tradition has the Ark being spirited away by the prophet Jeremiah, some forty years in advance of Nebuchadnezzar's troops, to await the coming of the Messiah.

The fate of the Ark, hidden or lost for all time, is one more archaeological mystery that, barring acts of God, I rather think will be successfully plumbed someday.

IV

BACK
TO
CANAAN

II

WHENCE CAME THE PHILISTINES

When Darwin wrote the Origin of Species *in 1859, he introduced the volume with a long disquisition on the breeding of domestic animals, as a way of leading his audience to the difficult terrain of natural selection through a familiar backyard (and barnyard) path. Today this material is as foreign to most readers as the subject it was meant to illustrate—and this is because we have, in the intervening century, come indoors. . . . The "ancients," as we generally refer to everyone from the time of Plato back to* Homo erectus, *lived and worked outdoors, and did most of their thinking there as well. Because we have come indoors in the last hundred years, much that was obvious to earlier generations of scholars about ancient [legends] is no longer obvious to us. That [the legends] are to a large extent stories about nature has passed in the last few generations from something that "goes without saying" to something that "cannot be said."*
—MOTT T. GREENE, *Natural Knowledge in Preclassical Antiquity*

JERUSALEM, THE ARAB QUARTER, AUTUMN A.D. 1991

At Ami Mazar's instruction, I have begun exploring the back lots and shops of the world's Mecca to the black-market antiquities trade. Ami is the nephew of the great-grandfather of Israeli archaeology, Benjamin Mazar. Jerusalem's antiquities trade is, in the strictest technical sense, legal, but it is a thorn in Ami Mazar's side.

"Israel is the only country in the Middle East that trades legally in antiquities," Mazar has explained. "Don't ask me why. We try now to change the law, but changing the law is very difficult. It has to go through a whole legislation system. It's a grave situation, but there it is."

According to Mazar, some government archaeologists and bureaucrats claim that by their legitimizing of the trade, controls can be exerted over what is traded. They claim that if something unique turns up in the shops, the government can purchase it. They claim that legitimacy prevents clandestine excavations and treasure hunt operations of the sort we have seen near the buried city of Thera.

The claims do not always hold up very well. In one of the shops my wife and I have found Minoan pottery from the Aegean and carvings from as far away as China. There are gold-inlaid Bronze Age daggers from Greece and Lebanon. For the archaeological mercenaries of the world, it seems all roads lead to Jerusalem.

Some of the smallest, dirtiest, and most obscure-appearing shops contain the largest treasures. I have seen whole shelves of Roman glass—and tossed among the vials, a beautiful earring, hand-sculpted to microscopic detail, hopelessly misplaced by a Roman woman when Herod Agrippa I and the emperor Claudius ruled over these parts. One antiquities trafficker slid aside a dust-covered bookshelf, revealing the room we had been standing in to be merely the antechamber to an even larger room cut into the very bedrock of the walled city. My eye was drawn first to a prominently displayed picture of the trafficker standing in this very same treasure-filled chamber, shaking hands with the now-legendary Israeli general Moshe Dayan. Philistine pottery lay all around me, and Gloria, my wife, found mixed in with it a crude imitation of a Philistine beer jug manufactured by one of the first Israelite tribes to settle in this area. It was genuine—from the time of Joshua and Judges. No need for making counterfeits here. No value in it either. During an interval of about twenty minutes, in only one shop, I twice saw sacks of artifacts being brought in by nomadic herdsmen, then traded for a mere pittance in cash. I could have had my pick from the sacks, and the shop owner offered to write me official cards with serial numbers and site identifications (despite the fact that the sites to which the objects belonged had never been identified by the herdsmen).

When prospective customers arrived and asked where a certain jug had come from, they were most frequently told Jericho. Why not? The name was familiar to most people, and such familiarity immediately increased its demand (hence its selling price) by a factor of ten. I have come to suspect that if all the Jericho jugs for sale in Jerusalem were gathered in one place, they would form a mound high enough to bury ten Jerichos. But the saddest cut of all is that these piles and piles of misidentified material once constituted a great deal of scientific information, now lost for all time. No one will ever know what a given bowl or jug could have revealed to us about a stratum at the site from which it was actually taken.

Personally, I object to the artifact trade only to the extent that it produces a loss of scientific information. Indeed, if not for the mislabeling of antiquities and the lack of documentation showing precisely when and where (to the precise century and inch) they came from, I could argue that worldwide dispersal into private collections is the best way to preserve the past. Historically, great museums and libraries have been located in the centers of our cities, which have always been prime targets in time of war. As we begin to enter an era of small-scale nuclear terrorism (made possible by easier access to highly refined heavy metals, and by a sinister little design called "the Sears and Roebuck eight-hundred-dollar car trunk atomic bomb" [batteries and uranium not included]*), we should perhaps be drawing some relevant

*This particular little nightmare (part of what got me and one of my associates christened "the Pablo Picasso and Salvador Dali of nuclear destruction") was so named because almost everything needed to build it could be ordered through the 1990 Sears and Roebuck catalog, and it could have fit easily into the trunk of a car. My brief participation in brainstorming sessions on counternuclear terrorism, just prior to the Gulf War, had the distinction of being the only point in my life at which the objects I dreamed up gave me nightmares. I began

lessons from the fates of Alexandria, Rome, Dresden, and Hiroshima. Time and again, the labor and knowledge of centuries has been assembled in one place, where it could be destroyed in an instant.

Over cups of the strongest coffee I have ever tasted, I explained to the owner of the secret chamber that all we really needed was more cooperation between archaeologists and collectors, so that a worldwide computer net, listing the locations of artifacts from all the major time periods, could eventually be established. "For if instead we establish laws and strict penalties against possession," I said, "the collectors will all go underground, and uncountable pieces of archaeology's most challenging jigsaw puzzles will disappear from view without our ever knowing they existed."

I meant every word of what I had said, and the dealer, who watched my eyes the whole time, seemed to sense this. He came to trust me, and another bookshelf slid aside, revealing a bearded Babylonian sphinx standing three feet tall at the shoulders. There were fragments of a gold and lapis eagle from the city of Ur, and a crushed golden cup decorated with winged bulls. From the Temple of Nineveh he handed me the right half of a copper helmet. It looked more like a mask than a helmet, with eye slits, lips, and a full beard.

"Look upon the face of Naram-Sin," the dealer said, "grandson to Sargon of Agade." He next opened up a tiny cardboard box. A three-inch-wide patch of woven gold and silver thread lay upon a bed of cotton. "A fragment from the robe of Thomas Lawrence," my host explained. My eyes narrowed suspiciously. We both laughed. And then I pointed to the boxes of cuneiform tablets and cylinder seals lying in the sphinx chamber.

"My latest acquisitions," the dealer explained. He began reading cuneiform to us, but I assumed it was all a show, intended to induce me to buy one of the tablets or seals, whose prices were astronomical and whose authenticity I was coming more and more to doubt . . . until he mentioned the name of a city and a god few people had heard of yet: "Mashkan-shapir . . . Nergal . . ."

Elizabeth Stone and Paul Zimansky had expressed fears that American bombs might have obliterated Iraq's ancient treasures, but the growing accumulation of Babylonian antiquities in Jerusalem suggested that the realities of the postwar economy might prove to be a greater threat to archaeological sites than Tomahawk cruise missiles. Iraq's economic base was rendered virtually nonexistent, and for many, the only easy way of

---

distancing myself from the sessions after only two weeks, and returned to archaeology and antimatter rockets.

putting food on the table was to mine the half million archaeological sites running the length of the Tigris and Euphrates rivers, to transport the most marketable objects across Jordan, and to procure whatever goods could be traded for them.*

It occurred to me that in Jerusalem the market for Mesopotamian antiquities was flooding, amounting to a kind of foreign aid program from Israel to the people of Iraq. But my thoughts on the subject were interrupted by a commotion outside. Three German tourists had just been stabbed, and the commotion grew louder and closer each passing second. With a wall of solid rock behind us, and in front of us a shop door that opened directly into the trouble, there was nothing to do except hope for the best and prepare for the worst. The dealer had no guns (he was an Arab living in an occupied city, and to be caught with a gun would have, in any situation except the one we now faced, gotten him into more trouble than the gun was worth). The nearest thing to a weapon at hand was a Crusader-period sword. It was so badly rusted and blistered that I feared it might break on the first strike. But it was considerably longer and more intimidating than the assailant's knife, and though the blade was dull and might break, one well-placed strike would be all that was needed.

I was about to ask Gloria, "Do you want to handle this, or should I?" (my wife is more skilled with swords than I am) when a half dozen gun-wielding soldiers came swarming through the Damascus Gate and solved our problem.

The three of us shivered inwardly, and shared a nervous laugh of relief. The dealer poured fresh coffee and shook his head. "You really thought to kill him with that antique?"

"If I had to."

"You crazy archaeologist types. I wonder what you were like as children."

"Terrors."

"I'll bet."

〰〰〰〰

At four o'clock the sun was setting behind the Western Wall, and two rabbis, friends of Father MacQuitty, hurried me past a crowd of young soldiers into a thickly overgrown and mostly hidden depression at the foot of the Temple Mount. I watched the soldiers close ranks and block the

---

*To judge from Naram-Sin's helmet, which appeared to have been recently crushed under falling concrete, some of the objects in the sphinx chamber had been looted from a bombed museum.

entrance off for us, but my mind was not really on the soldiers that afternoon. The focus of my attention was three recently and clandestinely moved stones, forming a breach through which a reasonably thin man could squeeze beneath the Western Wall. I was promised a glimpse, at least, of what almost everyone except me believed to be Solomon's tunnel system. And if we were not caught, I was told, the rabbis intended to follow the labyrinth all the way to the lost vault of the Ark. I pushed my infamous green canvas bag in ahead of me and held a Grumman penlight between my teeth.* The penlight, like the bag, was badly worn, having been with me on every expedition since its 1983 flight aboard the space shuttle *Challenger*. It revealed a long, empty corridor that stretched beyond the range of my search beam into total darkness. At first it seemed to me to have been carved by centuries of water escaping from somewhere above, but as my eyes grew accustomed to the dim lighting, I could see that the entire left side of the chamber—more than thirty feet of it—was a wall of hand-hewn stone . . . No, not just a wall. More details emerged from the dark. There was a right angle at the top of the stone. It was actually a huge monolith turned on its side. Someone had cut it, dragged it, and buried it here at what must have been incredible cost, and for God only knew what reason. I was struck dumb with amazement. Was it granite? I wanted to know. Was it actually cut from one of the hardest of all available materials? That would have made whoever carried it here appear even stranger. I was about to move in for a closer look when my penlight, for the first time in eight years, gave out. I fumbled for and found the spare in my bag, but a voice behind me called, "Come out. The Arabs know you are here."

It was one of the soldiers, and she needed to say no more. (Her instructions, before we entered, had been simply this: "If we call you out, we don't care if you've found the Holy of Holies itself. You come out immediately.") Within twenty seconds I and my two companions had retreated from the opening, and less than two minutes later we had disappeared into the markets of the Old City.

Deep within the Jewish quarter we found an outdoor café and stopped for ice-cold colas.

"What did you see?" the older rabbi asked.

---

*Part of the infamy seems to arise from the fact that the bag is badly in need of a good dry-cleaning. It has survived the collapse of a rampart at the Great Wall of China (following a Beijing earthquake). It has been buried under and recovered from Bet-She'an, risen above the Shanghai flood, and come through a firefight and an "unscheduled energetic disassembly" (which is, in the wacky world of nuclear weaponry, just another way of saying, "Oops!"). I never carry a briefcase—just that cumbersome green bag, and it seems to bother some people that it is always at my side, like a faithful dog.

"The chamber was empty—but that wall! It's a slab, I think. A slab as big as anything I've ever seen in Egypt!"

The rabbi nodded. "What did I tell you? And not just one big slab. We've seen at least two others just like it."

I unzipped my green carryall and threw in the two penlights. Something caught the younger rabbi's eye. He reached in and pulled out a bag of Cheez Doodles.

"Corn oil," he read. "Oh, yes. American food. Sodium acid pyrophosphate . . . polysorbate sixty . . . red dye forty, and yellow five . . . my compliments to the chemist."

"They're not really for eating," I tried to explain. "I always bring them into caves, as a backup to my backup lights, if ever I find myself down to the last ounce of fuel in my Bic flick."

"You're making this up," the rabbi said.

"No. No. Not at all. They actually extend the life of the flick. A Cheez Doodle on the end of a tweezer will burn for at least two minutes. A bag of them can provide you with an extra hour of light if you're stuck underground." I then explained how this exact scenario had been played out in New Zealand, during my 1981 exploration of a cave in which a Maori princess, pursued by a vengeful husband, had spent a lifetime in hiding some two hundred years earlier. My guide and I ended up a quarter mile inside a mountain with all our spare lights hopelessly destroyed. A Cheez Doodle saved my life that day, revealing a ten-foot drop onto a field of stalagmites, pointing up like a hundred swords.

"One archaeologist has told me that he prefers Twinkies," I continued. "Mounted on a stick, they burn for a very long time. Evidently, the heat is always bringing new oil to the surface."*

"Twinkies and Cheez Doodles?" the rabbi said.

"Hey, if you don't need them for fuel, you can always eat them when you get out."

"Twinkies and Cheez Doodles?" the rabbi said again. "My, you really are a Philistine."

〰〰〰〰

Philistines, as defined by Jeremiah, were a remnant from the coasts of Crete. In Jeremiah 47:1–4 the prophet described an angry God's vengeance upon

---

*Spelunkers and archaeologists take note: In recent years, nutritional awareness has led to a change in the recipe for Twinkies. Cholesterol-rich coconut oil (the substance that once made Twinkies burn so brightly) has been replaced by less combustible cottonseed and soybean oils.

these "seashore enemies of the children of Israel. . . . Behold, waters are rising from the north, a torrent in flood, it shall flood the land and all that is in it, the cities and their people. . . . Yes, the Lord is destroying the Philistines."

Amos, a minor biblical prophet of the eighth century B.C., linked the second millennium B.C. migrations of Hebrews from Egypt and Philistines from Crete (Kaphtor) to a day when lands melted (an apparent memory of Thera) and the sea rose up to terrorize the Earth.

"And the Lord God of hosts," he wrote in Amos 10:5–7, "is he that toucheth the land, and it shall melt, and all that dwell therein shall mourn: and it shall rise up wholly like a flood. . . . It is he that calleth for the waters of the sea, and poureth them out upon the face of the earth. . . . Have I not brought up Israel out of the land of Egypt? And the Philistines from Crete?"

Genesis 10:14 identifies the Minoan island of Crete as the place "whence came the Philistines." The Bible refers to the people of ancient Crete as the "Kaphtor" (a probable adaptation of *Keftiu,* the Egyptian word for Cretans), whom history remembers as the only people Tuthmosis III and his viziers considered civilized enough to be worthy of a name other than "barbarians" or "abominations against Ra."

A Cretic, as defined by the *Oxford English Dictionary,* is someone "belonging to Crete." The dictionary further describes *creticism* as "Cretan behaviour, *i.e.* lying . . . cheating." It would appear, on a purely linguistic basis, that like serpents and comets, people from Crete have come down through Western tradition with a bad reputation. Their cultural descendants, the Philistines, have not fared any better.

The dictionary defines a Philistine as "One of an alien warlike people . . . who occupied the southern sea-coast of Palestine, and in early times constantly harassed the Israelites . . . a person who is lacking in or smugly indifferent to culture, aesthetic refinement, etc., or is contentedly commonplace in ideas and tastes."

For most people, as for my two rabbi friends, the latter definition of the Philistines is the more familiar one. How they and their ancestors came to characterize common tastes and uncultured behavior has never been one of archaeology's unsolved mysteries. Much that went into building the modern English language was inherited from the West's most widely read book. To the authors of the Bible, the Philistines were enemies and, seen in this light, were simply bound to receive a lot of bad press.

Now, in one of history's strangest role reversals, it is Trude and Moshe

Dothan, and Ami Mazar—Hebrew archaeologists of modern Israel—who are at the forefront of the emerging respectability of Philistine culture.

∿∿∿∿∿

According to Moshe Dothan, the giant slabs I saw under the Temple Mount date from shortly after the Philistine period, "probably around 970 or 980 B.C.," he says. "But that's just a guess, using the Bible as a guide. We don't know exactly." He believes they are foundation stones from Solomon's Temple, but their size is beyond reason: more than thirty feet long and weighing many, many tons. "We have these stones," he explains. "And tunnels. But all excavation has been halted. There has been great fury over further work, for it is a very controversial thing to be digging under the Dome of the Rock."

Moshe's first dig began shortly after he left his native Poland in 1938, three months before Hitler invaded. (Fool's luck, he calls it. "I've had that kind of luck all my life." Most archaeologists seem to have that kind of luck. It's like a God-given law.)* He settled in Jerusalem in 1939, at the age of seventeen, and began studying archaeology under Benjamin Mazar. He especially enjoyed learning the language of the ancient Bible, but Hitler's expansion through Europe and into North Africa cut his studies short. Israel, at that time, was a mandate of Great Britain, so Moshe volunteered for three and a half years in the British army. Up to that point, he had been unsure about spending the rest of his life digging into lost worlds, yet as he traveled through Europe sabotaging Hitler's bridges, he was continually stumbling upon the ruined bridges, aqueducts, and pillars of earlier civilizations; and by the time he returned to Jerusalem in the summer of 1945, his course was decided. He took up spade and textbook under his old mentor, Benjamin Mazar, and began working toward his Ph.D. Together, professor and student traveled up the river Jordan to Bet-She'an, site of the Philis-

---

*I am no exception to the law of "fool's luck." There is a colleague who, for reasons that will never be clear, devoted extraordinary energies to plunging this book's expeditions into total chaos. I ended up nicknaming him "my guardian angel" because, though I doubt it was what he had in mind, he saved my life—repeatedly. A disruption of my plans in Egypt caused Gloria and me to switch booking, at the last minute, off a ship that subsequently went down, drowning half its passengers. Some months later, in what was beginning to look like a coyote-versus-roadrunner cartoon carried a step further, another change of plans was forced upon Gloria and me, and we were diverted from being directly in the path of Saddam Hussein's tanks. And this wasn't the last such incident. These odd coincidences continued for more than two years. Fool's luck. Like Moshe Dothan, I've always had it, and my closest friends tell me I always will.

tines' last stand against the Israelites about 1010 B.C. A thousand years after Saul, David, and the Philistines, the mound of Bet-She'an towered 120 feet over the Jordan, and had been capped and surrounded by a Roman city, complete with a seven-thousand-seat theater, bath houses, temples, and rows of colonnaded shops. The earth beneath the Roman streets plunged backward through time into the Bronze Age, a period that was becoming increasingly fascinating to Moshe Dothan when yet another war interrupted his studies.

At four o'clock in the afternoon of May 14, 1948, his voice quivering with emotion, the Israeli leader David Ben-Gurion called out over the airwaves from Tel Aviv: "By virtue of . . . the resolution of the United Nations, we hereby proclaim the establishment of the Jewish state in Palestine—to be called Israel."

At precisely that moment, in the port city of Haifa, Sir Alan Cunningham, the last British High Commissioner of Palestine, gave a salute toward Jerusalem and sailed out into the Mediterranean. With this simple gesture, twenty-eight years of British rule ended. Partly out of sympathy for the victims of Hitler's attempts at what is now termed "ethnic cleansing" and partly out of guilt for such inexplicable behaviors as turning shiploads of Jewish refugees back to the Nazi slaughter, Britain, the United States, and other members of the United Nations drew up plans for the partition of Palestine into separate Arab and Jewish states. Arrayed against the borders of the newly born state, and drawing plans for the total erasure of those borders, were the tanks, planes, and artillery of five Arab nations, their leaders vowing to drive the Israelites into the sea, as the Israelites had driven the Philistines.

To the Arabs, the United Nations' decision was the dismemberment of their homeland, and during the year leading up to the decision, Jerusalem had been in a state of civil war, with Arab and Jewish bomb squads detonating explosives in each other's marketplaces. The slowly evacuating British forces, wishing only to get out as quickly and quietly as possible, declined to intervene. Meanwhile, in the overstocked weapons depots of a half dozen postwar nations, the emerging Israeli Army was able to purchase, at bargain basement prices, machine guns, mortars, rocket launchers, Flying Fortresses, and, in one of history's stranger turns, Luftwaffe fighter planes whose swastikas were overpainted with a field of white and the blue star of David.

Moshe joined the Israeli Army ("this time it was *our* army, not the British army"), and even during the yearlong War for Independence, he had managed to snatch opportunities to explore Bet-She'an and at least three other archaeological sites. Like most wars, it was long hours, days and even weeks of quiet, interrupted by seconds of chaos and terror. When the smoke began to settle in 1949, the Lebanese, Syrians, and Iraqis had withdrawn across their borders. The Egyptians were forced out of southern Israel into the Sinai Peninsula. The Jordanians annexed Old Jerusalem and all the land lying between the city and the Jordan River (this region, known as the West Bank, included most of the land allotted to Palestine's Arabs by the UN), and Israel found its borders expanded 21 percent beyond those originally drawn under the United Nations plan. That was the year Moshe met a fellow soldier and student of archaeology named Trude. Somewhere between strafings and discussions of Philistine beer jugs, they fell in love and were married.

Unlike her husband, Trude had known from earliest childhood that she wanted to be an archaeologist. She grew up in Jerusalem, where her mother (a painter) and her father (an architect) took her along on their numerous explorations of caves, tombs, and ancient roadbeds that seemed to be revealed every time someone dug a new basement. Unlike most archaeologists, she has nurtured a teasing and most refreshing sense of humor. Ask how old she is, and she will tell you, "Didn't your mother teach you never to ask a woman that question? I'm five thousand years, plus or minus." She has a warm, cherubic smile, reminding one of a happy child somehow continuing to live in an adult's body, never having allowed that six-year-old's sense of wonder to drain away.

"I had already started my studies at the university," she recalls, "when the War for Independence began. I was in the army, then, and I was also a student of Benjamin Mazar, who took me to a river site built by the Philistines. That is where I first became acquainted with the Sea People, who apparently came down from Crete sometime after 1600 B.C., bringing with them a veneer of Aegean [that is, Minoan] background. They've followed me ever since.

"Dealing with the unknown and suddenly seeing a pattern in it—that has always been archaeology's allure for me, as when I traveled south of the Gaza Strip in 1968, trying to locate the source of unusual artifacts that started turning up in Jerusalem's antiquities market. As it turned out, we found a city buried under the sand."

In the souvenir shops of the Walled City, Trude found the market suddenly flooding with Egyptian scarabs, carnelian necklaces, golden earrings, signet rings, and alabaster vessels—all typical grave offerings from the time of Joshua. She suspected that modern tomb robbers must have come upon a Late Bronze Age cemetery filled with untold treasures. Her theory was confirmed when a shop owner who understood her interest in the Philistine period called to tell her something very strange had been brought out of the desert.

Trude was led to four baked-earth coffin lids propped against a back room wall. Modeled in clay, four large-eyed faces stared at her with an uncomfortable intensity—"Spooky, but they had a way of growing on me."

She called them anthropoid coffins, because they seemed to her to be an attempt to imitate in pottery the gilded human-shaped mummy cases the Egyptians had been building since before the time of Abraham.

"Where are these coming from?" she asked.

"From Hebron," the shop owner said. "You know, Hebron around Jerusalem."

"This is impossible," said Trude, for the artifacts were full of pure, white sea sand. She could see immediately that they came from somewhere near the coast, somewhere near Gaza or Sinai. Unfortunately this was autumn 1967, just after the Six-Day War, in which Israeli war planes had won command over Egypt's airspace, allowing battalions of tanks to claim as occupied territories not only the Gaza Strip but the West Bank of the Jordan River, the Golan Heights, and the entire Sinai Peninsula. The new borders west of Jerusalem were still being contested by artillery duels and, except for military vehicles, were essentially a forbidden zone on the day Trude Dothan made her back-room discovery.

Barred from exploring the zone, she visited Moshe Dayan, the charismatic general with the eyepatch, who had been the prime architect of Israel's victory in the just-ended war. She knew that the general (now promoted to Minister of Defense) shared her keen interest in antiquities. "Unfortunately," she recalls, "I also knew he would probably go to Gaza and dig up the site himself. It was not exactly what we archaeologists would have liked to see, but [on account of our common interest] I knew that if I needed his help, he was a person I could turn to. So I sought him out, and I said, 'You know . . . these coffins . . . there is a site out there somewhere, and I have this terrible feeling that treasures we may never guess at will be looted and perhaps even melted for their gold value before science knows they exist. What have you heard? Do you know where they're coming from?' "

"No, I don't," the general said. "But I promise you, the moment I do know, I will call you."

Nine months passed. Trude was at home in Jerusalem when Dayan called. "I know where it is," he announced. "Tomorrow morning come to my office. I will meet you, and we will go to the Gaza Strip."

Leading a convoy of armored cars, the Defense Minister brought Trude and her husband to a place called Deir el-Balah, an unassuming-looking tilled field in the midst of sand dunes, about a mile from the Mediterranean coast. Sixty soldiers went with them, and as the Dothans walked among freshly plowed rows and began plucking pottery shards from between the furrows, Trude became aware of a tense and barely restrained hostility—not on the part of the farm owner (a Gaza lawyer), or among the local Bedouins, but among her own troops. Putting the tensions out of her thoughts, she saw for the first time the source of the coffins. Hundreds of fragments littered the field, which had been hastily tilled in a feeble attempt to hide evidence of clandestine excavations. At least a half dozen pottery coffins had been plowed into hundreds of pieces—too many to be effectively hidden. Almost complete clay and alabaster vessels were found scattered across the field, and a carnelian seal, worth more than its weight in gold to science yet cast aside by tomb robbers interested only in gold. One face of the seal showed an Egyptian king standing in his chariot. The other bore the hieroglyphic insignia of Ramses II, who (according to the Theran ash layer calendar) had lived about 1430 B.C., some two hundred years after Tuthmosis III.

As the Dothans rushed about snatching up every significant object they could carry, some of the soldiers began to lose their composure.

"We wasted our time on these archaeologists," one of them protested. "On archaeologists!"

"They became very angry," Trude recalls. But General Dayan ordered them into the field to help her bring back as many pieces as the armored vehicles could hold, and quickly, for in those days the security situation in the Gaza Strip was such that any military scouting force could remain safely in one place only for a few minutes.

The Dothans had time enough to begin scratching the surface (and, oh, what a strange and thrilling surface it proved to be), but nearly four years were to pass before the Minister of Defense allowed the Dothans to return—four years during which lootings from the site continued to disperse through a black-market antiquities trade which, more and more, was coming to be focused on Jerusalem. Moshe Dayan worked closely with the city's antiquities dealers and had much to do with legitimizing the trade that was to become the pollution of many an archaeologist's life. Much of the material from Deir el-Balah came directly into the Defense Minister's private collection (which, in time, Trudy persuaded him to donate to the Israel Museum, where it can be seen today).

This was also a time during which Trude made her acquaintance with

Spyridon Marinatos. He was the last remnant of the Leonard Woolley school of archaeology, "a character," she recalls. "A real character. And a man with vision. He was a very elegant man, very gentlemanly; and what I remember most is these huge blue, rather piercing eyes. He had contacted me at that time because I had been working on these coffins from Gaza, looking for possible connections with the Philistines. The faces on my coffin lids, with their large, open eyes, looked to him like the large-eyed golden death masks of the Mycenaean kings [dating from post-Thera Greece and Crete].

"He had a theory about what he called 'the people with the open eyes.' He believed there was a direct connection between the coffin makers of Deir el-Balah and the people who carved Greek death masks with large, piercing eyes, as in depictions of Alexander the Great (and the real-life Marinatos himself, I used to joke at that time). This was a very strange theory, and yet . . . he was a special kind of explorer. He did not do things in the very systematic, step-by-step way we are accustomed to working in today. But he had a vision. It's almost a second sight, what he had. It was an intuitive gift, and he was not afraid of strange theories (as in the case of his 'people

Spyridon Marinatos, discoverer of a lost Minoan world, saw a connection between the Aegean, Egypt, and Israel long before Minoan settlements were found (in 1991) on the western shore of the Galilee and (in 1993) at Avaris on the Nile. Shown here are two funerary representations of Marinatos' "wide-eyed people": one from a Deir el-Balah coffin lid, the other from the golden death mask of a Mycenaean king. He believed that Minoan ships might have spread customs as well as trade goods, and that similar death masks meant common origins.

with the open eyes' sharing a common cultural ancestry, or the idea that his lost Minoans and the buried city of Thera were the origin of Plato's Atlantis legend)."

As far as Marinatos was concerned, his theories did not always have to be proved correct. Whenever new evidence took down an old theory (even if it was his favorite and oldest theory), it usually erected a newer and even more thrilling one in its place, and therefore, being wrong was no cause for embarrassment or despair. Quite the contrary, nature's occasional surprises were cause for celebration. To Trude Dothan, Marinatos had a way of breaking away from the herd, of attacking a new puzzle in ways that often seemed to her like "going out backward," which she had never seen anyone do before.

"My husband and I were really there at the beginning," says Trude, "when Marinatos began revealing his lost city of Thera to the world. And he brought geologists down into the volcanic death shroud, showed them the ruins, put these scientists shoulder to shoulder with archaeologists. I would say this was the beginning of the overlap of fields, the beginning of this focus: What happened on Thera? How do we interpret it? How do we date it?"

Sadly, Marinatos' hoped-for marriage of diverse fields has been continually on the verge of divorce. I have spent much of the past five years traveling up and down the riverworlds, redating the entire Mideast on the basis of how far artifacts reside above or below the 1628 B.C. Thera-Tuthmosis III time probe. To me, the Theran ash layer is little different from an older and even more violent destruction layer that turned up in New Zealand in 1980. At that time, I was studying a 14.8 million B.C. layer of rocks coincident with a period of global cooling, widespread extinction, and the Ries Basin asteroid impact, when Luis Alvarez's team arrived and found a stratum fifty million years beneath me to be laced with debris from an explosive force more powerful than a thousand Theras. About 64.4 million B.C. (give or take a hundred thousand years) our solar system appears to have passed through one of the galaxy's many dust clouds, picking up a sudden influx of platinum-rich dust particles, a few large boulders, and at least two flying mountains of ice laced with carbonaceous rock. One of them was at least as wide as the World Trade Center's twin towers are tall when it impacted the Earth at about thirty miles per second. Within fifty thousand years a second object, slightly larger than Manhattan Island, dug out a two-hundred-mile diameter crater near the Yucatán Peninsula and blasted molten pieces of the Mexican countryside as far away as Ohio. The resulting ash layer, like the ash layer of Thera, can be read around the world.

While paleontologists, astronomers, and geophysicists have been debating the details ever since, the debates have never gone beyond legitimate scientific criticism. By contrast with the marriage between geology and archaeology, the marriage of paleobiology with astrobiology has been exceedingly smooth. There have been no character assassinations, no homicidal and suicidal threats. The difference, I suspect, arises from the fact that paleontologists and astronomers are forced every day to comprehend the meaning of a million years (and if such comprehension does not drive us mad, we tend to get along fairly well with each other, as long as we stick to dinosaurs and stay away from the anthropologists). By comparison, most of the archaeologists I have known think only in millennia or centuries, and this may explain why one Egyptologist, to whom five thousand years meant all the time in the world, did indeed make both homicidal and suicidal threats when I tried to argue that, if we have a reliable date on the Theran ash layer, Tuthmosis III must have lived more than 120 years before her textbooks said he did. Another archaeologist, raising his hands above his head (as if to show that he bore no weapons during a debate that was beginning to degenerate from name-calling to food throwing), summed it up best when he said, "After studying the answers to all those questions given by volcanologists and archaeologists, do you really think that time is the same thing for both of them?"

Christos Doumas, a student of Marinatos' who now lives at the Thera excavation, expresses strong reservations about ash layer dates provided by paleontologists, dendrochronologists, and geophysicists and believes that "scientific dating" should never be taken as gospel.

Trude Dothan agrees with him: "I would not say that to include geology brings the time lines into clearer focus. There is not much agreement [between the archaeologist's pottery clock and, for example, the paleontologist's Nile Delta ash layer and the carbon decay products that surround it]. So if we want to rely on the sciences, it really does not help us a lot."

Trude's husband prefers not even to talk about the Theran ash layer, especially about how it may fit in with the Exodus story. "Your 1628 B.C. time line . . . your redating of Tuthmosis III and my Philistines . . . I don't know. It is generating all sorts of ideas. And I want very much not to touch this thing. This is still a very blank page of history."

~~~~~~

During the drive back from Gaza, Trude had turned to the Minister of Defense and said, "Well, I'm ready to dig now."

A soldier laughed.

"You've got to be kidding," replied Dayan. "It's too dangerous. But I promise you, the moment you can go back, you will." That was in 1968. The year 1971 came and went, and while the Dothans continued to wait for their clearance to Gaza, the grave robbers came to Deir el-Balah—came repeatedly—and went away with earrings, golden amulets, and a beautiful jeweled death mask that turned up in the Jerusalem market. The Shah of Iran's festivities marking twenty-five hundred years of the Iranian Empire came and went, becoming the most expensive celebration of the twentieth century. Anwar al-Sadat came to power in Egypt, and the threat of yet another Arab-Israeli war came and refused to go away.

"Finally," Trude recalls, "in March 1972 permission was granted for the start of our first field season. At Deir el-Balah we were housed in a military garrison, and soldiers accompanied us at all times.

"We searched for weeks without finding anything because it had all been robbed and plowed." And where the land had not been plundered and planted, it was buried under high, shifting dunes of white sand (the same fine sand, Trude soon realized, that had clung to the treasures she discovered in the Jerusalem markets). Luckless, the Dothans hired the local Bedouins to assist in their search for a pocket of the unplundered past, but the tribespeople kept leading them to empty depressions in the ground.

Defeatism was beginning to take hold when Moshe Dayan called their attention to an elderly Bedouin who was walking among the dunes with a stick, using it like a divining rod.

"So, on this day," says Trude Dothan, "we went to the old man with a small group of soldiers, and I knew from the moment I met him that he was the chief tomb robber. What I could not have known was that in time he would become my best friend. His name was Khammad, and he turned out to be one of the most fascinating men I have ever known. He was a dignified and deeply religious man, and when he promised to guide us with his expertise, I somehow knew he would not betray us.

"I remember him leading us over the dunes, every so often probing the earth with his long stick, yet after two weeks of searching, we had found nothing. It was really most frustrating. Usually, when we went to a site, we had a very good idea where to dig. We knew how to work. But there, at Deir el-Balah, we were working hand in hand with the plunderers.

"So finally, after two totally uneventful weeks, Khammad and his people came to me, and we had a talk. My husband and I had, up to that point, been hiring the Bedouins as cheap labor. They now wanted us to improve the arrangement. They wanted us to find them some real money, to begin paying them real working wages. After long negotiations we agreed to the new arrangement. And that was the end of our problem.

"Two days later Khammad came to me and said, 'Well, suppose I told you that one day soon we will find a tomb. One day it is coming. Tomorrow it will be coming.' "

In the morning Khammad went out with his divining rod to find the tomb. He held the staff before him, as if he were Merlin . . . or Moses. It was an impressive display. But, Trude told herself, he knows in advance where the tomb is. He knows because he has robbed. He knows because he has already worked the site for so long. He knows everything about this place!

At the foothills of the high dunes Khammad paused, jabbed his staff into the ground, and announced, "Here will be a tomb."

Beneath the dune lay the original ground surface, dark and fertile, readily distinguishable from the white sands. Clearing the soil away, the Dothans got their first view of an unplundered coffin: a wide-eyed face with baked-clay hands folded peacefully on the lid. At first viewing, it had seemed completely undisturbed, but further excavation revealed that the lid had caved in under the weight of the dune and that the white sands had spilled into the cavity, pressing its contents in place, fossilizing them for all time.

As the sand pit was widened to allow careful removal and photographic mapping of the treasures, Khammad became visibly uneasy. He was the

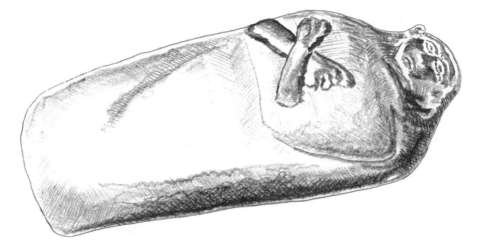

The anthropoid coffins of Deir el-Balah are almost identical to those found at a Philistine site in Bet-She'an. This burial practice might have been borrowed from the Egyptians during a time when the Philistines' ancestors, among them the Minoans, had settled peacefully along the Nile. When she discovered the coffins, archaeologist Trude Dothan found the crossed arms and faces with open eyes "a little spooky at first, but they grow on you."

leader of the local Bedouins, and Trude could not help noticing the dejected expressions on their faces as the first objects of gold came out of the grave. The men were hurriedly jotting figures down in notebooks. Trude guessed that they were privately estimating the price each artifact would have brought on the Jerusalem market and balancing this against what she was paying them to give up their treasure hoard. Khammad had decided that their tomb-robbing days were over, and Trude knew that they would follow him so long as he remained a committed member of her team. But as he watched a small fortune about to disappear into some museum, it was clear to the Dothans that Khammad was beginning to have second thoughts. It began to look as if this might become the last Gaza tomb they would ever be allowed to find. And there was nothing in the world they could do about it.

As more and more coffin pieces and the sands that had slipped through them were removed, the first bones came into view. "A truly dazzling sight," Trude recalls. "Inside lay the skeletons of a man and a woman, touching each other in a pose so romantic that we nicknamed them Romeo and Juliet."

Glistening in the sand beside the skulls were twenty gold pendants, mingled with crescent-shaped ingots, pairs of earrings, and braided gold loops with fruit-shaped drops. One bony hand still bore a golden seal ring.

Game over, Trude thought. She looked at Khammad, and he was looking more distressed than ever. And then an anthropologist removed the skulls from the grave, set them down on a rock, and photographed them. He turned them upside down in his hands, poking at their teeth, examining muscle scars. "Well," he said, "she is twenty. That one is twenty-six."

Trude remembers: "Without our planning it that way or expecting it, the announcement won Khammad over—instantly. Everything changed. He looked at us and then at the skulls, and suddenly he saw something that fascinated him: that we could know she died at the age of twenty. And from that moment on, I think he saw how we worked and what sorts of information could be dredged from the past. In that single moment he really became part of our group.

"He died a few years ago, and I miss him terribly. He was one of the sharpest men I ever met. And in time his tribe found a whole settlement concealed beneath the dunes. Considering the opulence of the grave gifts we had found, we expected the settlement to yield up buildings with expensive wall frescoes and stone columns, much like Spyridon Marinatos' buried city of Thera. But there were just plain mud-brick walls and small rooms, very similar to what one sees in little Arab villages today. And I said to Khammad, 'I don't believe that people living in such humble, mud-brick houses could have been the same wealthy men and women buried in the cemetery.'

"Khammad laughed. He said, 'Look at my house. It is made of mud-bricks like these. Sometimes I even live in a tent.' Yet there was no doubt that he was a wealthy man.

"Those were very romantic days, the likes of which I have never had before or after. We tried burrowing sideways into the settlement [slicing it open like a layer cake], but the continuation of the buildings ran under a sand dune that towered about forty-three feet over our heads."

The explorers had time enough to see wonderful Minoan and Mycenaean vessels (which had come to Israel by trade from the people known in the Bible as Kaphtorim). In a layer immediately above the Mycenaean pots, the Dothans saw a Philistine settlement, but from the trickle of sand already beginning to drift down, they realized that they had reached their limit, that further digging could bring the whole dune down upon them, a turn of events that might have become confusing to archaeologists three thousand years hence: skeletons with twentieth century A.D. fillings in their teeth, and cameras at their sides, splayed out on the floor of a Philistine home. But a war interrupted them, and perhaps saved them.

~~~~~~

On October 6, 1973, while the Israelis were celebrating the holiest day in the Jewish calendar, Egypt and Syria launched what was to go down in history as the Yom Kippur War. The Dothans had returned home to Jerusalem for the Jewish New Year when Egyptian troops crossed the Suez Canal into Sinai and Gaza, and Syrian tanks and infantry stormed into the northern territories. From across the borders Soviet-made rockets rained down upon Israeli air bases, destroying a large portion of the air force while it was still on the ground. So devastating was the defeat that Defense Minister Dayan broke down into tears on public television.

Two days later Israel struck back—struck back furiously—and somehow drove the Syrians back within twenty-five miles of Damascus. Equally miraculous, and in an ironic reversal of the Exodus route in the time of Tuthmosis III, Israeli tanks rode west over the red sands between Egypt's Bitter Lakes and the Reed Sea, entrenching themselves within sixty miles of Cairo and the Pyramids. The global aftershocks of oil embargoes and political negotiations, of wars and rumors of wars, kept the Dothans from Khammad and the Gaza dune field until 1977.

When Trude returned, she found the site totally undisturbed. Thanks to

Khammad, the foothills of the dunes were exactly as she had left them, still awaiting her spade. The dunes themselves, being a relatively young geologic feature, had to be removed before exploration of the lost settlement could continue.

"We called in construction and sand removal experts for advice," says Trude. "Heavy earthmoving equipment was the only practical answer. But the cost of such a massive undertaking was far beyond our budget."

One of the construction experts had pointed out to Trude that the pure white sand was valued for making fine glass. Meanwhile, the lawyer who owned the land had become interested in the fertile soil beneath the dunes. He liked the prospect that after the dunes were removed and the archaeologists had completed their work, his fields would be expanded. With the landowner's backing, Trude persuaded the courts to declare the area a "free sand zone." During the year that followed, trucks carted the dunes away, clearing a path for the archaeologists.

More unplundered tombs were uncovered. The Dothans did not understand why they always found more than one body inside each coffin; one afternoon, when a lid was removed and three skeletons were found tangled together, the mystery grew even deeper. They were two men and a woman, all buried at the same hour. They had survived into their thirties and forties and were therefore rather elderly by the standards of their time. For some reason the coffin held no burial gifts (was this a sign of poverty, or scandal and disgrace?). The skulls of the men were undamaged, but the woman's forehead appeared to have been smashed.

During an open-air breakfast, as they sipped cups of the bittersweet coffee Khammad's son had prepared, the archaeologists speculated about the matter: Perhaps two brothers and a sister had perished during an epidemic, and the large hole in the woman's skull was the result of a failed attempt to save her with trepanation (an ancient form of brain surgery). Someone raised the more gruesome possibility that the crowded coffins reflected a custom of killing and entombing wives and other family members with the deceased to accompany them in the next world. Ritual burial was always hard to interpret. In one pit, Gaza's enigmatic contemporaries of Joshua and Saul had interred a decapitated puppy with its head placed between the hind legs. Trude could not even guess what this was supposed to signify. But she did have definite ideas about the three skeletons whose tomb lacked burial gifts. She wondered if the damage to the woman's skull might be the result of a Bronze Age love triangle that ended in a double murder-suicide or perhaps a double murder in which the killer, having been tried and found guilty, was buried with his victims.

Elsewhere in the site, at a level slightly above the coffins, Trude found a whole stratum filled with shards of typical Philistine pottery. "This," she recalls, "unmistakably demonstrated the beginning of the end of Egyptian domination [during the thirteenth and fourteenth centuries B.C.] and shed new light on the people who had built the human-shaped coffins. That the graves bear Egyptian signet rings and scarabs, and that the coffins themselves are essentially clay imitations of Egyptian mummy cases, bespeak a people deeply steeped in Egyptian culture and religion. The period in which they lived was one of intensive trade with Egypt and of great ethnic changes and political upheaval. It was the time of the last flowering of the Egyptian New Kingdom before its decline to the point where the Bible scorned it (in II Kings 18:21) as a 'bruised reed.'

"After defeating the invading northern People of the Sea (among whom the Philistines played a dominant role), Pharaoh Ramses III [about 1300 B.C.] appointed their remnants as mercenaries in his own strongholds. Defeated in war, the Philistines borrowed from the culture of their conquerors, including their burial customs."

The Dothans had seen a similar adoption of foreign burial customs in the Jordan Valley, in the Egyptian-Philistine strata of Bet-She'an. There, the coffins' faces had been decorated with the same "feathered crowns" identified on the Karnak walls of Ramses III's Mortuary Temple as belonging to the Sea People. A comparison with the Philistine coffins of Bet-She'an convinced Trude that the slightly older Deir el-Balah coffins must have been the prototypes for those found at Bet-She'an.

And where, in this picture, do we find the Hebrew tribes?

"If it were not for the Bible," says Moshe Dothan, "nobody would even start to think about it. We know that the Israelites probably did not enter Canaan in only one large migration (as the Book of Exodus would have it) but instead arrived in small groups. Some of them settled here permanently. Others probably moved on. It was not a very simple picture, with uniform and simultaneous changes sweeping over the entire country. Rather, this period was characterized by the rise, the fall, the comings and goings of Canaanite, Egyptian, Philistine, and Israelite cultures, and all of these things tended to overlap.

"About 1450 B.C. [based on Theran ash layer dates] we begin to find prodigious quantities of pottery connected with the Israelites (most notably low-quality but distinctive imitations of Philistine beer jugs). We know, therefore, that they became deeply entrenched in the land around the time of Ramses II and the Deir el-Balah coffins, and that by the time Ramses III defeated the Philistines [in approximately 1300 B.C.], the enmity between our two cultures was already more than a century old."

JERUSALEM, TIME PRESENT

An irony: Though Israel is the only place in the world where the market for ancient artifacts is aboveground and legal, Israeli archaeologists are the only ones I have encountered who, without exception, do not decorate their living rooms with antiquities, who do not even keep the smallest souvenirs from their own sites (even as their next-door neighbors convert broken Roman columns—which are to be found everywhere—into backyard planters and coffee tables). Instead of Egyptian and Philistine limestone reliefs, the Dothans' and Mazars' homes display modern works of art, occasional reproductions of an ancient work, and photographs—volumes and volumes of photographs.

Trude opens a book to photos of a procession of men painted on an Egyptian wall. They are the ancient inhabitants of Crete, people ancestral to the Philistines. "Well," she explains, pointing, "here, in these hidden tombs, you have what are called the emissary drawings, such as those found in the tomb of Rekhmire [foreign minister to Tuthmosis III]. And we see in these drawings [and read in the hieroglyphs that accompany them] people arriving from many nations. And among them are Keftiu. Keftiu are the Minoans."

She pulls out more photos, including a partly reconstructed painting on plaster showing a leaf and a vine. Even in fragmentary form, faded and photographed in poor light, it is clearly a Minoan design. It looks as if it might be a close-up of one of Thera's frescoes. But it is not from Thera. I know that, because Trude is grinning.

There it is, that easy, mischievous smile. "This is something quite new. *Minoan,* and it is on the Sea of Galilee."

"My God . . ."

"That's what I said when I saw it for the first time. I looked at this one leaf, painted against a whitewashed background, and it looked just like the hand of one of the painters from Thera."

"What was it? An outpost of some kind?"

Trude shakes her head. "A palace or whatever. We don't know enough about it yet. The excavation is just beginning."

The big question is: Was the site built before or after Thera exploded? If before, we can only wonder what its builders, or its builders' descendants, must have thought watching from the Galilee as their civilization went up like smoke from a furnace on the western horizon. If after, they would have come as remnants of that fallen empire, as refugees somehow accepted by the surrounding Canaanite population. The latter seems to me less likely.

Good land was scarce. A relatively powerless competitor would not readily be accepted here, leading me to suspect that the site predates the 1628 B.C. explosion.

There had always been indications of contact between Canaanites and Minoans, but they were vague, limited to the occasional storage jar of distinctly Cretan design. But never has there been evidence of a full-blown Minoan settlement even along the coast, and finding an outpost of the world's first naval empire so far inland as the Galilee is even more surprising. It suggests a very large Minoan presence.

"My big question, Charles, is did they come here to settle or to do battle?"

"Maybe they intended to do neither. Maybe it was like much of the world today. Representatives of an economic superpower came out to do their business, to seek out new trade, and some of them liked what they found in new and exotic lands and decided to stay here."

"Yes," says Trude. "I can see them coming along with their trinkets and other trade goods."

"Then you have little neighborhoods of people from other countries springing up."

"They would have tended to settle around the trade routes, wherever the ships [and caravans] stopped. But the Minoans of the Galilee . . ." Trude trails off into thought. "We have detective work ahead of us," she says at last. "We do not know that a trade route actually stopped at the Galilee site. We cannot know if these people were colonists, sailors, or traders."*

---

*Detective work is always full of surprises, especially when we try to reconstruct life amid vanished species and vanished civilizations. During the winter of 1994, Nanno Marinatos, the archaeologist daughter of Spyridon Marinatos, called to say, "I have wonderful news for you." A second Minoan outpost, apparently dating to the time of the Thera explosion, or perhaps a little earlier, had been unearthed at Avaris, Egypt. In pre–Eighteenth Dynasty times, Avaris was the capitol of Egypt's Hyksos oppressors (and, according to Israelite oral traditions, was the city from which the Exodus began). During only one year of excavation, three major buildings at Avaris were already rivaling the best Minoan architecture of Thera and Crete. Over plates of octopus and sweet shrimp at one of Manhattan's all-you-can-eat sushi bars, Nanno described four hundred crates filled with frescoes from the walls of Avaris, depicting beautiful landscapes, royal processions, hunting scenes, and Minoan gymnasts leaping over the backs of charging bulls.

"They were living in luxury even in the heart of Egypt," I said, grinning.

"Not likely a military outpost," Nanno agreed.

"They had the world's first navy, but it begins to look as if their real power was in trade," I said. "Their real power was economic." We laughed at the strangeness of it, and anyone passing by could have guessed that someone's detective work was coming along nicely.

She returns her attention to photos of foundations and broken plaster, and no matter how hard she tries to convince herself otherwise, she continues to see the hand of a nameless artist from Marinatos' lost city of Thera in the fresco fragments. She reads the bits of plaster and mud brick the way one reads a book. It is as if, in the way that books allow, she slips into the mind of the fresco painter.

The first Book of Samuel (13:19–23) claims that an apparent remnant of the Minoans (among them the Philistines), while still relative newcomers to the land of Israel, somehow managed to establish a monopoly over the metal industry during the time of Saul. At the Philistine city of Ashdod (a few miles north of Deir el-Balah), seals inscribed on metal ingots lend support to the claim, and attest to the literacy of the Sea People. They were, at that time, using a script almost identical to the lost (that is, almost impossible to decipher) Minoan Linear A script known from Late Bronze Age Crete and Cyprus. Far from being uncivilized brutes deserving (in biblical accounts) of utter extermination, the Philistines, like their contemporary migrants the Israelites, brought with them a military prowess, sophisticated technology, and a cultural heritage already more than a thousand years old.

"The hard facts are seldom able to alter perceptions that have become ingrained over centuries," says Trude. "From the perspective of the Bible, the Philistines were a nation of uncouth barbarians, and remain so to the present. . . . Whether or not they can ever shed their negative image, they have, at least, emerged from the web of myth onto the stage of history."

And for that, we (and presumably the Philistines) can thank the Dothans.

BET-SHE'AN, TIME PRESENT

It is one of the hottest places on Earth. And it is a humid heat, a heat so unbelievably potent that discomfort is sometimes replaced by fascination. Here, in the upper Jordan Valley, we stand at the bottom of a scratch in the Earth's surface. We are located far enough below sea level that if a major earthquake were to breach the natural dams in the south, the inrush of the Red Sea would drown this whole valley from horizon to horizon, in a flood as deep as most of the world's skyscrapers. Because we stand below sea level, the atmosphere piles up more thickly than on the earth above. It traps more of the Sun's infrared radiation, like a greenhouse. Downriver the Dead Sea is evaporating at a furious rate. The land around it is a desert, for the warm, moist air rises in a vertical column, then spreads out like the head of a mighty anvil. The Sun's anvil stores the heat and moisture greedily, to give them forth later, as its uppermost decks spread north and brush over the hills of Bet-She'an. After only an hour's exertion in the hills, one comes away feeling like a snail with a beard. The air is so dense that it actually shields us from the Sun's ultraviolet rays, and downriver, where the Jordan meets the Dead Sea and the valley runs more than a thousand feet deeper, it is virtually impossible to get a sunburn.

Ami Mazar tells me that the remains of some twenty-five cities lay under my feet, each built upon the ruins of its predecessor. The earliest dates back to about 5000 B.C. (this places the settlement of Bet-She'an about three thousand years behind Jericho and fifteen hundred years ahead of the Pyramids). The Philistine rulers of Bet-She'an are said to have displayed the bodies of Saul and Jonathan on the city walls.

The mound upon which we stand is the residuum of all those twenty-five layers of civilization, piled more than 120 feet high. In the distance, the river Jordan can be glimpsed through treetops and morning haze, and the air, though shimmering with heat, brings to us the sweet fragrance of orange groves.

Excavation teams have changed dramatically since the time of Leonard Woolley at Babylon, Spyridon Marinatos at Thera, and the Dothans at Deir el-Balah. There are no more Hamoudis or Khammads. The face you meet in the pit is most likely to be that of an American housewife on vacation. Because budgets are dwindling and paid workers are so expensive, and because lost civilizations hold a special fascination for many people, the new generation of Israeli archaeologists has been able to tap into a volunteer workforce.* Presently, Ami's team is sinking shafts and trenches into the peak of the mound. On the floors of the shallowest trenches, the volunteers are standing in the time of Ramses III and the Sea People. The deeper slices plunge through Thera into the Middle Bronze Age, where circular, rock-lined pits have been uncovered just twenty feet from the walls of a temple. To Ami Mazar, they were sacrificial pits because a layer of black ash and lamb bones were found at the bottom. To me (or at least to that paleontological side of me), they might just as well have been barbecue pits, for all I care. The important thing about them is that the pits would have been an

---

*At this writing, and for the foreseeable future, lists of sites requiring volunteers are announced every January in the *Biblical Archaeology Review*. "It is very romantic to many of our visitors," says Ami. "Archaeology, the search for our origins—people from all over the world are coming to this place. And while there are all these global divisions and conflicts, Israeli archaeology is taking on an international atmosphere. That's one of the nice things about it."

effective trap for air- and rain-borne debris. The bottom of one pit predated the Thera explosion only by about a hundred years, and somewhere between the lamb bones and the Philistines, Thera's ash cloud must have deposited at least a few of its very distinctive grains. I would have liked to look for them (one more snippet of evidence linking Thera and Tuthmosis III to the year 1628 B.C. and thus allowing us to more precisely date Bet-She'an), but the archaeologists were searching for something else. If indeed any ash layer existed, they erased it completely on their way down to the lamb bones.

Ami points up from the pit to a horizon of mud bricks, barely discernible in a slice his volunteers have made through the Canaanite Period: "This site's major importance was during the Late Bronze Age. At that time, Bet-She'an became a stronghold of the Egyptian government. The Egyptians conquered this country."

"Now, under whose reign was this?"

"Tuthmosis the Third," comes the reply. "And for about three hundred fifty years afterward, this country was part of the Egyptian Empire.* Though the local people continued to live as Canaanites, the Egyptians were controlling the country, exploiting it."

"Tuthmosis the Third," I say. "You know what I am thinking?"

"Yes, I'm afraid I do."

"Somewhere in the Tuthmosis the Third layer, it's got to exist. We've identified a Theran ash layer in the Nile, to the south. And we have it to the north, in Turkey and Cyprus. There simply has to be an ash layer in Israel. Has to be. It was certainly blowing this way."

"No one has found Theran ash in Israel. But maybe that is because we didn't look very well, or know what to look for."

"I would look for microscopic shards of volcanic glass mixed in with a layer of dirt three inches, perhaps six inches thick. If we're looking at an ancient Bet-She'an street, the ash would have been trampled in with the preexisting ground surface, and it would have continued to be mixed in with any new sediments accumulating on top of the road—"

Ami stops me. "Let us not even touch this subject," he protests. "Because it is very, very highly debated amongst us. And the raising of the Thera

---

*Ami Mazar also notes that during some part of the tumultuous century immediately following Tuthmosis III, a series of destructions and rebuildings occurred throughout the country. The destructions are of unknown origin, and there appears also to have been a general decline of settlement during that same period (as if some city-dwelling communities had taken up an archaeologically invisible, nomadic existence).

dates is very new to us, as you know. I have read your report, and I think it is still too early. I wouldn't even try to redate *Egypt* on the basis of Thera—much less Israel."

"All right. We'll stay out of that argument for now."

"I'd like it if you do. It's really a minefield, because I am digging in the same period to which you and a handful of volcanologists have brought such calamity."

He shifts attention to the four Canaanite temples of Bet-She'an. An earlier team from the University of Pennsylvania found the temples built one on top of the other, each more elaborate than its predecessor, and each from a different century—"from the eleventh century, twelfth century, thirteenth century, and fourteenth century B.C.," Ami explains. He notes that the oldest temple was built about a hundred years after the reign of Tuthmosis III, and in my mind (using Thera, 1628 B.C. as a compass setting), I immediately push the first Canaanite temple back to the fifteenth or sixteenth century B.C., filing the recalibration away for future reference.

"Four successive temples," Ami says. It reminds me of what Elizabeth Stone and Paul Zimansky have said about ancient Iraq: that the more chaotic a land is, the greater the need to create a few stable, mystic places and to focus on them for many centuries. "So I assumed," Ami continues, "that if there were four temples, one on top of the other, there should be a fifth one somewhere below."

He points to a one-room shrine: mud-brick benches covered over with white plaster. The walls of the shrine are long gone. There is bench space for perhaps a half dozen priests. This is where it began. This is where the site of Bet-She'an first became sacred.

"You see," says Ami, "This continuity of culture is very common in the Canaanite Period. Temple One is an unusually small shrine. Its date, too, is interesting: sixteenth century B.C. [seventeenth century B.C., I recalibrate]—very close to the date of . . ."

"Yes?"

"Thera."

I have difficulty restraining a smile. And Ami, quite unexpectedly, announces, "Thera, of course, by the conventional date, was destroyed around 1500 B.C." He has met me almost halfway, pulling the "conventional date" back nearly a hundred years, and my grin widens uncontrollably.

"That's not the conventional date," I point out. "Convention places Thera between 1400 and 1450 B.C., and usually closer to 1400."

"Don't raise it anymore," Ami says.

"I won't."

"Good."

So we stand upon the anvil of the Sun, Ami and I, poised between horizons of civilization now so utterly lost that the very sunlight under which the temple builders toiled has already passed the Pleiades and the great Horsehead Nebula in Orion. According to a legend of the Mycenaeans (the Greek conquerors of post-Thera Crete who absorbed the culture of the Minoans), a Canaanite princess was brought from Israel to Crete by the god Zeus, who appeared to her in the form of a beautiful chestnut bull. She mounted his back, and he raced off with her across the winds of the sea, vowing his eternal love. Her name was Europa, and in addition to giving her name to a continent, and to a Jovian moon, she brought forth King Minos through divine birth and became part of the Minoan-Canaanite Creation myth.

"You know," Ami says, "the Bible contains a lot of material that evolved from common wisdom—the accumulated wisdom of the ancient Middle East: Minoan, Canaanite, Egyptian, and Babylonian.

"And when I look at the first chapters of the Book of Genesis, to me, the Hebrew becomes a condensed, telescopic view of the entire history of Earth from the Big Bang."

"Yes, Ami, the Big Bang. *That* odd parallel, I have always wondered about!"

"And there it is, the astronomers' Big Bang, in Chapter One. Because what does it say? It says, in the beginning there was nothing. There was total void. And then there was light."

"A whole lot of light, if our astronomy serves us correctly. Looking out from our solar system, we have discovered a universe that appears to be scattering to all points of the celestial compass, one that is blowing up even as we watch it. It is as though our galaxy and all its neighbors were caught up in the aftermath of a stupendous detonation. And when we trace the explosion backward in time, we discover that the event we call the Big Bang did not occur somewhere in space. It occupied all of space, bursting forth from every direction at once as photons of light. For more than three hundred thousand years, the universe was bathed in a bright yellow glow."

"I think, Charles, that the intuition of the Babylonians was more advanced than we have assumed. The story of the Creation reflects their level of scientific thought, which in some essentials is very close to what we know today. How did the ancients know it, without studying archaeology, paleontology, and astrophysics? I don't know, and neither do you. But the fact is that they had this observation."

For my own part, the similarities are just another of nature's odd coincidences. Interesting to think about, but not necessarily Earthshaking.

Then, again, for all I know, there's so much more I don't know. For all I know, Ami Mazar may be on to something. If the parallels between parts of the biblical Creation story and the Big Bang do not seem particularly dramatic, can anyone read the Creation myth of the ancient Maya (who, like the Babylonians, were accomplished mathematicians and astronomers) without wondering if humanity is somehow doomed to keep repeating itself?

The Bible opens with the words, "In the beginning, God created the heaven and the earth. And the earth was without form and void; and darkness was upon the face of the deep." The Mayan Popol Vuh echoes: "Before the world was created, calm and silence were the great kings that ruled. Nothing existed . . . the face of the earth was unseen. There was only motionless sea, and a great emptiness of sky. . . . It was night, silence stood in the dark."

Genesis: "And a wind from God moved over the surface of the waters. And God said, Let there be light. And there was light."

Popol Vuh: "Flatness and emptiness, only the sea, alone and breathless. . . . In the darkness the Creators waited. . . . Then let the emptiness fill! they said. Let the light break on the ridges, let the sky fill up with the yellow light of dawn!"

Genesis: "And God said, Let there be a firmament in the midst of the waters, and let it divide water from water."

Popol Vuh: "Let the water weave its way downward so the earth can show its face."

Genesis: "And God said . . . Let the dry land appear: and it was so."

Popol Vuh: "Earth! the Creators called. They called only once, and it was there, from a mist, from a cloud of dust, the mountains appeared instantly."

Genesis: "And God said, Let the earth bring forth grass, herb yielding seed, and fruit tree yielding fruit after its kind . . . and it was so."

Popol Vuh: "At this single word the groves of cypresses and pines sent out shoots."

Genesis: "And God said, Let the waters swim abundantly with moving creatures that have life, and let birds fly above the earth in the open firmament of heaven. And God created the great crocodiles, and every living creature that moves."

Popol Vuh: "The Creators often asked, Will this silence reign under the trees forever? Suddenly there were the animals: deer, birds, jaguars, snakes."

Genesis: "And God said, let us make Mankind."

Popol Vuh: "Let our glory be a man walking on a path through the trees! the Creators called."

Beginnings have always been shadowy and strange, have always intrigued us. By the standards of most people the temples of the Maya, the Babylonians, the Philistines, and the Canaanites are so old that they go back almost to the dawn of time itself; but just the other night, about an hour after sunset, I watched the *Mir* space station racing overhead and glimpsed archaeological time scales shifting paleontological. The artifact will be going around and around up there for the next million years, untouched by wind and rain. Its hull will still be shiny long after Ami and I and the towers of Manhattan have crumbled to dust, long after the Sphinx has ceased to be even a memory.

I wonder what the Canaanite temple builders and mythmakers of seventeenth century B.C. Bet-She'an would have thought if there had been a way for them to know that here, on the other side of Year Zero, people would be drawing aluminum and titanium from the rocks and hurling contraptions of flying fire into the frontiers of space. I wonder what they would have thought if they could have looked ahead to Christmas Eve, A.D. 1968, and seen that the Creation stories of their time were so enduring and so pervasive that when men first went to the moon, they were compelled to send a most unexpected message to "all the people back on Earth."

From *Apollo 8,* a quarter million miles away, an echo from the past crackled down:

"In the beginning, God created the heavens and the earth. . . ."

12

SECRETS
OF THE
SCROLLS

If by "God" one means the set of physical laws that govern the universe, then clearly there is such a God. This God is emotionally unsatisfying . . . it does not make much sense to pray to the law of gravity.
—CARL SAGAN, to *U.S. News & World Report,*
   December 23, 1991

Dear God, if you're there—and I hope you are—not tonight. Not tonight!
—CHARLIE'S PRAYER, as he prepared for a night's
   sleep after touring one of Russia's
   Chernobyl-class reactors

Maybe this world is another planet's hell.
—ALDOUS HUXLEY

THE QUMRAN SETTLEMENT, A.D. 68

On the northwest shore of the Dead Sea, within sight of the Jericho mound, a small spring had long sustained a community whose foundation stones, baths, and underground water conduits would still be standing as the second millennium drew to a close.

Although the floor plans of kitchens, dining areas, stables, and work areas were going to suggest to future historians that the place was a scrollery (a forerunner of the medieval monasteries in which multiple copies of the Bible would be tediously handwritten), no one would ever really be sure who the Qumran occupants were or exactly what they were doing here.

About this time, a Roman historian and naturalist, known throughout the world as Pliny the Elder, began exploring the Jordan Valley and the Dead Sea. A few miles south of Jericho, where Qumran lies, he came across what he described as a spring-fed settlement of deeply religious men, whom he called the Essenes. Before his death in A.D. 79 (beneath the volcanic death cloud of Vesuvius) Pliny had time enough to finish a best-selling book detailing his adventures and, from his home in Pompeii, to send forth the only eyewitness account the world would have of the Qumran people:

> On the west side of the Dead Sea . . . is the solitary tribe of Essenes, which is remarkable beyond all the other tribes in the whole world, as it has no women and has renounced all sexual desire, has no money, and has only palm trees for company. Day by day the throng of refugees is recruited to an equal number by numerous accessions of persons tired of life and driven thither by waves of fortune to adopt their manners. Thus through thousands of ages (incredible to relate), a race in which no one is born lives on forever.*

*This English translation of Pliny, *Natural History 2,* can be found in H. Rackham, Loeb Classical Library (London: Heinemann/Cambridge, Mass: Harvard University Press, 1969). The apparent similarity between Pliny's Essenes and medieval monasteries may be more real

In clifftop caves near Pliny's ageless tribe, texts from diverse places were being placed in terra-cotta jars and sealed behind false rock faces for safe-keeping. At the lowest reaches of the Jordan Valley, in air that was always hot and dry, the Israelite book depository would last almost forever. But whoever created the archive was clearly not intent on inventing an Israelite version of the Library of Congress, with texts arranged by subject and author. The scrolls were deposited haphazardly and in random order, as if someone were hastily attempting to conceal whatever texts could be scooped up and ferried to the caves. Scrolls written in Greek were placed alongside those written in Hebrew. Multiple copies of the Book of Isaiah were mingled with land deeds, marriage contracts, bills of divorce, scriptural commentaries, and at least one private letter. They were random snapshots of life in and around the Jordan Valley about the time of the Hebrew revolt against Rome. This was the end of the Second Temple Period, the rout of Solomon's kingdom.

~~~~~~

Israel's First Temple Period ended in 586 B.C., when the Babylonian king Nebuchadnezzar captured Jerusalem, put the torch to every building that stood atop the Temple Mount, and condemned large numbers of survivors to slave labor in Babylon. Twenty-five years later Nebuchadnezzar was dead and the descendants of Israelite captives were running the local Babylonian markets. Under an easing of old restrictions, educated Jewish scribes began assembling and editing hundreds of ancient documents to form the books of the Old Testament. Thirty-five years later, almost on schedule, a new cycle of chaos and decline swept through the Tigris-Euphrates world. Babylon soon lost its grip on Israel, and with riots, economic collapse, and famine spreading throughout the land, there was nothing of value to keep the Hebrews in ancient Iraq. So they migrated home, bringing with them the first official synthesis of the biblical texts.

In 444 B.C., as the Second Temple was dedicated, a new empire was being born in the west and beginning to spread east. Four centuries later, as the Psalms of Solomon were being written, Rome annexed Crete, Cyprus, Turkey, and Jerusalem. From 37 to 4 B.C., the Roman puppet king

---

than apparent. This was the time of the first generations of Christianity, and the soon-to-be-born holy Roman Catholic Church would, like the Essene community, eventually move toward a tradition of living on through a constantly recruited priesthood of celibate men, leading one to wonder if the Essenes, as described by Pliny, were in some way ancestral to the Roman Catholic Church.

Herod the Great more than doubled the size of the Temple Mount, adding a fortress and giant pools.

In A.D. 18 Joseph Caiaphas became high priest of Jerusalem, and approximately eleven years later (according to the New Testament) John the Baptist and Jesus were put to death. In A.D. 39, as the emperor Gaius Caesar Germanicus (Caligula) proclaimed himself a living god and decreed that the priests of the Temple Mount would henceforth worship him as king of the Jews, James the Righteous (also known as James the Just) began preaching the faith that eventually became known to the world as Christianity. Like the Temple priests, James (according to accounts given in Scripture, and by the second-century church historian Hegesippus) wore priestly robes and was allowed to enter the innermost and holiest of the Temple's chambers. The authors of the New Testament called him "the brother of the Lord," leaving scholars to ponder for all time whether Christianity's founders had meant him to be the spiritual or actual brother of Jesus. It is clear, however, from the New Testament epistle that bears his name, that James did not view himself as helping found a new religion but as a Jewish reformer. To him, Christianity was simply a fulfillment of Judaism, and that was why he continued to pray daily in the Temple and to defend the laws of Moses, even as King Herod Agrippa I (who had been instrumental in the condemnation of Jesus as a heretic) wrapped himself in a silver robe, declared that the Temple priests would worship him as the Messiah, and promptly died of a mysterious wasting disease.

By A.D. 66 James, the reformer, had gained favor even with the Pharisees, leaders of the synagogues who were becoming prototypes for the coming age of rabbinic Judaism. But the Sadducees, an aristocracy of high priests left over from the Herodian dynasty of Roman puppets, saw him as a threat to their power and began drawing their plans against him. Proclaiming that the "Righteous One" had gone astray, they justified their intentions by invoking a passage from Isaiah 3:10–12: "Say of the righteous, that it shall be well with him. . . . Alas! . . . for according to the deserving of his hands shall be done to him."

Convincing themselves that Isaiah had thus prophesied the death of the "Righteous One," they surrounded him one afternoon while he prayed in the Temple, then threw him off a parapet and bludgeoned his dead body with clubs. Publicly they claimed merely to be bringing Isaiah's prophecy to fulfillment, fully expecting their claim to be the end of all arguments.

Among the scrolls hidden by the Qumran community are references to at least two saviorlike figures. Both were martyred. One of them, possibly James the Righteous, was identified as the "Teacher of Righteousness."

In Jerusalem the priesthood's justification for the murder of James became everything but the end of the argument, and A.D. 66 went down in history as the year of the Judean revolt. Among its first casualties were the high priests of the Temple Mount, executed as pro-Roman collaborators. Initially the Hebrews were victorious. They drove the Romans completely outside Jerusalem's gates; but the reaction from Rome was very swift, very brutal, and utterly devastating. A huge expeditionary force was dispatched under Vespasian. About the same time the Qumran scrolls were being sealed in caves, Jerusalem was under siege, the Second Temple would soon be burning, and Israel's Hebrews were preparing for a return to tribal nomadism and guerrilla warfare.

The Qumran people, knowing that Roman legions were poised for an assault on the Jordan Valley, could have anticipated nothing except death if they remained behind. Whether they fled into the mountains and survived or remained at Qumran and died, none of them would ever return to the caves—ever again. They simply inserted hundreds of scrolls—both sacred and mundane—into clay storage jars and cast them into futurity like messages in bottles. Just how far into futurity those bottles would sail, the preservers of Qumran could scarcely have dreamed.

〜〜〜〜

In A.D. 1947, as Arabs and Jews picked up weapons and resumed their four-thousand-year war against each other, a young Bedouin shepherd, pasturing his goats on a strip of scrub grass amid sand-blasted foundations and collapsed walls of the Qumran settlement, noticed a cave entrance exposed in a cliff. While the goats grazed, he idly tossed stones through the opening, using it as the bull's-eye in a game of target practice, and was surprised to hear the hollow crash of pottery vessels breaking inside. Hoping that he had come across something of value to the antiquities dealers, he climbed inside to have a look around. But he found no hidden Philistine tomb. No gold. No lost treasures of Solomon's Temple. There was only a collection of nondescript, baked-clay jars containing hand-scrawled papers that he feared were worthless. Just the same, he stuffed as many pieces of paper as he was able to carry into a single jar and brought it to Bethlehem. The antiquities dealers he met in the marketplace responded enthusiastically, and he was relieved to learn that the scrolls were indeed valuable after all. Three dealers were actually eager to bid against each other, and the shepherd settled the bid for fifteen dollars.

For nearly twenty centuries the ruins of the Qumran community had been crumbling in silence. Now, after all those years, first-century Israelites were about to speak, now that a shepherd boy had discovered the Dead Sea Scrolls. During the next few years 800 scrolls were recovered from the caves (of which 127 are now known to be biblical texts), but as more than one scholar has since observed, the Dead Sea Scrolls have always had an affinity for the shadows. After two millennia of burial, more than half of the scrolls' secrets were to be kept under lock and key by a self-appointed priesthood of fewer than a dozen scholars who claimed exclusive rights to the documents. Nowhere in the world had such a convention ever been written into law. There was simply no precedent for making proprietary claims on ancient historical documents. Legally, they had been in the public domain for a very long time; yet nearly a half century was to pass without translations or even photographs of the texts being released, while some of the scroll fragments, upon being removed from Qumran to the relatively humid elevations of Jerusalem, began crumbling into thick brown powder. They, and whatever had been written upon them, were lost forever. In spite of this, members of the "priesthood" justified their proprietary rights with claims that only they were competent to curate and translate the scrolls. Inevitably some of them began to grow old and die and, alarmingly, to bequeath their proprietary rights to a chosen dynasty of scholarly heirs.

It was only a matter of time before such seemingly inexplicable behavior began attracting conspiracy theorists. The most popular villain was the Vatican, allegedly hoping to hide the secret teachings of Christ. Others speculated that the scrolls reflected badly on the early Christians, or the Jews, or both. But sometimes things are simply as they appear. The truth leaked out every time a scroll team member labeled anyone else who wanted to examine "their" materials an "incompetent." It was simply a game of egos.

As people accustomed to descending into the Earth by inches, sometimes using toothbrushes and paintbrushes, archaeologists are a notoriously patient, if not an entirely saintly, lot. But even the patience of a saint has its limits. For most of us, even the approach of the Dead Sea Scrolls fiasco's half century mark did not inspire thoughts of mutiny. The limits of self-control were not exceeded until mid-1990, when scroll fragments began surfacing in the antiquities market and one of the scroll editors suddenly blurted to a reporter that Judaism was, to him, "a horrible religion." For an encore he called himself an "anti-Judaist." How a man who even harbored such views came to be in charge of translating and interpreting ancient Jewish documents was anybody's guess. For most of us, it was the final indignity, the proverbial last straw.

By the autumn of 1991 the mutiny was well under way, led by *Biblical*

Archaeology Review publisher Hershel Shanks and Huntington Library director William Moffett. They were, in a sense, the Fletcher Christian and George Stewart of archaeology. Rallying to the call of "Free the scrolls and you will free the scholars!" Shanks released a computerized reconstruction of scroll documents, and Moffett followed up by opening his library's vaults and releasing copies of scroll photographs taken years earlier as a precautionary measure against the likely destruction of the originals in a part of the world about as politically and militarily stable as uranium 235.*

Within weeks of the mutiny the first publicly released translations of the hidden scrolls began shedding new light on the concept of "Messiah." A five-line fragment referred to a Hebrew prophet—perhaps Jesus or James, but not necessarily so—who shared an ancestry with King David and was put to death. The decay rates of certain radioactive elements in the scrolls themselves told us that they had been written over a three-hundred-year period beginning near 250 B.C. and ending about the time Rome crushed the Jewish revolt of A.D. 66–68. This meant that during or even prior to the time of Jesus, writers belonging to an apparently Jewish sect believed in a Messiah who would suffer and be (or had already suffered and been) put to death.

"We've known for a long time that there are connections between ideas contained in the scrolls and Christianity," said University of Chicago archaeologist Michael Wise, who had participated in the translation. "However, this particular idea—the idea of a dying Messiah—is new and explosive. [It suggests that] this was not an idea unique to Christianity."

But is this revelation really so new and explosive as it appears? I don't think so. The Book of Isaiah contains several references to the "suffering servant," and the fact that Herod Agrippa I, the emperor Caligula, and even the (second century A.D.) Hebrew general Simon Bar Kokhba, near the ends of their lives, had proclaimed themselves Messiah, tells us that there

---

*As it turned out, the only time a bomb ever got close to the scrolls was the day I arrived (in September 1991) at their depository in Jerusalem's Rockefeller Museum. Someone sent a hand grenade skidding down the hall in my general direction, and it fizzled. A dud; it wasn't even exciting. Just the same, authorities evacuated the museum, and I never did get to see the scrolls. Though officially listed as an attack by Arab terrorists, the grenade looked suspiciously like a trainer: no scoring for shrapnel, and possibly nothing more dramatic inside than a firecracker. No Arab terrorist would have risked his life to throw a practice grenade. There was, at that time, a great deal of bitterness over the Dead Sea Scroll mutineers (and America's growing reluctance to guarantee fourteen billion dollars in loans to Israel), and I came to suspect that a certain renegade archaeologist of my acquaintance had sent the grenade as "a message to the American."

השׄמׄים והארץ ישמעו למשיחו

מתיר אסורים פוקח עורים זוקף כ[פופים]

או ירפא חללים ומתים יחיה ענוים יבשר

[. . . The hea]vens and the earth will obey His Messiah,

He will release the captives, make the blind see, raise up the do[wntrodden.]

then he will heal the sick, resurrect the dead, and to the poor announce glad tidings.

The 2,000-year-old Dead Sea Scroll text known as 4Q521 (that is, fragment 521 from Qumran Cave Four), discovered in A.D. 1946 but hidden from the world's scientists and historians until the mutiny of 1991, suggests that the Hebrew scribe who composed it was intensely Messianic. (Translation by Wise and Eisenman, based upon reconstructions by Strugnel-Qimron.)

were a lot of people walking the Earth in those days who thought themselves to be God-like.* We must also remember that Hebrew tradition owed much to Egypt, where the pharaoh was himself a messianic figure, destined to die, venture into the underworld, face judgment and resurrection, and ultimately bring salvation to the souls of his people. In Greek tradition, the minor god Prometheus (whose golden statue adorns New York City's Rockefeller Center) brought mankind fire and the art of healing,

---

*Herod Agrippa I had grown up in the same imperial household as Caligula, and it is therefore no coincidence that Caligula, when he claimed that he had metamorphosed into a living god, associated himself with the Messiah whose coming the Jews had been prophesizing for centuries. From time to time, Caligula expressed regrets that the prophecies also indicated he would die young and hated by his own people (one of the few points on which the emperor turned out to be entirely correct). Of course, his habit of riding into Roman pubs on horseback and demanding drinks for both himself and his horse did not exactly endear him to people. Caligula was eventually stabbed to death by his own guards. His horse, meanwhile, was appointed to the Roman Senate (which, if anything like the modern American Senate, might have produced an improvement).

and for this Zeus chained him to a rock and each day sent an eagle to peck out his liver. The idea that a suffering servant of man might be persecuted— might even die for man's sake—was an old story by the time the Dead Sea Scrolls were written.

This by no means diminishes the scrolls or what can be learned from them. In fact, comparison with other snippets from the jigsaw puzzle of history enhances them. The Qumran texts are almost a thousand years older than the Aleppo Codex, which was written in Israel about A.D. 900 and was, until A.D. 1947, the oldest-known Hebrew manuscript containing the full text of the Bible. What has surprised most scholars is that in spite of the millennium of hand copying that separates them, the Aleppo Codex and the books of the Old Testament unearthed at Qumran are virtually identical. One of the oldest of the Dead Sea Scrolls, dating to about 200 B.C., is the Book of Isaiah. Only thirteen minor variations from the Aleppo text (and from its modern descendants) have been identified. During the approximately three and a half centuries separating the Qumran Book of Isaiah from the version originally compiled in Babylon, one might be tempted to posit even fewer changes than during the eleven centuries separating Qumran and Aleppo. If this assumption is correct, then the modern Hebrew Bible is, in most essentials, the same Bible one would have found in Babylon about 550 B.C.

Evidence supporting this assumption can be found in Jerusalem's silver scroll—which is, next to the writing on the 750 B.C. wall of Deir Alla, the earliest fragment of biblical text presently known. The scroll of beaten and inscribed silver was contained in a one-inch-diameter prayer amulet. Construction workers found it in A.D. 1981, when they accidentally broke into a grave site under New Jerusalem's cinema district. The verses on the scroll are literally microscopic, and are indistinguishable from the priestly benediction in Numbers 6:24–26. The biblical fragment, dating between 700 and 500 B.C., is almost an exact contemporary with the Babylonian compilations and reads: "The Lord bless thee and keep thee: The Lord make his face shine upon thee: The Lord lift up his countenance to thee, and give thee peace." The modern biblical version reads: "The LORD bless thee and keep thee: The LORD make his face shine upon thee: The LORD lift up his countenance to thee, and give thee peace."

More than twenty-five hundred years . . . yet we witness no mutation of the text. The DNA of moths living on the Hawaiian Islands has undergone vastly greater change in that same time frame.* It is testimony to the

---

*Two native species of moths (*Hedylepta maia* and *H. meyricki*) have evolved in Hawaii to feed exclusively on banana plants. Bananas were first brought to the islands by Polynesian settlers about sixteen hundred years ago. The new moth species can be no older.

amazing fidelity of transmission, through dozens of generations of pious scribes, once the old song stories and oral histories were committed to writing. It may also be an indication of the piety with which the song stories were themselves preserved, during the centuries before the Silver Scroll, Babylon, and Qumran.

~~~~~~

When I first arrived at Qumran, in the autumn of 1991, James Tabor (an archaeologist from the University of North Carolina) was probing the nearby cliffs with a ground-scan radar, hoping to reveal undiscovered caves in the rock face—caves that might contain more scrolls.* The Jewish-Christian minister Zola Levitt was also there, and a Jesuit friend who sometimes waxes agnostic.

"It's a very rich area," Tabor told Levitt. "From a hundred years before the time of Jesus through about a hundred fifty years after, Jewish religious refugees were coming out into this desert. And they left pottery . . . and scrolls."

He then asked Levitt what he thought a modern archaeologist might do if he found a hidden cave and all the right signs were present: silt from campfires coating the walls, Roman pottery shards eroding out of the cave entrance, and a dozen other promising hints of human habitation from just the right period.

"And," Tabor added, "imagine we have two thousand years' worth (about nine to twelve feet) of bat feces extending down to the original cave floor. We are modern archaeologists. What do we do?"

Levitt shrugged.

"We don't just take our team in with masks and start digging and sweating," said Tabor. "Rather, we move in with our scanners, peer through the layers of dung and debris, and image the pots and scrolls. We can determine precisely how deep they are, and in which corners they are located, before anyone actually starts digging.

"This," Tabor explained, "is the new lazy man's way to do archaeology."

---

*Such scanning technology has been evolving in leaps and bounds ever since. As this book goes to press, the first scanning deep-ocean robots are coming on line. We will first be using them in a region of the Atlantic where temperatures are only two degrees above freezing and there is no oxygen. Ships sinking into these waters might just as well have been flies sinking into amber. There are fourteen hundred of them, from all ages, and we now have the ability to "CAT scan" a Portuguese galleon, feed all the data into a computer, and "walk" through its interior with a video screen and a joystick before actual physical exploration begins.

∿∿∿∿

Hinting at what may yet be discovered, in one cave alone (Cave Four, the one nearest the Qumran ruins) eight hundred scroll fragments were excavated during the 1940s, and very little of that material could be seen until the mutiny of 1991.

Robert Eisenman, a mutineer from California State University at Long Beach, was also at Qumran. He recalled the personal agony that had surrounded the discovery of two copper scrolls in 1952: "There was a huge political struggle to get them open. There were Hebrew letters punched into the metal. What they revealed, when they were finally opened, was that we had the Temple treasure list here [the Ark of the Covenant was not on it]. Now, one might say: What did that matter? [Why the huge political struggle?] It mattered because, what it implied was that we had people here, at Qumran, who were not [according to the most widely accepted theory] a pious group of Essenes retiring from the Jerusalem mainstream, but a more activist, militant group, probably connected with the Jewish uprising against Rome. And the Temple treasure list we found in Cave Three probably related to attempts by the revolutionaries to sequester the Temple treasure in the face of the coming Roman armies.

"I feel that this movement, the movement that produced and hid the scrolls, is the messianic movement in Palestine. This literature is tremendously messianic. At the beginning we thought we were seeing documents having to do with *two* Messiahs. But [with the recent freeing of the scrolls] we are also seeing descriptions of a single, God-like Judeo-Christian Messiah.

"And if you read your Josephus, the Roman historian of this period who wrote *The Jewish War*—'[The Jews'] chief inducement to go to war was an oracle found in their sacred writings, announcing that at that time a man from their country would become Monarch of the whole world'*—you will

---

*To provide some idea of the supernatural belief system, or backdrop against which the Dead Sea Scrolls were written, it is worth noting that Josephus (as translated by G. A. Williamson) went on to say that: "[This Jewish prophecy] they took to mean the triumph of their own race, and many of their scholars were wildly out in their interpretation. In fact, the oracle pointed out the accession of Vespasian; for it was in Judea he was proclaimed Emperor. But it is not possible for men to escape from fate even if they see it coming. . . . Anyone who ponders these things will find that God cares for mankind and in all possible ways foreshows to His people the means of elevation, and that it is through folly and evils of their own doing that they come to destruction. Thus the Jews, after pulling down the [Roman] Citadel of Antonia [from the Temple Mount], made the Temple [of Solomon]

see that Josephus believed that what most moved the Jewish people to revolt against Rome was the prophecy that a world ruler would come out of Palestine. This tells us that the revolution against Rome, which we often think of as a political revolution, was not so political but religious and messianic. The people were moved by a messianic prophecy.

"I think we have here, at Qumran, the literature of the messianic movement. The last stages of this literature [up to the time of the Roman counterstrikes] are parallel to that community known in the New Testament as the Jerusalem Church. This is also the Jerusalem community of James the Righteous, sometimes called the brother of Jesus."

Conspiracy-of-silence theories aside, one can easily see how such conclusions might expose a few raw nerves or how a handful of would-be scroll monopolists would have wanted those conclusions, if proved correct, associated eternally with their own names.

One of the clandestinely released scroll fragments reads: "A shoot shall come forth from the root of Jesse [the father of King David]." The fragment appears to be a quote from Chapter 11, verse 1, of the Book of Isaiah: "And there shall come forth a rod out of the stem of Yishay [Jesse], and a branch shall grow out of his roots." This, by itself, is interesting but not particularly startling. Other copies of the Book of Isaiah have been found among the Dead Sea Scrolls, and we have become accustomed to reading ancient texts holding true to the versions received by us today. The reference to "the root of Jesse" is in keeping with the biblical specification that the Messiah, when he came, whoever he might be, had to come out of the family of David.

At our home base, inside Jerusalem's Christchurch compound, my friend Reverend Jill had been reacquainting me with the books of Isaiah, Matthew, and Luke.

"To truly understand Isaiah," she instructed, "you must go back to the original Hebrew. Only then can you understand it as the people who originally wrote it understood it. Only then can you read Isaiah as if you were living among the desert tribes, reading the words of the prophets."

One of the most interesting parts of Isaiah, the Reverend pointed out, was the prophecy of a child who would be called Immanuel (meaning "gift from heaven"). To the Christians, this became a reference to the virgin birth of Jesus, but this interpretation rested upon how one crucial word from Isaiah 7:14—*almah*—was translated. In the original Hebrew, *almah* re-

---

square in spite of the warning in their prophetic books that when the Temple became a square the City and the Sanctuary would fall."

ferred to a young woman who might or might not be a virgin. There was, in the Hebrew language, a specific word for virgin—*bethulah*—but the word was never used in the original version of Isaiah, which simply read: "Behold, a young woman shall conceive and bear a son. . . ."

"Here is what happened," said Reverend Jill. "The word *virgin* was inserted by the Christian compilers of the King James Version of Isaiah as a means of proving that the Hebrew Bible foretold the coming of Jesus and that God's covenants with Abraham and Moses should be replaced with a new and everlasting promise. The men who rewrote and misused that passage must have forgotten that Jesus was a Jew, that all of our roots were distinctly Jewish, and that Jesus' real message was one of brotherhood and mutual tolerance. Yet ours has often been the most intolerant of all religions, and down through the centuries this single mistranslated passage of Isaiah would be used even by the followers of Hitler as a means of dividing people—of separating "them" from "us"—and destroying "them." Think of what it must have been like to be a Jew in such times: to look upward, sometimes with dying eyes, and see your own Bible being misused against you."

No one was ever going to make a clearer argument for why we must go back to the original Hebrew if we hope to understand the origins of Christianity and Islam (both offshoots of Judaism) and why we must be very careful about our translators (especially since at least one of the would-be Dead Sea Scroll monopolists has expressed neo-Nazi sentiments). When I first arrived in the Holy Land, Christianity was to me just another branching lineage, almost like a new species, carrying with it the Old Testament "backbone" of its ancestral stock, much like the mitochondria and fossil oceans that run in our veins. But now my fascination deepened as I scanned copies of the newly released scroll fragments and came away with the unmistakable feeling that these were the traditions and history of a real people. I wondered what new astonishments the mutineers were about to uncover, as my eyes ran down columns of the first direct Hebrew-to-English translations and found snippets of sentences that were at once familiar yet somehow strange: "And they will put to death the prince of the congregation . . . and with piercings a priest shall give order. . . . He will release the captives, raise up the do[wntrodden]. . . . Then he will raise up the dead, and to the poor [he will] announce glad tidings."

University of Chicago mutineer Michael Wise was picking his way through the scrolls in much the same way Spyridon Marinatos had picked his way, an inch at a time, through the ash covering the lost city of Thera.

"Now," said Wise, "we know from other Dead Sea Scrolls that the words *prince of the congregation* refer to a messianic son of David figure. They are speaking of a messianic figure who is evidently being put to death, and the notion of woundings or piercing. The author appears to be referring to the 'suffering servant' of Isaiah Fifty-three. We don't have evidence that this very important chapter of the Hebrew Bible was understood messianically among the Jews at the time of Jesus—apart from this pierced Messiah text."*

Wise produced a second scroll fragment that began: "The heavens and the earth will listen to [obey] His Messiah."

This introductory passage of Dead Sea Scroll 4Q521 was immediately familiar to anyone who knew Matthew 28 (in which Jesus is quoted as saying, "All power in heaven and on earth has been given to me"). It was succeeded by lines in which the Messiah is described releasing captives, raising the dead, and bringing glad tidings to the poor.

"Do you know Isaiah Sixty-one: One?" Wise asked.

I consulted my copy of the Hebrew Bible and found Isaiah's prophecy of an anointed one† who would work spectacular signs before the Day of the Lord: "The spirit of the LORD GOD is upon me; because the Lord has anointed me . . . to [raise] up the downtrodden, to proclaim liberty to the captives . . . to announce good tidings to the poor."

"Now," said Wise, "look at the Gospel of Luke, Four:Eighteen."

The lines Jesus read to his followers at the synagogue in Nazareth, describing the signs he had worked, were almost identical to Isaiah 61:1, except for one glaring difference, a difference that it shared with our Dead Sea Scroll text.

"Isaiah Sixty-one, verse One, says nothing about this Anointed One raising the dead," I pointed out.

"That is correct," said Wise. "Indeed, in the entire Hebrew Bible there is nothing at all about a messianic figure raising the dead. Yet in Matthew and Luke we find the reference to raising the dead linked to glad tidings for the poor. The two phrases are linked as signs of the Messiah: The dead are raised up, and the poor have glad tidings preached to them—precisely as in our Dead Sea Scroll!"

Here again was a point over which wars had been fought and were still being fought.

*The "pierced Messiah text" is also known as 4Q285—that is: Cave 4, Qumran document number 285.

†Note that the Greek word *Christos* ("Christ") is a translation of the Hebrew *Meshiach* ("Messiah"), which means "Anointed One."

"Earth-shattering," said one of the archaeologists. "What's at stake, I think, is that from the scrolls the Jews will learn that messianic Judaism is an acceptable form of Jewish faith, because here it is—at Qumran."

And the Christians? What will the Christians in the West learn? Perhaps they will finally see that the real roots of their faith are out here in the desert, with these people, and that in the beginning, Christian, Muslim, and Jew were one.

Would only that they had seen it sooner, our history might have been far less bloody.

13

GOD,
THE
UNIVERSE,
AND
EVERYTHING

Biology is the production of order from chaos; evolution is chaos with feedback.
—BISHOP AND PELLEGRINO'S SECOND LAW

Where in this stormy universe could a man lay hold on faith? Perhaps God Himself was in creation, proceeding to unity through the infinity of nuclei in the infinity of matter. Perhaps God was everywhere . . . but only because God was Everything.
—PEARL BUCK, *Command the Morning*

WESTCHESTER COUNTY, NEW YORK, TIME PRESENT

Much of our ascent to ownership of the Earth has been simply this: new ways of using the rocks. The children of Mitochondrial Eve knew how to draw the best flint cores from limestone cliffs, and for a quarter million years there was the Stone Age. About six thousand years ago, Egypt's *adj mer*—the canal builders—learned how to draw copper from the rocks with a flame, and in rapid sequence there came the Copper Age, the Bronze Age, and the Iron Age. History's most intensive effort at separating metals from the rocks began when a modern government set out to murder all the Jews in Europe, then formed an alliance with another government attempting to apply its technology toward a similar slaughter in China. A result of the effort (known to history as the Manhattan Project) is perched before me, far beneath the dome of the New York Power Authority's Indian Point 3 plant. One would not know, to look at the enormous maze of buildings that surrounds it, that the actual core of the nuclear reactor is a relatively small affair. Standing barely thirty feet tall and slightly more than ten feet wide, it could easily be contained inside one of the stone monoliths beneath the Temple Mount. More than half the surrounding equipment—which covers an area even larger than the Temple Mount—is designed to contain the nuclear fires. The rest is basic nineteenth-century technology: the simple conversion of water into steam. The steam spins turbines. The turbines spin magnets about coils of wire. The magnets send streams of electrons coursing through the wire, and the resulting surge is pumped south toward New York City.

The uranium 235 in the reactor is refined to a purity of less than 4 percent. Using the world's most sophisticated refining facilities (again, larger than the Temple Mount), one can, as a matter of everyday routine, refine the metal to better than 90 percent purity (what we call bomb grade). In their purest form, walnut-size slugs of uranium 235 are harmless enough to sleep with under your pillow without fear of getting sick, and they shine even more brilliantly than silver or platinum. Nuggets of bomb-grade uranium are deceptively beautiful. Judging from appearance alone, one is challenged to believe that heavy, neutron-emitting metals can unleash such incredible powers. I could easily hold in both hands two eight-pound lumps of uranium, refined to better than 90 percent purity. Their combined volume would occupy a space slightly smaller than a grapefruit, and to make an atomic bomb, all I would have to do is clap the lumps together in my hands. It would, of course, be a very crude bomb, killing me and whoever happened to be standing near me, splattering some red-hot uranium and a lot of neutrons about, and generally creating an ugly but rather localized mess.

If I change the geometry of the lumps and, instead of simply clapping the two halves together, place one at the bottom of a ten-story drainpipe running down the side of my apartment building, add a little amercium* from some smoke detectors, and drop the second nugget down the pipe, I can blow up my entire building and break all the windows in town. From that point forward, designing a "better" nuclear weapon is just a matter of increasing its efficiency. Add plastic explosives, precision-timed detonators, and about eight hundred dollars' worth of equipment obtainable at the local shopping mall, and I can build a bomb small enough to fit in the trunk of my car and powerful enough to knock down every building in town. Given a couple of hundred thousand dollars more, I can vaporize the town and break windows in the next county.

These are the realities of the world we are creating. The same fire that warms and lights our homes can, at any moment, be turned against us. This is how it has been since *Homo erectus* times. The difference is that now we are handling much stronger fires. Our rocks contain traces of metals forged in the hearts of supernovae. They are the ash of stars that lived and died when our solar system was dust. Refined, arranged in specific geometries, and tweaked in just the right way, the primordial ash of Creation can be made to echo, billions of years later, the last shriek of an exploding star. If we have retained as much dryopithecine savagery as Bronze Age and Iron Age Scriptures suggest, then the shriek may yet manifest as brief reincarnations of distant suns burning hellishly in the centers of our cities. If we are wise, and perhaps even if we cling to some of the lessons contained in those

---

*Amercium (element no. 95) is, like kerosine thrown on a fire, a very good accelerant. Its one drawback is that, if not measured and configured in just the right way, it may cause the bomb to behave unpredictably (i.e.: beware of "premature disassembly").

very same Iron Age Scriptures ("Thou shalt not kill," would be a good start), the shriek will be harnessed, for decades to come, as a warm, steady glow in the reactor core. Born of Auschwitz, Pearl Harbor, and Hiroshima, Indian Point 3 (especially in this era of American and Russian nuclear disarmament) stands to become the ultimate realization of Isaiah's beating swords into plowshares.

There seem to be no limits to what the human mind is capable of dreaming and producing. But the one thought that stands foremost in my mind, as I study the cyclic collapse of past civilizations and dream of new machines for the advancement of our own civilization, is that as we begin forging the keys to the universe, we must be very, very careful that those same keys do not also open the gates to hell. At Brookhaven National Laboratory, nuclear physicist Jim Powell and I have, since 1984, been drawing up plans for a new era of refining in which atoms that did not exist on Earth until human beings conceived them will be (and at this writing are being) created. Chief among these is antihydrogen, which differs from ordinary hydrogen in having a negatively charged proton and a positively charged electron. These antimatter twins of hydrogen atoms can be produced by very brief and violent collisions in atomic accelerators. They will, very soon, be routinely collected and stabilized in special traps—essentially magnetic bottles—vacuum sealed and supercooled to temperatures so low that, unless there are other electronic civilizations amid the stars, will render the coldest places in the universe of human origin.*

As an energy source, antimatter, when combined with matter, is a thousand times more powerful than uranium. By detonating antimatter bombs barely larger than amoebae in a magnetic field, Powell and I can accelerate a spaceship up to 92 percent lightspeed. We have discovered that there are no major technological barriers to this dream. Actual voyages to the nearer stars should become possible by the middle of the twenty-first century, within the lifetimes of many people walking the Earth today. Propelled by microscopic suns of our own creation, we humans are about to forge new keys and throw open the doors to the galaxy, far sooner than even most science fiction writers anticipated.

What was it that God asked Job? "Hast thou commanded the morning?

---

*Given the number of stars in the galaxies and what we know about planet formation, the probability of intelligent life existing elsewhere in the universe, even in our own galaxy, is a statistical certainty. Given the immense distances between stars, it is also probable that there is no one else within a thousand light-years of us. That is probably just as well for us, for given the lessons of recent human history, if we did have neighbors, we'd probably be getting ourselves into another war about now.

Canst thou send lightnings? Hast thou seen the doors of deepest darkness? Canst thou bind the chains of the Pleiades, or loosen the cords of Orion?"

Yes, I believe we can, if we are wise, and abide by nature's laws, and pay very close attention to what we are doing. And yes, a few of us have indeed seen the doors of deepest darkness: Fred Haise tells me that when he was outward bound for the moon and looked back at the rapidly diminishing Earth, he was most impressed by the contrast between Earth and sky, by a darkness that is to this day, impossible for him to describe, a darkness such as he has never seen anywhere on Earth. And I remember when one of our space probes was hurtling through the outermost fringes of Saturn's rings, and a stream of new images and numbers—more than any mind could instantly absorb—was flooding in, and we scientists, overwhelmed by the flood, were trying to figure out what it all meant. The essayist and fantasy writer Harlan Ellison was there. He was watching us. He knew, perhaps better than any of us, what it all meant. He wrote:

> Norman Ness, from the Goddard Space Flight Center, principal investigator on the magnetic field team, explains how the Voyager passed through Saturn's bow shock wave at 4:50 P.M. when Titan was inside the magnetic field envelope of the planet. He speaks of the solar wind, the flow of ionized gas given off by the Sun that hisses through the Solar System. There is no poetry in the words . . . only in the way he speaks of it. Norman Ness barely realizes he has looked on the face of the Almighty.

∿∿∿∿

We sit, now, upon the solar system's largest collection of rocks, whirling around a very average star, in a galaxy containing some hundred stars for every man, woman, and child on Earth. Six hundred billion stars . . . and we live in a universe containing at least as many galaxies.

Given so many worlds and suns, so many opportunities for the origin of life, so many throws of the dice, it is almost inevitable that someone on the far rim of the Milky Way is asking the same questions we human beings have been asking for more than eight thousand years: Where did we come from? How does our universe work? How did it get here? Will it last forever or die? And where in it do we belong?

Space probes and electron microprobes suggest that we are members of a species whose blood runs back through the origin of the Earth . . . to carbon and phosphorus and sulfur—all seething in the hearts of dying stars . . . to the cauldron of Creation that preceded the Big Bang, when time itself was timeless.

By the time of the dryopithecines—only a few seconds ago, by a rock's standard of time—the numbers of nerve cells and the complexity of synaptic connections in the savage brain had been rising for millions of years. When the australopithecine "Lucy" stood upon the shores of Lake Turkana, the columns of her neocortex were already more alive with neurotransmitters, neuromodulators, and electrical charges than anything yet seen upon the Earth. As if following some true compass, the australopithecine lineages were diversifying into newer, even larger-brained tribes. Sooner or later one of those diverse branches was bound to reach a neural threshold.

By the time Mitochondrial Eve appeared, the threshold had been reached and exceeded. Out of inanimate carbon, hydrogen, phosphorus, and sulfur had emerged consciousness. Carbon and calcium knew fire and flint, and most important, it knew itself. Knowing itself, and lonely, it amassed in the riverworlds. With consciousness and massing behavior, both promise and dread entered the world. Anything that thinks can build. Anything that builds can destroy.

"The first thing that must be asked about future man," Charles Darwin said, "is whether he will be alive, and will know how to keep alive, and not whether it is a good thing that he should be alive."

"The first thing I'd like to ask about future man," Father Mervyn Fernando* once told me, "is whether the evolutionary process stops with our present civilization or are there further stages ahead? If you look back to the origins of man, if you are really paying attention, you can get bearings and sights to peer into the future. As the diversity of life on Earth increased, some organisms became increasingly complex. Increasing complexity brought increasing consciousness, and now one conscious organism has spread from pole to pole, over the oceans and under them, enclosing the Earth in a single thinking envelope. And the envelope is becoming more and more unified. We see this happening before our very eyes."

The way Father Fernando viewed civilization, five billion people, seemingly mindless of what was actually happening, were creating a world mind connected by a network of satellites, telephones, and fax machines. The planet was acquiring a nervous system, and there was no telling what shape it would eventually take. What surprised me is how closely the priest's vision of man's ascent and ultimate fate agreed with the views of NASA scientist Jesco von Puttkamer.

*Father Fernando, a Jesuit, is the director of Sri Lanka's Institute for Integral Education. Along with his colleague Arthur C. Clarke, Father John MacQuitty, and the explorer Robert Ballard, he is a chief instigator of this book.

"The origin and persistence of consciousness are the key to our evolutionary vocation," said von Puttkamer, who was part of the rocket team that put Scott Carpenter into orbit. "And it may be that we are already creating our next evolutionary stage."

For more than a decade von Puttkamer had been speaking about what he liked to call "the soul in the machine."

"It is an attempt," he told me, "to explain why machines, in my opinion, as they get more complex, do become more responsive to humans. They become more sophisticated. They become more automatic. They become more independent. The robot spacecraft *Voyager 2,* as it flew past Saturn, was in a sense very independent from humans. One command beamed up from Earth triggered a whole chain of commands. So if you look at the inanimate matter all these automatic systems are made from—"

"Oh, my God!" I remember cutting him off in mid-sentence; then we both trailed off. The soul in von Puttkamer's machine had raised something new in my mind and was raising a chill of gooseflesh on my arms.

"Do you mean," I said at last, "that as biology has done with carbon and phosphorus and sulfur, so, too, are we doing with silicon, plastic, and steel? A computer's circuits, and the nerves bundled in my neocortex, move pulses of electrons around in organized fashion. Are you suggesting that by creating artificial intelligence, we may actually be creating our next evolutionary offshoot, that we are creating life?"

"In a sense, yes. The computer chip is still inanimate matter, but it is obviously more than sand."

"But what is it that is more?"

"An ever-increasing complexity," said von Puttkamer, "which shows itself in terms of a consciousness." As he followed the neural networks of modern computers backward through time, back through the ancestral and already primitive brain of *Voyager 2,** back past the human brains that had created the machines, and into the Earth itself—all the way back to atoms of carbon or silicon—he began to see that there must have been a continuing chain of increasing degrees of consciousness, which started with the simplest form of matter. "It has to start somewhere, and I figure it starts at the electron. The electron could actually be the unit of consciousness, meaning that human brains are simply the electron's way of reaching increased complexity."

*It is worth noting that when *Voyager 2* flew past Neptune and Triton in 1989, the laptop word processors owned by journalists reporting, from NASA, on the spacefaring robot's discoveries, were a quantum leap ahead in computer intelligence over the robot launched into space only twelve years earlier.

"You make it sound as if electrons do this by design," I said.

"And who's to say there is no grand design?"

"I try not to view nature that way."

"You wear the badge of Darwin with too much pride, Charlie. Be careful. It can blind you. A good scientist leaves all possibilities open, even the possibility that there is a grand design. Look at us, for example. We are building more and more complex machines, and we really don't know why. We are just doing it. Somewhere we think it's the right thing to do, almost as if we were following some deep-rooted instinct. We build *Voyager 2* and plan to follow it with a whole generation of more sophisticated spacefaring robots. We are forever reaching on, and if somebody asks us why we are doing it, basically we have no answer. We are just supposed to be doing it."

I never was able to stop thinking about those electrons . . . the way von Puttkamer had spoken about them. How is it that certain nerve fibers arranged in certain ways allow us to think? A lot of matter from diverse places (salts from Antarctica, calcium from a Triassic pond) was assembled, not very long ago, into the first diploid cell of the yet-to-be-born, from which unfurled genetic blueprints for a gridwork of brain cells. And although the electrons coursing through the neural grid are the basis of every thought we have, they somehow produce a mind that, as it asks questions and designs computers to help answer them, feels quite separate from the cells themselves. The electrons are working in our best interest, supposedly. But when I imagine what I would look like if all the organic molecules in my body could be made invisible, so that I could look in a mirror and see only the paths of freely moving electrons, I know that the outline of my entire body would be there in every detail, brightest at the brain and spine; even the nerves in my fingertips and eyelids would show up as streams of electrons. As von Puttkamer would have it, the electrons, being among the very first particles to come out of the Big Bang, waited more than twelve billion years for planets like Earth to form and then to sprout life, waited for moments such as this. Perhaps it is really the electrons who are thinking these words. Perhaps our bodies are little more than vessels serving their interests, and as we set forth to design increasingly advanced artificial brains, it is possible to believe that the *sine qua non* of our existence is to build larger, faster electron vessels, perhaps even to eventually clear the decks for them, as the dinosaurs once cleared the decks for us.

Viewed in this light, the history of the universe has been a tremendous waiting game. In the beginning, electrons emerged as the perfect parasites, fresh and hot from the Big Bang, awaiting only the arrival of the perfect host. If this is true, it is not I but the electrons coursing through my brain who have peered into the core of Indian Point 3, who are looking back across all time and asking, "Where did we come from?" These were precisely the thoughts running through my mind as I sat with Arthur C. Clarke at his computer console in Sri Lanka. He was one of the creators of Father Fernando's membrane of human thought, which now so completely surrounds the planet. Through one of his satellites Clarke had recently established a link with black hole theorist Stephen Hawking, half a world away at Cambridge University.

We knew that the space between the galaxies was constantly expanding, so that the galaxies themselves were moving away from each other. The farther away they were from each other, the greater the expansion and the faster they were being pushed apart. When we ran the cosmic movie backward in time, we came to a moment, more than twelve billion years ago, in which all the matter in the universe was superimposed. The entire mass of our galaxy was compressed, along with many other galaxies, into a space smaller than a virus. If you have ever wondered about the environment inside a black hole, just track the universe back to an instant before the Big Bang and have a look around.

The American astronomer Carl Sagan had recently suggested (through that same satellite link) "that a key, unanswered and perhaps unanswerable question is: Where did that [primordial black hole] come from? What was before that? And if it was made from nothing, who made it and who made the maker? And of course, there's an infinite regress behind that."

Arthur C. Clarke was "doodling with his computer," producing a particularly colorful visualization of an infinitely regressing mathematical formula called the M-Set, "which," he explained, "leads us toward an understanding of the infinite."

The computer screen displayed a strange geometric shape, something Euclid had never imagined, from whose surface spirals and webs sprang. It was defined by an equation of only two terms, $Z^2 + C$.

"That's all there is to it," Clarke said. "Yet that simple equation can carry on over and over. I can crank it round and round and plot the result on the screen. And I can tell the computer to go to any spot, recompute that small area to a higher degree of precision, and then blow it up on the screen. So we can essentially use the computer as a microscope."

I showed Clarke how I had applied this same concept of increasing magnification to human evolution, starting at about 20 million B.C. (as illustrated in Chapter 4). In the first illustration the branching pattern of the dryopithecine "tree" resembled a candelabrum. A second illustration zoomed in on a small corner of the first, and a third on a small corner of

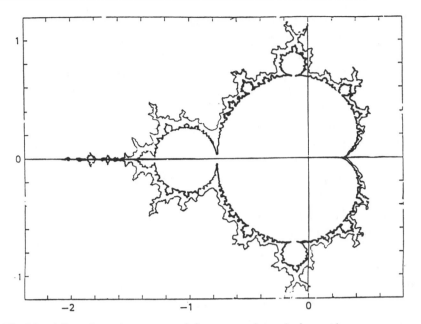

The Mandelbrot Set, when converted from a mathematical equation to computer graphics, provides a visual introduction to the fractal universe—which, even if nothing more than a mathematical aberration, is an analogy for a universe that has no beginning or end, whose edge is nowhere and whose center is everywhere. (Graphics from the computer of Arthur C. Clarke.)

the second, and so on, until after five levels of magnification, we had resolved a seventy-five-year span in a single family.

"Two things impressed me," I said. "First, while zooming in on a single family lineage, I had magnified the original picture in the series more than two hundred sixty thousand times, which meant that in the final frame we were looking at a portion of a page four hundred miles wide. Second, no matter what level of magnification I chose, I kept encountering the same candelabrum shape. Some of the finer details might have altered from frame to frame, but the overall pattern was the same, as if nature had a habit of repeating itself."

Arthur smiled. "You ain't seen nothing yet. Using my computer as a microscope on the M-Set, I have found what look like black holes. So what I'm going to do now is zoom into its horizon, increasing the magnification manyfold."

He pressed a button, and a maze of tiny spirals at the hole's edge filled the screen. Seconds later we were descending upon spirals within the spirals. He had magnified the hole about a thousand times, meaning that the picture we had begun with was more than two thousand feet across. Exploring the spirals, he quickly found a second black hole, different in its finer details but essentially the same as the first. He zoomed in on it and near its outer boundary found another.

"Now, my third hole looks just like the others, but this is on a far greater magnification. The original is now, I think, about ten million miles wide."

We penetrated farther into the M-Set, finding another black hole, and another, and another. At last, Clarke stopped. "I'm rather proud of this one," he said. "Because on this scale, that original little picture you saw is the width of the orbit of Mars.

"Each time we move forward into the M-Set, we're being drawn into mathematical infinity. We're being sucked into it. Now, the thing that fascinates me about this is that it *is* infinite in detail. Unlike your candelabrum phylogeny, which runs into a dead end when you reach a twentieth-century family, the M-Set goes on forever and ever.* Now, I would like to ask Stephen Hawking this question: Is the real universe also infinite in detail, like the M-Set, or is there a limit, as with phylogeny? I mean, we know we can zoom down into a single carbon atom in a human hair, and we know that we will find subatomic particles of descending size and increasing detail, right down to the quarks, but does it continue forever and ever, or is there a basement to the real universe?"

"We will discover new structures when we look at the universe on smaller and smaller scales," Hawking said. "But in the case of the universe, there seems to be a limited scale. It is called the Planck length, and it is about a million billion billion times smaller than an inch. This means that there is a limit to how complex the universe can be. It also means that the universe could be described by a theory that is fairly simple, at least on scales of the Planck length. I just hope that we are smart enough to find it."

So, looking down as far as he could, Hawking saw a limit to smallness. The universe of the particle physicist was more like the universe of the paleontologist than that of the mathematician. And that meant the universe might be within human grasp, that we might eventually discover the complete theory of everything.

---

*The candelabrum phylogeny becomes more detailed if we continue tracing its progression into the future, and if we focus not just on one lineage but on the diversification of the virtually immortal DNA molecule. Its center becomes the origin of life and its edge the end of the universe itself.

"If we can come to a true understanding of how the universe works," Hawking had often said, "we shall all, philosophers, scientists, and just ordinary people, be able to take part in the discussion of why it is that we and the universe exist. If we find the answer to that, it would be the ultimate triumph of human reason . . . for then we would truly know the mind of God."

The next obvious question begged asking: "Are we smart enough to find the theory of God, the universe, and everything?"

"Well, I wonder," said Clarke. "Because after all, we are still fairly primitive organisms. And the universe is very old. And I just don't know. I would like to think so, but then there is a feeling: When we have found it, then what? Where do we go from here?

"I can imagine the Old Testament God finding himself (or herself) in very much that same situation. According to one emerging view of the Creation, once God set the laws of the cosmos into motion—atomic mass, atomic weight, the bonding habits of carbon, hydrogen, phosphorus, and sulfur—no one could tamper with them again, not even God. So God had to create the universe just to see what would happen. Even God did not know."

"The question," said Hawking, "of whether God is bound by the laws of science is a bit like the question: Can God make a stone that is so heavy that he cannot lift it? I don't think it is very useful to speculate on what God might or might not be able to do. Rather, we should examine what he actually does with the universe we live in. All our observations suggest that it operates according to well-defined laws. These laws might have been ordained by God, but it seems that he does not intervene in the universe to break the laws; at least, not once he had set the universe going. However, until recently it was thought that the laws would necessarily break down at the beginning of the universe. That would have meant that God would have had complete freedom to choose how the universe began. In the last few years, however, we have realized that the laws of science may hold even at the beginning of time. In that case God would have had no freedom.

"I should probably add that I use [the word] *God* in the same sense that Einstein did. It is really the reason why the universe is, as it is, and why the universe exists at all."

"That limits God today to the role of innocent bystander," objected Father Fernando. He was not so sure that as Hawking had observed, there was no evidence of God intervening in the universe to break its laws. The priest had encountered cases in which charismatic prayer had produced what could only be described as medical mysteries, if not miracles. During the autumn of 1991, in Jerusalem, a scientist of his acquaintance had been approached by a charismatic healer from India who asked permission to pray over him, without ever giving his name or even asking for a charitable donation. The scientist, who happened to be an agnostic, experienced the spontaneous reassembly of a knee so undeniably damaged as to be inoperable, a total loss.*

"I do not think God is just a bystander," Father Fernando continued. "I suspect that despite the depressing difficulties of the present, mankind is being called to a great and grand destiny."

"And that evolutionary destiny is the discovery of the universe," added Father MacQuitty. "It may be that mankind is being called upon now to go freely and consciously into the deserts of space. I feel that when you look into space, when you look above and beyond to what it is that we are really all contained in—*the infinite*—you begin to understand loneliness. And perhaps more so than in the deserts of the East, God is within the loneliness of the cosmos. In the generations to come, people will go forth into a revealing universe, into a newer, stranger sort of womb than we have ever known before. That is what the Earth has really been headed for, during all of its four and a half billion years."

〰〰〰〰

Arthur C. Clarke had said that he was certain the human species, if it survived the next few years, would go forth to colonize the solar system and then go on to send ships out to the stars.

I'm not so sure. The operative phrase is, "if it survived the next few

*On another occasion Father Fernando asked me what I thought about the Near Death Experience, in which people all over the world, snatched by modern medical science from the moment between dying and death, recounted stories of leaving their bodies and moving toward a light. I told him that I tended to regard it as a hallucinatory state brought on by extreme physical stress (having experienced it myself and found the encounter with the light to have been very dreamlike and not, as others had reported, "more real than here"). What interested me most was that if the world's religious leaders were to take these reports as really meaning anything, they would have to come to terms with the fact that Christians and Jews, Muslims and Hindus all meet the gods and prophets in whom they have come to believe. "If there is something more to human consciousness than streams of electrons," I told the priest, "if the Near Death Experience is teaching us that man has a soul, then our souls are telling us that all of the world's religions are equally valid." And we came to agree that if *we* were planning the universe, that would be a sensible way of setting it up. "As for the paradox of all religions—even those with contradictory views—being simultaneously correct," said Father Fernando, "well, and why not? Isn't that child's play in a universe that already manifests every photon of light as both particle and wave at the same time?"

years." As a paleontologist and space scientist I was fairly certain that so long as we could avoid nuclear war, our civilization would have enough staying power to enter the age of antimatter propulsion. But when I surfaced from cosmic and geologic time scales into the shallows of archaeologic time, I began to see too many cyclic collapses of civilizations.

Over and over again I have watched how the salting of fields and the fouling of riverworld nests led first to environmental collapse, then to economic collapse, and ultimately to the total breakdown of social order.

Seeing past failures, and trying to draw from them signposts for the road ahead as ozone holes begin to open in the sky and acid rain falls upon those life-giving top few inches of the ocean, I have found it easy to believe that humans must learn to act as guardians of an Earth now threatened with permanent, grand-scale derailment by our intervention.

But such views, no matter how well intentioned, are rooted in the old sin of pride and exaggerated self-importance, suggests paleontologist Stephen Jay Gould. "We are one among millions of species, stewards of nothing. By what argument could we, arising just a geological microsecond ago, become responsible for the affairs of a world 4.5 billion years old, teeming with life that has been evolving and diversifying for at least three quarters of that immense span? Nature does not exist for us, had no idea we were coming, and doesn't give a damn about us."

Gould validates and perhaps surpasses everything Clarke, Hawking, and I have been talking about. If one finds twenty million years of candelabras and the mere size of the universe belittling, then seen against our planet's time scale and its own naturally occurring geologic upheavals, we are truly as nothing. All the megatonnage in our nuclear arsenals barely equals the explosive force of the volcano that caused a false global winter during the summer of 1627 B.C. The shock totally disrupted human civilization, but life itself went on. Even after the asteroid bombardment of 64.4 million B.C., which might have been the final *coup de grace* for the dinosaurs, life went on. The survivors simply radiated and diversified into new directions, including snakes and dryopithecines.

Against time's vastness, man will simply exist here for a short term and then, like most species, very probably be extinct in the next geologic period. We are just another species passing through, and this fact has lately (and wrongly) become the basis for arguments that if we maim or pollute, if we make another human being's life miserable or kill the last humpback whale, it won't matter a million years from now, or even a hundred. Arguing that the world can easily afford a few less humpbacks (at our immediate scale of human lifetimes) because they will inevitably become extinct, if by nothing else than through evolving (at geologic time scales) into something else, is not much different from arguing that none of the Iraqi or Israeli children caught in the crossfire of war against Saddam Hussein really matter because in a hundred years they will all be old or dead anyway.

Paleontologists, we people who dwell in the deeps of time, must be the first to stand against such arguments. In a way that is difficult for outsiders to understand, we develop a perspective that, while outwardly reducing man's importance, paradoxically amplifies a feeling for human time scales, and the special miracle that is even a single human life, or even the life of an insect. But as I have said, this feeling, at first glance, seems difficult and paradoxical to all but a few dozen of us who have spent more time than anyone else on Earth thinking about time. The Hebrew scribes who compiled the Bible about 550 B.C. had access to ancient texts which, according to Babylonian scientists, suggested that the world was nearly five hundred thousand years old. The Hebrews shortened the Earth's past to thirty-five hundred years (before their time), thus banishing all hints of geologic time scales. As a basis for a book dictating the rules of human behavior, it was the smart thing to do. In a reduced time frame the death of a single child, the tribulations of a single tribe, took on added and immediate significance, even to a nonscientist.

What the biblical writers probably never anticipated was that more than two thousand years later, scientists would rediscover and surpass the older, Babylonian chronology.

"I love geological time," says Gould, "a wondrous and expansive notion that sets the foundation of my chosen profession, but such immensity is not the proper scale for my personal life.

"We have a legitimately parochial interest in . . . the happiness and prosperity of our children, the suffering of our fellows. [At geologic time scales] the planet will recover [even] from nuclear holocaust, but we will be killed and maimed by the billions, and our cultures will [at our immediate scale of human lifetimes] perish.

"I have a decidedly unradical suggestion to make about an appropriate environmental ethic; one rooted in the issue of appropriate human scale versus the majesty . . . of geological time.

"Christians call this the principle of the golden rule; Plato, Hillel, and Confucius knew the same maxim by other names. I cannot think of a better principle based on enlightened self-interest. If we all treated others as we wished to be treated ourselves, then decency and stability would have to prevail.

"I suggest that we execute such a pact with our planet. She holds all the cards and has immense power over us—so such a contract, which we desperately need but which she does not at her own time scale, would be a blessing for us and an indulgence for her. We had better sign the papers while she is still willing to make a deal. If we treat her nicely, she will keep us going for a while. If we scratch her, she will bleed, kick us out, bandage up, and go about her business at her planetary scale."

∿∿∿∿

"I think the human species will go on to colonize the solar system, and then the stars," Clarke had said.

"It is our destiny," Father MacQuitty had said, "to know the universe."

Stephen Hawking, who had spent a lifetime getting to know the universe, by penetrating deeper than anyone since Einstein into its physical laws, noted that knowledge alone should not be confused with learning: "I don't think that physics tells us how to behave to our neighbors."[*]

If the universe really is our destiny, will we bring religion with us? I wondered. And can religion be equated with learning (humanity's growing pains, perhaps?) and science with knowledge? And if so, which is more important?

Gould regarded science and religion as being of equal dignity, and seemed to be echoing Hawking. "The two should not conflict," he said, "because science treats factual reality, while religion struggles with human morality. I do not view moral argument as a whit less important than factual investigation."

Father Fernando agreed with Stephen Jay Gould. Scientific knowledge and religious learning were equally important. But he expressed fears that our technological world was turning away from spirituality and becoming increasingly unbalanced: "So much knowledge," he warned. "So little wisdom."

"And you, Charles?" Father MacQuitty asked. "Where do you see humanity going?"

"I'm not as optimistic as the others," I said. "I've known Karnak, Jericho, Bet-She'an, and Babylon too well to think we've got that much of a shot. There is a long, difficult road ahead, and it diverges here, near the border of the third millennium A.D. A thousand years from now we will either be an archaeological curiosity that our own descendant starfarers look back to and ask, 'How did they ever accomplish so much with so little?' or

[*]"Ah," quipped Arthur C. Clarke, "but physics may determine who our neighbors are and on what planets they live!"

we will be another in a long line of vanished civilizations—mysteriously advanced for our time and full of promise—just another lost Eden, romanticized and overglorified forever. It's the doors of heaven and Earth or the gates of hell, the universe or nothing. That is the choice man is coming to."

"Then you were wrong about something," MacQuitty said.

"What do you mean?"

"You once told me that if this is how far we've come, we have not come very far. But that is not the whole story, my friend, is it?"

"No," I conceded. "That is not the whole picture. I'm afraid the real story is this: Whatever we're coming to, we're almost there."

AFTERWORD:
Where
Are They
Today?

SCOTT CARPENTER, one of the original *Mercury 7* astronauts, has the peculiar distinction of being the first (and at this writing the only) human ever to penetrate both inner and outer space. On May 24, 1962, he became the second American in orbit. Five years later he became director of the Navy's Deep Submergence Project, and today he works closely with Jacques Cousteau to improve the ocean environment.

AGATHA CHRISTIE was already one of the world's best-known mystery writers when, on December 3, 1926, she herself mysteriously disappeared. Her car was found in such a position as to indicate, according to police reports, "that some unusual proceeding had taken place: the car being found well off the main road with its [removable canvas] roof buried in some bushes." A black shoe, a heavy coat, and a brown, fur-lined glove were left at the scene—the last items one would voluntarily discard on a snowy night. Murder was suspected, and as the days wore on, more and more of the suspicion began to fall on her husband, Colonel Archibald Christie. A string of clues revealed that Colonel Christie had been keeping a mistress named Nancy Neele, for whom he was planning to divorce his wife. After ten days, with, according to a family physician, "Mr. Christie's nerves completely spent," owing to the possibility of a murder trial and the gallows looming in his future, Mrs. Christie suddenly surfaced from seclusion, where she had been sending forth snippets of potentially incriminating evidence under the assumed name of Miss Neele. In a brilliant turning of the tables on her tormentor, it was Agatha Christie who then demanded a divorce from the colonel, instead of the other way around. Mr. Christie granted it immediately, and without raising the slightest fuss, he abandoned all claim to future royalties on his wife's detective novels. Wishing to "get away from it all," Agatha Christie soon joined Leonard Woolley at his excavations, where she found the inspiration for the characters and settings in *Death on the Nile* and *Murder in Mesopotamia,* and where she met her next husband, Max Mallowan. As near as any historian can tell, Mallowan never strayed on his wife, never even gave it a passing thought.

ARTHUR C. CLARKE lives on the island of Sri Lanka (which is, next to New Zealand, the most beautiful place on Earth), where he supports educational institutions, advises scientists and political leaders all over the planet, writes about five books simultaneously, and occasionally moves mountains.

MOSHE DAYAN, the charismatic Israeli general who helped to establish Jerusalem as a center of both the aboveground and black-market antiquities trades, donated his own priceless collection of artifacts (including a set of anthropoid coffins from Gaza's Philistine Period) to the Israeli National Museum, where it remains under the curatorship of Trude and Moshe Dothan.

The DEAD SEA SCROLLS were finally released in their entirety in 1992, and new scanning techniques that allow archaeologists to see through the ground promise to reveal more hidden chambers at Qumran, and perhaps more scrolls.

TRUDE and MOSHE DOTHAN continue to explore the lost worlds of the Philistines and their Minoan cousins, with much of their emphasis beginning to focus on the Minoan site at Kandy, on the Sea of Galilee.

FATHER MERVYN FERNANDO continues as director of Subodhi, the Institute of Integral Education, Sri Lanka. Arthur C. Clarke is one of the institute's chief supporters. During the civil disturbances of the 1980s and early '90s, while universities remained closed, Subodhi and the Arthur C. Clarke Center (located across the river from each other and connected by a bridge) were among the very few educational institutions permitted by both the government and the revolutionaries to remain open.

QUEEN HATSHEPSUT appears to have ruled Egypt (as one of the coregent pharaohs of the Oppression) during the decades immediately preceding the explosion of Thera (in 1628 B.C.). Her stepson and successor, Tuthmosis III, never did accept the concept of a woman god-king and, after engineering her murder, continued to persecute her even in death by desecrating her mummy case. In an attempt to prevent her resurrection in the afterlife (where, according to Egyptian tradition, she might be reborn as the one anointed on Judgment Day to lead the souls of her people to Paradise), he chiseled away the golden sculpture of her face. The sarcophagus (or mummy case) was constructed in multiple layers, each enclosing a smaller replica and the smallest enclosing the mummy itself. One of the multiple layers, with most of its face gone but with much of the gold still remaining around the cheeks and forehead, was discovered early in the twentieth century and eventually made its way to the Cairo Museum. The other layers have disappeared into history. There is a persistent rumor in Egypt that Hatshepsut left behind an evil curse on anyone who despoiled her grave or attempted to remove its contents from the shores of the Nile. Tuthmosis III was apparently the first to feel the curse, in the form of crop-ravaging ash and other catastrophes, or "plagues" brought on by the passover of Thera's volcanic death cloud. In A.D. 1910, the British Egyptologist Douglass Murray is said to have been among the next to feel the curse. An American treasure hunter approached him in Cairo, offering for sale one of several portions of Hatshepsut's sarcophagus. The American died before he could cash Murray's check. Three days later Murray's gun exploded, blowing off most of his right hand. What remained turned gangrenous, requiring amputation of the arm at the elbow. En route to England with the sarcophagus, Murray received word via wireless telegraph that two of his closest friends and two of his servants had died suddenly. Upon arrival in England, he began to feel superstitious and left Hatshepsut's case in the house of a girlfriend who had taken a fancy to it. The girlfriend soon came down with a mysterious wasting disease. Then her mother died suddenly. After the funeral, her lawyer delivered the sarcophagus back to Murray, who promptly unloaded it on the British Museum. The British Museum already had more sarcophagi than it needed. Not so the American Museum of Natural History in New York, so a trade was made for Alberta dinosaur skeletons. Before the deal was complete, the British Museum's director of Egyptology and his photographer were dead. Queen Hatshepsut was beginning to lose her charm. The curators packed the sarcophagus into a wooden crate and saw it lowered into the hold of a ship. And on April 10, 1912, the ship sailed—Southampton to New York. Hatshepsut, according to legend, departed for America on the *Titanic* (although many items were not logged at all or were logged under "disguise," a thorough investigation of the *Titanic*'s cargo manifest fails to confirm Hatshepsut's presence, and it is noteworthy that Egypt's oral traditions state merely that the sarcophagus went down on *a* ship that sailed for New York in 1912, widely believed but not necessarily known to be the *Titanic*). As for the Alberta dinosaurs, their shipment was evidently delayed until a substitute sarcophagus could be sent.

In 1916 Charles and Levi Sternberg packed two duck-billed dinosaurs into crates, saw them lowered into the hold of a London-bound steamer, and sent them east. About halfway across the Atlantic, the ship was torpedoed by a German submarine. Eventually the same convective spasms that have created the Mid-Atlantic Ridge and sent whole continents adrift will carry the steamship, like a parcel on a conveyor belt, eastward to the edge of Western Europe. She will complete her voyage, about 250 million years after leaving the dock at Manhattan. By then the vessel will be an iron-oxide fossil sandwiched between layers of sedimentary rock, and if there are sentient beings on this planet at that time, their paleontologists will be forced to ask how humans and dinosaurs came to be fossilized together on a Victorian-era steamship.

STEPHEN HAWKING, the Cambridge, England, physicist and sometimes chaotician who saw physics as a window on "the mind of God," continues to open up new windows.

JERICHO, the town described in the Book of Joshua as having been defeated by Hebrew armies carrying the Ark of the Covenant before them, and whose land was said to forever thereafter be cursed, was ceded, along with the Dothans' Deir el-Balah site in Gaza, to the Palestinians as part of the 1993 peace accord between Israel and the PLO. Even as the accord was signed, hard-liner Palestinians claimed that it betrayed the sacrifices they had made during years of chaos and defiance, while hard-liner Israeli factions claimed that Gaza and other occupied territories were promised to the descendants of Abraham in the Bible. A few days later, on September 24, 1993, an Israeli farm worker was stabbed to death just hours before the Yom Kippur fast began. The Islamic Resistance Movement claimed responsibility, and Israeli voices vowed revenge. Separately yet with one voice, extremist factions on both sides were reserving their right to hate each other forever.

KARNAK, the palace center from which Hatshepsut and Tuthmosis III ruled, was covered more than thirty feet deep in the broken pottery shards and dust of later civilizations. At the uppermost levels of ancient stone pillars, on the more recent ground surface of the Christian Era, the devout chiseled away the faces of Egyptian gods and kings because they were "pagan" and "blasphemous." In the earth, dozens of feet below, carvings on the older, Eighteenth Dynasty ground surface were protected from the Christians simply by being beyond their reach. Then, during the mid-twentieth century, archaeologists carted away all those dozens of feet of protective earth, and tourists could be heard to ask how the faces of ancient Egypt, "all the way up there," came to be obliterated by chisels. In October 1992 the whole idea of exposing to daylight whatever had survived the Christian destructions began to seem, at best, a questionable tactic. One of the more radical Islamic factions began calling for the destruction of Egypt's pharaonic monuments as "pagan" sites. A small group of chisel-wielding men was stopped from attacking Queen Hatshepsut's obelisk, but not before they had caused some minor damage to a panel proclaiming her divinity. Two days later an earthquake rumbled inexplicably up the Nile Valley and devastated Cairo, rumbled through a thick layer of sedimentary deposits long known to be one of the few geologically stable regions along the entire length of the riverworlds.

THOMAS EDWARD LAWRENCE, the British soldier and writer who studied under Leonard Woolley before becoming known to the world as Lawrence of Arabia, died in a motorcycle accident in 1935. Few would remember that he began his career as an archaeologist.

LILITH, Adam's first wife, was, according to ancient Babylonian and modern Hebrew tradition, cast out of Eden before man violated the forbidden tree and gained knowledge of good and evil. Without such knowledge she presented no challenge to the Creator(s) and, unlike Adam and Eve, never had her immortality taken away. According to legend, she still walks the Earth today.

MAX MALLOWAN was already famous when Agatha Christie began joining him at excavations on the Tigris and Euphrates rivers, where he was at once "thoroughly bowled over by her." As marriage became inevitable, Mallowan asked her if she would mind spending her future with someone whose profession was "digging up the dead," and she replied, "I adore corpses and stiffs." But religious differences, in those days, were designed to keep them apart. Agatha decided that she would be converted to Roman Catholicism on her deathbed so they could be buried together. When the church decreed that it would never recognize Mallowan's marriage to a Protestant under any circumstances, he married her anyway, and together they explored the past. But at times it seemed to be the future that fascinated Agatha most, perhaps because of the desert nights' incomparable view of the galaxy.

In her letters she expressed to Max the realization that her own life was minuscule against the scale of archaeological time frames, and that man's existence from the beginning of archaeological time was minuscule against the age of the Earth. It became difficult for her to imagine God being pleased with a world in which species and civilizations rose and fell brutally, wastefully—in which every plague or flood, every famine or volcanic eruption, was attributed to "God's will." But all the suffering of countless generations would not have been the product of a tragic fate if God were in fact not a *cause* in our past but an *effect* in our future. If God were truly a creation of our distant descendants, then the Babylonian, Minoan, and Hiroshima destructions did not matter. "The creation of God is what we are moving to," she wrote.

The Roman Catholic Church viewed such thoughts as blasphemous, and when it was announced that even if Agatha converted "tomorrow," she and Max would never be allowed burial together on hallowed ground, Mallowan left his faith in a fit of fury. Agatha Christie died on January 12, 1975, after which Max Mallowan's health began a rapid decline. In August 1978, he followed her.

SPYRIDON MARINATOS, who, along with Mallowan, Woolley, and Lawrence, stands out as one of archaeology's more brilliant figures, collapsed on the main road of his buried city on October 1, 1974, and died there. He was buried in one of the buildings near the place where he fell. The city itself stands to become an archaeological treasure even more valuable than the tomb of Tutankhamen (which could easily be contained in one of the houses along its main road). At this writing, only a single city block has been excavated, yet the buried city spreads nearly a mile in diameter, and beneath its streets and shops lie the ruins of another volcano-ravaged city, almost a thousand years older. Virtually all excavation came to a halt with Marinatos' death. The archaeologists who were being groomed by Marinatos to carry on his work were ousted in a horrible spasm of piracy and political infighting that continues to this day and rivals the Dead Sea Scrolls scandal. By 1991 hundreds of priceless fossil impressions of wooden structures had crumbled to dust, Theran artifacts began appearing on the Jerusalem antiquities market, and the city's funding (supported by ticket sales to tourists) had been pilfered so utterly that there was no money to maintain a protective rain shelter over the ruins, much less explore them further. At this writing the future of the lost city is still in doubt ("Atlantis is in danger of disappearing for the second time"), and some archaeologists can be heard to remark acidly that no one will ever really know what happened the day Thera exploded until certain people in the Greek political arena either die of old age or are swallowed by the next Thera upheaval.

BENJAMIN MAZAR, the great-grandfather of Israeli archaeology, spends his ninth decade exploring, as always (with most of his attention now focusing on the tunnels and monoliths beneath the Temple Mount). His nephew, Ami Mazar, carries on the family tradition at Bet-She'an.

NOAH'S ARK, and the stories associated with it, probably date back to actual events of the Babylonian flood period about 2800 B.C., in which valued farm animals were hastily loaded onto barges when the entire Tigris-Euphrates plain and all its cities disappeared under stormwaters. About 3,000 years later, early Christian tradition began placing the final resting place of the legendary Ark on Mount Ararat, in northern Turkey. During the Middle Ages, at least two monasteries were built on Ararat, in the shapes of boats. They became destinations of religious pilgrimages. During the late twentieth century, expeditions to Ararat began hauling pieces of wood down from the mountain's snowfields. With the exception of one obvious hoax involving a piece of late twentieth century firewood, all recovered pieces of "Noah's Ark" could be carbon dated to the time of the monasteries.

ELIZABETH STONE and PAUL ZIMANSKY, the American archaeologists who have spent much of their life together exploring the Tigris-Euphrates region, have collected enough material to keep them writing for years about city life in the time of Abraham and Lot. Like most good detective stories, the clues in hand raise at least as many questions as they answer, so Stone and Zimansky must await the death of Saddam Hussein's regime before they can return to Iraq and learn what really happened the day Mashkan-shapir burned down.

TUTANKHAMEN's golden death mask can be seen on display in the Cairo Museum. As in the case of Queen Hatshepsut, mythology has it that there is an evil curse associated with the pharaoh's tomb and its contents. When the mask and other treasures from "The Tomb of King Tut" arrived at New York's American Museum of Natural History in 1977, there was much public interest in the rumors of a curse. At that time, Charles Pellegrino was helping to pay his way through college by working as a radio disc jockey. From WMCA (on what was then the "Sally Jesse Raphael Show") he gave the following report: "On the day the tomb was opened, strange things began happening. A python swallowed [Howard] Carter's pet canary. Word spread that the python was the king's messenger of revenge. Then one day in April 1923, before the king's sarcophagus had been discovered, [Lord] Carnavon's barber nicked a mosquito bite with his razor. It became infected, and Carnavon, who had financed the excavation, died. At the moment he died, all the lights in Cairo went out, giving more credence to the talk of a curse. Carnavon's son, the current earl, remembers the death and blackout well since he was there at the time. This sweltering July 14, here in New York for an NBC special, he replies that he does not believe in the curse but that he will not disbelieve it either. He said that at the hour his father died, another strange happening occurred: 'At the identical moment, a little fox terrier bitch I had given my father sat up in the basket where she was sleeping in the housekeeper's room and suddenly let out howls like a wolf. Foam flecked her lips. She gave a final howl, and died.' " After the broadcast ended, Pellegrino whispered to a friend, "Carnavon! What a load of crap." And then all the lights in New York went out, triggering the longest night of looting and mayhem in the city's history.

TUTHMOSIS III, the apparent pharaoh of the Thera explosion and the Exodus, and the first Egyptian conqueror of the eastern Mediterranean world, died of natural causes at an advanced old age. Despite a labyrinth of false walls and deathtraps, robbers had entered his tomb by 1000 B.C. and removed almost everything of value, including his golden death mask. Gemstones and sheets of gold were stripped away from the wooden sarcophagus, and little more than the mummy itself and objects of wood and alabaster remained. The same fate befell every pharaoh except Tutankhamen, who still sleeps in his golden casket beneath the Valley of the Kings. Elsewhere in the valley, thirteen desecrated royal mummies, including Tuthmosis III, were eventually sheltered in the tomb of Amenhotep II. This tomb, too, was broken into. The greater part of whatever art treasures remained were spirited away and dismembered for their gold and jewels. Most of the funeral equipment—delicate alabaster containers filled with oil and perfumes—had been drained of their contents and smashed against the walls. And then, for nearly three thousand years, the tomb was miraculously forgotten. When archaeologists entered it in A.D. 1898, they found the mummies intact. Much had been taken, it was true, but the shelter had escaped the total destruction seen in other tombs. The body of Tuthmosis III himself still lay within its own (albeit plundered) sarcophagus, seeming to have been spared the final indignity of being stripped completely and burned for fuel, but within a year modern tomb robbers, doubtless with the cooperation of the guards, removed almost all that had survived. The pharaoh's funerary wrappings were cut and torn, and the body within was searched and stripped of rings and other jewelry. What remained of the pharaoh was removed to the Cairo Museum, where, today, a visitor can look upon the face that (if the prophet really existed) must have looked upon the face of Moses.

From the desert sands around Cairo, archaeologists of the nineteenth century A.D. recovered an obelisk inscribed with the name Tuthmosis III. After burying Hatshepsut's obelisks at Karnak and declaring the very concept of a woman monarch blasphemous, the pharaoh erected the huge limestone needle in his own honor, intending it to carry his name into our time, where he would forever eclipse the queen of the Nile. It now stands in Central Park, New York, clearly proclaiming the name Tuthmosis III in Egyptian hieroglyphic art, for all the world to see. But New Yorkers, as a rule, do not read hieroglyphs, and as the stone slowly dissolves under layers of moss, acid rain, and pigeon droppings, and as visitors to the park pose beneath it for photographs, children may be heard to ask, "Mommy, why do they call that rock Cleopatra's Needle?"

LEONARD WOOLLEY, the English archaeologist and explorer who became mentor to Max Mallowan and Thomas Lawrence, continued exploring, producing books, and generally "enjoying life to the fullest" until his death at age eighty in 1960. During the early 1990s, as the U.S.-Iraq War loomed, CIA analysts in the United States discovered that for all their satellites and sensors, it was impossible for them to determine whether or not the desert soil would support the massive array of fighting machines they planned to send across the Tigris-Euphrates plain. Putting their satellite data aside, reconnaissance experts went to the library of Congress and looked up Leonard Woolley's eighty-year-old archaeological surveys. It was a turn of events that would have made Lawrence smile wryly: still spying after all these years.

Selected Bibliography

Ager, D. *The New Catastrophism: The Importance of the Rare Event in Geological History*. Cambridge: Cambridge University Press, 1993.

Asimov, I. *Asimov's Guide to the Bible*. New York: Avenel, 1981.

———. *In the Beginning: Science Faces God in the Book of Genesis*. New York: Crown, 1981.

Anati, E. *Palestine Before the Hebrews*. New York: Knopf, 1961.

Baigent, M., and R. Leigh. *The Dead Sea Scrolls Deception*. New York: Summit Books, 1991.

Baker, R. R. "Environmental Conditions Inside a Burning Cigarette." *Analytical Calorimetry*, 4 (1977).

Barrow, J. D. *Theories of Everything*. London: Oxford University Press, 1991.

———. *The World Within the World*. London: Oxford University Press, 1991.

Boyle, C., et al. *Time Frame*. New York: Time-Life Books, 1990. (This superbly researched series of books, which covers global events from the dawn of man into the space age, is probably the most invaluable reference work of its kind.)

Cann, R. L. "The Search for Eve." *Science*, 256 (1992), p. 79.

Carter, H. *The Tomb of Tutankhamen*. London: Century Hutchinson, Ltd., 1983. (Originally published in installments between 1923 and 1933.)

Clarke, A. C., S. Hawking, and C. Sagan. *God, the Universe and Everything*. London: BBC Television, 1990. (Transcripts available: educational television of the "must view" kind.)

Coe, M. D. *Breaking the Maya Code*. New York: Thames and Hudson, 1992.

Christie, A. *Come Tell Me How You Live*. London: Collins, 1943.

———. *An Autobiography*. London: Collins, 1977.

Crick, F. *The Astonishing Hypothesis: The Scientific Search for the Soul*. New York: Scribners, 1994.

———, et al. "Mind and Brain," *Scientific American* (special issue), 267 (September 1992).

Culotta, E. "A New Take on Anthropoid Origins." *Science*, 256 (1992), pp. 1516–17.

Deiss, J. J. *Herculaneum: Italy's Buried Treasure.* New York: Harper & Row, 1985.

Dening, G. *Mr. Bligh's Bad Language: Passion, Power and Theatre on the Bounty.* New York: Cambridge University Press, 1992.

Diamond, J. "A Pox on Our Genes." *Natural History* (February 1990), pp. 26–30.

Dothan, M. "High Loop-Handled Cups and Relations Between Mesopotamia, Palestine, and Egypt." *PEQ,* 335 (1953).

Dothan, T. "Gaza Sands Yield Lost Outpost of the Egyptian Empire." *National Geographic* (December 1982), pp. 739–68.

———— and M., *People of the Sea.* New York: Macmillan, 1992.

————, and S. Gitin. "Ekron of the Philistines." *Biblical Archaeology Review,* 16 (1990), pp. 26–36.

Doumas, C., et al. *Thera and the Aegean World.* 3 vols. Athens: G. Tsiveriotis, Ltd., 1978, 1980, 1991.

————. *The Excavations at Akrotiri, Thera,* vol. I, *Delta House;* vol. II, *West House.* Athens: in press.

Eldredge, N. *The Miner's Canary: Unraveling the Mysteries of Extinction.* New York: Prentice Hall, 1991.

————. *Fossils: The Evolution and Extinction of Species.* Washington, D.C.: Smithsonian Books, 1992.

————, and I. Tattersall. *The Myths of Human Evolution.* New York: Columbia University Press, 1982.

Esse, D. L., et al. *Methods of Investigation on the Dead Sea Scrolls and the Khirbet Qumran Site: Present Realities and Future Prospects.* New York: Annals of the New York Academy of Sciences [9301], 1993.

Fagan, B. M. *Return to Babylon: Travelers, Archaeologists and Monuments in Mesopotamia.* Boston: Little, Brown and Co., 1979.

Francheteau, J., R. Hekinian, and R. D. Ballard. "Morphology and Evolution of Hydrothermal Deposits at the Axis of the East Pacific Rise." *Oceanologica Acta,* 8 (1985).

Gans, C., and G. C. Gorniak. "How Does the Toad Flip Its Tongue: A Test of Two Hypotheses." *Science,* 216 (1982).

Gardner, M. "The Great Pyramid," in *Fads and Fallacies in the Name of Science.* New York: Dover, 1957.

Gibbons, A. "Mitochondrial Eve: Wounded, but Not Dead Yet." *Science,* 257 (1992), pp. 873–75.

Gould, S. J. *Ever Since Darwin.* New York: W. W. Norton, 1977.

————. *The Mismeasure of Man.* New York: W. W. Norton, 1982.

————. "The Golden Rule—a Proper Scale for Our Environmental Crisis." *Natural History* (September 1990), pp. 24–30.

Grant, M. *The History of Ancient Israel.* New York: Charles Scribner's Sons, 1984.

Greene, M. T. *Natural Knowledge in Preclassical Antiquity.* Baltimore: Johns Hopkins University Press, 1992.

Greenhut, Z., and R. Reich. "The Tomb of Caiaphas." *Biblical Archaeology Review,* 18 (1992), pp. 28–57.

Hammer, C. U., et al. "Dating of the Santorini [Thera] Eruption." *Nature,* 332 (1988), pp. 401–02.

Han, T. M., and B. Runnegar. "Megascopic Eukaryotic Algar from the 2.1-Billion-Year-Old Negaunee Iron Formation, Michigan." *Science,* 257 (1992), pp. 232–35.

Hawking, S., et al. "The Beginning of the Universe," in Texas/ESO-CERN Symposium on Relativistic Astrophysics, Cosmology, and Fundamental Physics. New York: Annals of the New York Academy of Sciences [647], 1991.

Heilprin, A. *The Eruption of Pelée.* Philadelphia: J. B. Lippincott Co., 1908.

Hitti, P. K. *History of the Arabs.* New York: St. Martin's Press, 1970.

Hood, S. *The Minoans.* London: Thames and Hudson, 1971.

Hsu, K. *The Mediterranean Was a Desert.* Princeton: Princeton University Press, 1983.

Hughes, M. K. "Ice Layer Dating of Eruption at Santorini [Thera]." *Nature,* 335 (1988), pp. 211–12.

Jacobsen, T. *The Treasures of Darkness.* New Haven: Yale University Press, 1976.

Keller, W. *The Bible as History.* New York: Morrow, 1981.

Kunstel, M., and J. Albright. *Their Promised Land: Arab and Jew in History's Cauldron—One Valley in the Jerusalem Hills.* New York: Crown, 1990.

Le Strange, G. *The Land of the Eastern Caliphate.* Cambridge: Cambridge University Press, 1905.

Levy, S. *Artificial Life: The Quest for a New Creation.* New York: Pantheon, 1992.

Lloyd, S. *Foundations in the Dust.* London: Oxford University Press, 1947.

Mallowan, M. *Early Mesopotamia and Iran.* New York: McGraw-Hill, 1965.

———. *Nimrod and Its Remains.* 2 vols. London: Collins, 1966.

———. *Mallowan's Memoirs.* New York: Dodd, Mead, 1977.

Marinatos, N. *Art and Religion in Thera: Reconstructing a Bronze Age Society.* Athens: Andromedas I, 1984.

Marinatos, S. *Some Words About the Legend of Atlantis.* Athens: Athens Museum, 1969.

———. "On the Chronological Sequence of Thera's Catastrophes." *Acta* (1971), pp. 403–06.

———. "Thera: Key to the Riddle of Minos." *National Geographic,* 141 (1972), pp. 702–26.

———. *Excavations at Thera, 1968–1974.* Athens: Athens Museum, 1975.

Meyers, R. J. "An Instance of the Pitfalls Prevalent in Graveyard Research." *Biometrics,* 19 (1963), pp. 643–50.

Michael, H., and G. Weinstein. "Radiocarbon Dates for the Destruction Level of Akrotiri, Thera." Philadelphia: Temple University Aegean Symposium, 1977, pp. 27–30.

Nelson, R. *Popol Vuh: The Great Mythological Book of the Ancient Maya.* Boston: Houghton Mifflin, 1976.

Pang, K. D. "The Legacies of eruption." *The Sciences,* 31 (1991), pp. 30–35.

Pellegrino, C. R. "The Trouble with Nemesis." *Evolutionary Theory,* 7 (December 1985), pp. 219–21.

———. *Unearthing Atlantis: An Archaeological Odyssey.* New York: Random House, 1991.

———, and Joshua Stoff. *Chariots for Apollo: The Untold Story Behind the Race for the Moon.* Originally published: New York: Atheneum, 1985. Reprinted: Blue Ridge Summit, Pa.: TAB Books, 1987.

———, and J. A. Stoff. *Darwin's Universe: Origins and Crises in the History of Life.* Originally published: New Zealand, 1982; New York: Van Nostrand Reinhold, 1983. Reprinted: Blue Ridge Summit, Pa.: TAB Books, 1986.

Platon, N. *Zakros: The Discovery of a Lost Palace of Ancient Crete.* New York: Charles Scribner's and Sons, 1971.

Prichard, J. B. *Ancient Near Eastern Texts Relating to the Old Testament.* Princeton: Princeton University Press, 1955.

———, ed. *Harper Atlas of the Bible.* New York: Harper & Row, 1987.

Rathje, W., and C. Murphy. *Rubbish! The Archaeology of Garbage: What Our Garbage Tells Us About Ourselves.* New York: HarperCollins, 1992.

Raup, D. M. *Extinction: Bad Genes or Bad Luck?* New York: W. W. Norton, 1991.

Redford, D. B. *Egypt, Canaan and Israel in Ancient Times.* Princeton: Princeton University Press, 1992.

Rightmire, P. *The Evolution of Homo Erectus.* Cambridge: Cambridge University Press, 1990.

Roaf, M. *Cultural Atlas of Mesopotamia and the Ancient Near East.* New York: Facts on File, 1990.

Romer, J. *Testament: The Bible as History.* New York: Henry Holt and Co., 1988.

Ross, P. E. "Trends in Molecular Archaeology." *Scientific American,* 266 (1992), pp. 115–25.

Saggs, H.W.F. *Civilization Before Greece and Rome.* New Haven: Yale University Press, 1989.

Shanks, H., and D. P. Cole, eds. *Archaeology and the Bible,* vol. I, *Early Israel.* Washington, D.C.: Biblical Archaeology Society, 1990.

Sola, S. M., and M. Kohler. "Recent Discoveries of *Dryopithecus* Shed New Light on Evolution of Great Apes." *Nature,* 365 (1993), pp. 543–45.

Stanley, D., and H. Cheng. "Cores of Santorini [Thera] Ash Layer in the Nile Delta." *Science,* 240 (1987), pp. 497–500.

Stone, E., and P. Zimansky. "Mashkan-shapir and the Anatomy of an Old Babylonian City." *Biblical Archaeologist* (December 1992), pp. 212–18.

Stone, E. C. "The Tell Abu Duwari Project, Iraq, 1987." *Journal of Field Archaeology,* 17 (1990), pp. 141–61.

Steindorff, G. *When Egypt Ruled the East.* Chicago: University of Chicago Press, 1965.

Sullivan, D. G. "The Discovery of Santorini [Thera] Minoan Tephra in Western Turkey." *Nature,* 333 (1988), pp. 552–54.

Symons, G. J. *The Eruption of Krakatoa and Subsequent Phenomena.* London: Report of the Krakatoa Committee of the Royal Society, 1888.

Tabor, J. D., et al. "BAS Dead Sea Scroll Research Council." *Biblical Archaeology Review,* 18 (1992), pp. 53–66.

Templeton, et al. "Human Origins and Analysis of Mitochondrian DNA Sequences . . . 'African Eve.' " *Science,* 255 (1992), pp. 686–87, 737–39.

Thomas, G., and M. Morgan Witts. *The Day Their World Ended: The Destruction of St. Pierre and All Its 30,000 Inhabitants.* New York: Stein and Day, 1969.

Tianyuan, L., and D. A. Etler. "New Middle Pleistocene Hominid Crania from Yunxian in China." *Nature,* 404 (1992), pp. 404–07.

Warren, P. *The Aegean Civilization.* Oxford: Elsevier, 1975.

Wells, E. *Hatshepsut.* New York: Doubleday, 1969.

Wilford, J. N. "From Space, It's Clear: All Roads Go to Oman's Lost City of Frankincense." *The New York Times,* June 25, 1991, pp. A-1–14.

Wilson, A. C., et al. "Debate: Is an African 'Eve' the Mother of Us All?" *Scientific American,* 266 (1992), pp. 66–83.

Wood, B. G. "Did the Israelites Conquer Jericho? A New Look at the Archaeological Evidence." *Biblical Archaeology Review,* 16 (1990), pp. 44–53.

———. "The Philistines Enter Canaan." *Biblical Archaeology Review,* 17 (1991), pp. 44–52.

Woolley, L. *Excavations at Ur.* New York: Barnes and Noble, 1954.

———. *Dead Cities and Living Men.* New York: Philosophical Library, 1956.

Index

Page numbers in *italics* refer to illustrations.

DR. CHARLES PELLEGRINO wears many hats. He has been known to work simultaneously in crustaceology, paleontology, preliminary design of advanced rocket systems, and marine archaeology. He has been described by Stephen Jay Gould as a space scientist who occasionally looks down and by Arthur C. Clarke as "the polymathic astro-geologist–nuclear physicist who happens to be the world's first astropaleontologist." He was, with James Powell, Harvey Meyerson, and the late Senator Spark Matsunaga, a framer of the U.S.-Russian Space Cooperation Initiative (which included, among its designs, an International Space Station and joint Mars missions). At Brookhaven National Laboratory he and Dr. James Powell coordinate brainstorming sessions on the next seventy years; projects currently under design by Powell and Pellegrino range from a global system of high-speed Maglev trains (New York to Sydney in five hours) to relativistic flight (Valkyrie rockets) and the raising of an archaeological site (the wreckage of a Portuguese galleon and the mud that contained it) completely intact from a quarter mile under the Atlantic Ocean.

In the late 1970s Dr. Pellegrino and Dr. Jesse A. Stoff produced the original models that predicted the discovery of oceans inside certain moons of Jupiter and Saturn. While looking at the requirements for robot exploration of those new oceans, Pellegrino sailed with Dr. Robert Ballard, worked with the deep-sea robot *Argo,* and traced the *Titanic* debris field backward in time to reconstruct the liner's last three minutes. He has since, with James Powell, developed an economically viable means of raising and displaying the *Titanic*'s four-hundred-foot-long bow section.

Through his work on ancient DNA (including a recipe involving dinosaur cells that may be preserved in mouth parts and in the stomachs of ninety-five-million-year-old amberized flies), Pellegrino hopes one day to redefine extinction. His hope—and his recipe—became the basis for the Michael Crichton novel/Steven Spielberg film *Jurassic Park.*

ABOUT THE TYPE

This book was set in Galliard, a typeface designed by Matthew Carter for the Merganthaler Linotype Company in 1978. Galliard is based on the sixteenth-century typefaces of Robert Granjon, which give it classic lines yet interject a contemporary look.